Atomic and Molecular Physics

Introduction to
Advanced Topics

Atomic and Molecular Physics

Introduction to Advanced Topics

Editors
Rajesh Srivastava
Rakesh Choubisa

Narosa Publishing House

New Delhi   Chennai   Mumbai   Kolkata

Editors
Rajesh Srivastava
Department of Physics
Indian Institute of Technology Roorkee
Roorkee

Rakesh Choubisa
Department of Physics
Birla Institute of Technology and Science Pilani
Pilani

NAROSA PUBLISHING HOUSE PVT. LTD.

22 Delhi Medical Association Road, Daryaganj, New Delhi 110 002
35-36 Greams Road, Thousand Lights, Chennai 600 006
306 Shiv Centre, Sector 17, Vashi, Navi Mumbai 400 703
2F-2G Shivam Chambers, 53 Syed Amir Ali Avenue, Kolkata 700 019

www.narosa.com

Printed from the camera-ready copy provided by the Editors.

ISBN 978-81-8487-169-2

Published by N.K. Mehra for Narosa Publishing House Pvt. Ltd.,
22 Delhi Medical Association Road, Daryaganj, New Delhi 110 002

Printed in India

Preface

Department of Science and Technology (DST), Government of India, New Delhi through its Science and Engineering Research Council (SERC) sponsors series of Schools in the different fields of research in Science. The main objective of these schools has been to expose the research area and motivate the students planning to join research and to increase the level of academic standard of those who are beginners to the field. A DST sponsored school on the *Theoretical Techniques in Atomic and Molecular Collision Physics* was recently organized by us at Birla Institute of Technology and Science (BITS), Pilani.

In general the level of theoretical research has gone very high. Since the research in Atomic Physics has build up from an early age of fundamental science the complexity in this field in particular has become analytically and computationally very challenging. Usually students receive very little coverage of Atomic Collision Physics and the numerical methods in their Physics curriculum at graduate level. The course of School organized at BITS, Pilani was designed to provide the students the basics in Atomic Collision Physics and introduce them to the frontier research areas so that they can take up research in this field in an easy way. Also this course is very useful to start research in experimental Atomic and Molecular Physics as the students can have after this course a very good theoretical background. This course is also helpful to start research in interrelated research areas like Laser physics, Astrophysics and Plasma and Fusion research where such a background of theoretical Atomic Collision Physics is an integral part.

In view of the material covered in the school which is not available in a single book, we decided to publish the lecture notes in the present book form. We expect that the book will serve the requirements of students in future. The course was designed to cover important topics of Atomic Collision Physics *viz*. Atomic structure calculations, Photoionization of atomic systems, Electron-atom collisions, Ion-atom collisions, Collisions involving exotic particles, Ultracold atoms and Bose-Einstein condensation as well as Atomic data and Plasma diagnostics.

It is pleasure to acknowledge the help and full cooperation of the Vice Chancellor, the Director and their Administrative staff as well as faculty and research students of the Physics Department of the BITS, Pilani in holding this School. We are thankful to DST, New Delhi for full financial support to hold this School at Pilani as well as for publication of this book. We are grateful to Prof. P. C. Deshmukh (IIT Madras), Prof. Kailash Rustagi (IIT Bombay), Prof. E. Krishnakumar (TIFR, Mumbai), Dr. K. P. Subhramanian (PRL, Ahmedabad) and Dr. Amitava Roy (DST, New Delhi) for their valuable suggestions in organizing the School as well as all authors for their contributions in this book.

Rajesh Srivastava
Rakesh Choubisa

Contents

Hartee-Fock Self-Consistent Field Method for Many-Electron Systems

Pranawa C. Deshmukh[1*], Alak Banik[2] and Dilip Angom[3]

[1]Indian Institute of Technology Madras, Chennai
[2]Space Applications Centre, Ahmadabad
[3]Physical Research Laboratory, Ahmadabad
email: pcd@physics.iitm.ac.in[*]

1. INTRODUCTION

It is well-known that the first model of the electronic structure of atoms based on the idea of quantization of the angular momentum was proposed by Niels Bohr in the year 1913. Bohr's model was based on the planetary model of Kepler, but the Bohr-Kepler orbits are indefensible since an orbit requires a simultaneous description of position and momentum of the electron which correspond to mutually incompatible observables. The Bohr-model thus had to give way to formal quantum theory based on the Schrodinger equation and the Heisenberg's principle of uncertainty.

The Schrodinger equation for the Hydrogen atom has exact analytical solution, but an atom consisting of two or more electrons poses a formidable challenge. As pointed out by Professor G. E. Brown, if one is looking for exact solutions, "having no body at all is already too many"; even the analysis of the vacuum state requires approximations! Quantum elementary particles such as in a many-electron system are fundamentally identical, and one cannot track the temporal dynamics of each electron separately when it is in the company of another. The two electrons are indistinguishable; a 'two-electron' system can therefore be best described essentially only as a 'two-electron' wavefunction, sometimes called a

a fundamental particle! The fundamental particle in this case also is of course the individual single electron itself, since the two-electron geminal can be written as a product of single-electron wavefunctions, but the form of the product function must respect the indistinguishability of the two electrons! Electrons being fermions, the geminal wavefunction must be anti-symmetric. The Hartree-Fock (HF) method [1,2] employs an extension of this idea in which the many-electron wavefunction is written as an anti-symmetric product of single-electron wavefunctions. The method nevertheless belongs to the family of 'single-particle-approximations', also called as the 'Independent Particle Approximation' (IPA), since the many-electron wavefunction is nonetheless expressible as a linear superposition of products of single-particle wavefunctions subject to the condition that the superposition is anti-symmetric.

In the HF IPA, each electron in an atom/ion consisting of the N electrons is considered to experience a potential determined by the central field nuclear attraction and a static average potential determined by the remaining (N-1) electrons.

The N-electron Schrodinger equation to be solved is $H^{(N)}\psi^{(N)} = E^{(N)}\psi^{(N)}$, and its construct expresses a "Catch-22" situation. The two-electron Coulomb interaction term in the N-electron Hamiltonian,

$$H^{(N)}(q_1,q_2,...,q_N) = \sum_{i=1}^{N}\left(-\frac{1}{2}\nabla_i^2 - \frac{Z}{r_i}\right) + \sum_{i<j=1}^{N}\frac{1}{r_{ij}}$$

$$= \sum_{i}f(|\vec{r}_i|) + \sum_{i<j}v\left(|\vec{r}_i - \vec{r}_j|\right) , \tag{1}$$

$$= H_1 + H_2.$$

can be defined only in terms of the electron charge densities that generate the electron-electron interactions, but the charge densities themselves must be expressed only in terms of the electron wavefunctions, which are to be obtained from the solution to the N-electron Schrodinger equation - but that requires the Hamiltonian to be set up in the first place!

It was D. R. Hartree (1897-1958) who came up with the first idea of getting Self Consistent Field (SCF) solutions to a many-electron problem as a strategy to break the "Catch-22" state. D. R. Hartree was helped by his father, William Hartree, in solving the numerical problems involved in solving the SCF problem [3,4]. It is no wonder that with his numerical skills, D. R. Hartree designed a large differential analyzer, in 1935, prototype for which was a small-scale machine built from pieces of children's Meccano - which actually solved useful equations concerned with atomic theory in 1934! When John Eckert set up the ENIAC, Hartree was asked to go to the USA to advice on its use. Hartree showed how to use ENIAC to calculate trajectories of projectiles. Hartree predicted at Cambridge in 1946 that: "It may well be that the high-speed digital computer will have as great an influence on civilization as the advent of nuclear power." How truthful Hartree's vision has turned out to be!

Hartree's original method made use of the IPA in which the N-electron wavefunction was written as a product of one-electron wavefunctions. It was the extension of this idea that has come to be known as the Hartree-Fock method in which the Pauli's exclusion principle got formally incorporated. It accounts for the fundamental identity of the electrons that makes a many-electron wavefunction anti-symmetric. This description of the many-electron system is in accordance with the Fermi-Dirac (FD) statistics.

The 'spin' of the electron plays an essential role in determining the fermion character of the electron. The 'spin' is an intrinsic property, just like mass and charge, of the electron; it corresponds to the electron's intrinsic angular momentum. It has no classical analogue. The 'rule' that particles with half-integral spin observe Fermi-Dirac statistics (and those with integral spins observe the Bose-Einstein statistics) is well-known, but it is based on very deep and difficult principles. With reference to it, Feynman remarks [5]: "It appears to be one of the few places in physics where there is a rule which can be stated very simply, but for which no one has found a simple and easy explanation. The explanation is down deep in relativistic quantum mechanics.....". The electron spin emerges naturally from the relativistic (Dirac) equation, but it is included in the N-electron non-relativistic Schrodinger equation on an ad-hoc basis in the HF SCF method.

The one-electron 'spin-orbital' $u(q)$ for an electron is a function of four coordinates represented collectively by $q = (\vec{r}, \zeta)$ wherein $\vec{r}$ is the electron's position vector made up of the three space coordinates, and ζ represents its spin coordinate. The atomic electron's spin-orbital is described by the central field quantum numbers n, l, m_l, m_s. The spin-orbital is expressed as the product, $u_i(q_j) = \psi_{n_i, l_i, m_{l_i}}(\vec{r}_j) \times \chi_{m_{s_i}}(\zeta_j)$, of the 'orbital' part with the 'spin' part. The subscripts in this notation denote the 'good quantum numbers' of occupied quantum states, and the arguments denote the coordinates. In the Dirac notation, the spin-orbital is $\langle q_j | i \rangle = \langle \vec{r}_j, \zeta_j | n_i, l_i, m_{l_i}, m_{s_i} \rangle$.

The N-electron anti-symmetric product of N single-particle spin-orbitals is conveniently written as a determinant, known as the Slater Determinant (SD), named after John Slater:

$$\Psi^{(N)}(q_1, q_2 \ldots q_N) = \frac{1}{\sqrt{N!}} \begin{vmatrix} u_1(q_1) & u_1(q_2) & \ldots & u_1(q_N) \\ u_2(q_1) & u_2(q_2) & \ldots & u_2(q_N) \\ \vdots & \vdots & u_i(q_j) & \vdots \\ u_N(q_1) & u_N(q_2) & \ldots & u_N(q_N) \end{vmatrix}. \tag{2}$$

The SD explicitly manifests the Pauli Exclusion Principle, since the determinant would vanish if any two rows were to be equal. It is also manifestly anti-symmetric, since the determinant would change its sign every time the parity of the permutations of the N identical electrons is odd.

2. THE SELF CONSISTENT FIELD THEORY

The SCF strategy consists of using some 'guess' wavefunctions to construct the Hamiltonian, and then solve the Schrodinger equation for this Hamiltonian:

$$H^{(N)}(q_1,..,q_N)\psi^{(N)}(q_1,..,q_N) = E^{(N)}\psi^{(N)}(q_1,..,q_N). \tag{3}$$

One then inquires if the solutions to the Schrodinger equation yield the same wavefunctions that one had guessed [1,2]. Comparison of the two sets of wavefunctions is then tested to lie within a desired numerical convergence criterion. If the convergence fails, the trial functions are varied and the process iterated upon till self-consistency is attained to yield numerical solutions. The basic methodology of the HF SCF scheme is very well described in a number of text books, such as [1,2].

Now, a variation of the one-electron spin-orbitals is required to attain the SCF, but the variation is subject to the constraints of (i) normalization of the spin-orbitals $\langle u_i | u_j \rangle = 1$ for $j = i$, and (ii) the orthogonality $\langle u_i | u_j \rangle = 0$ for $j \neq i$; i and j stand respectively for the 'collective complete set of good quantum numbers' of the ith and j^{th} occupied single-particle states.

The Hartree-Fock strategy to seek self-consistent-field solutions to the N-electron Schrodinger equation is inspired by a very powerful principle that is well-known in fundamental physics, namely the principle of variation. The SCF solutions are obtained by employing the variational principle, that the correct solutions would be those as would make the expectation value of the N-electron Hamiltonian in the N-electron Slater determinant wavefunction an 'extremum'. Accordingly,

$$\delta \langle \psi^{(N)} | H^{(N)} | \psi^{(N)} \rangle = 0 . \tag{4}$$

As mentioned above, the variation is implemented subject to the constraints

$$\langle i | j \rangle = \delta_{ij} . \tag{5}$$

It is clear from Eq. 1 and Eq.4 that we need for our analysis $\langle \Psi | \Omega | \Psi \rangle$, with $\Omega = H_1$ and $\Omega = H_2$. It may be noted here that the Slater determinantal wavefunction can be conveniently written as

$$\psi^{(N)} = \frac{1}{\sqrt{N!}} \sum_{P=1}^{N!} (-1)^p P \{u_{\alpha_1}(q_1) u_{\alpha_2}(q_2)...u_{\alpha_N}(q_N)\} \tag{6}$$

where the summation is over all possible $N!$ number of permutations amongst the N completely identical electrons and p is the parity of the permutation P.

To obtain the variation in the expectation value of the Hamiltonian referred to in Eq.1, one first observes that

$$\langle \psi^{(N)} | H | \psi^{(N)} \rangle = \sum_{i=1}^{N} \int dV u^*_{\alpha_i}(\vec{r}) f(\vec{r}) u_{\alpha_i}(\vec{r}) +$$

$$+\frac{1}{2} \sum_{\alpha_i} \sum_{\alpha_j} \left[\begin{array}{l} \int\int dV_1 dV_2 u^*_{\alpha_i}(\vec{r}_1) u^*_{\alpha_j}(\vec{r}_2) \frac{1}{r_{12}} u_{\alpha_i}(\vec{r}_1) u_{\alpha_j}(\vec{r}_2) \\[2ex] -\delta(m_{s_i}, m_{s_j}) \int\int dV_1 dV_2 u^*_{\alpha_i}(\vec{r}_2) u^*_{\alpha_j}(\vec{r}_1) \frac{1}{r_{12}} u_{\alpha_i}(\vec{r}_1) u_{\alpha_j}(\vec{r}_2) \end{array} \right]$$

$$(7a)$$

in which the operators $f(|\vec{r}|)$ and $\dfrac{1}{r_{ij}} = v(|\vec{r}_i - \vec{r}_j|)$ have been introduced in Eq.1.

This result can be written in a compact form using the Dirac notation:

$$\langle \psi^{(N)} | H | \psi^{(N)} \rangle = \sum_{i=1}^{N} \langle i | f | i \rangle + \frac{1}{2} \sum_{j=1}^{N} \sum_{i=1}^{N} \left[\langle ij | g | ij \rangle - \langle ij | g | ji \rangle \right] \qquad (7b)$$

The first of the two-center integrals in Eq.7a is called as the 'Coulomb integral', and the second is the 'Exchange integral'.

Using Lagrange's method of variational multipliers [6] λ_{ij}, the condition (Eq.4) of the 'extremum', subject to the constraints described by Eq.5, is then expressed by the following relation:

$$0 = \delta \left\{ \begin{array}{l} \langle \psi^{(N)} | H | \psi^{(N)} \rangle + \\[1ex] +\sum_{i=1}^{N} \lambda_{ii} \int dV\, u^*_i(\vec{r}) u_i(\vec{r}) + \\[1ex] +\sum_{i\rangle j} \delta(m_{s_i}, m_{s_j}) \left[\lambda_{ij} \int dV\, u^*_i(\vec{r}) u_j(\vec{r}) + \lambda_{ji} \int dV\, u^*_j(\vec{r}) u_i(\vec{r}) \right] \end{array} \right\} \qquad (8a)$$

At this juncture, an important approximation, namely the 'frozen orbital approximation', is introduced in the HF SCF methodology. According to this, variations in the single particle orbitals are made one at a time, which is to say that the other $N-1$ orbitals are considered 'frozen' during the consideration of the variation in each orbital.

Within the frozen orbital approximation, allowing for a variation in only the kth orbital and in none other, Eq.8a gives:

$$0 = \left[\int dV_1 \{\delta u^*_k(\vec{r}_1)\} \left(\begin{array}{l} f(\vec{r}_1)u_k(\vec{r}_1) + \\ + \sum_j \left[\int dV_2 \dfrac{u^*_j(\vec{r}_2)}{r_{12}} \left(\begin{array}{l} u_k(\vec{r}_1)u_j(\vec{r}_2) \\ -\delta(m_{s_k},m_{s_j})u_k(\vec{r}_2)u_j(\vec{r}_1) \end{array} \right) \right] + \\ + \sum_j \delta(m_{s_k},m_{s_j})\lambda_{kj}u_j(\vec{r}_1) \end{array} \right) \right. $$
$$\left. + \int dV_1 \{\delta u_k(\vec{r}_1)\}(\ )^* \vphantom{\int} \right.$$
$$\left. + \left\{ \sum_j \delta(m_{s_k},m_{s_j}) \left[\lambda_{kj} \int dV_1 \left(\delta u^*_k(\vec{r}_1)\right) u_j(\vec{r}_1) + \lambda_{jk} \int dV_1 \ u^*_j(\vec{r}_1)\left(\delta u_k(\vec{r}_1)\right) \right] \right\} \right] \tag{8b}$$

The necessary and sufficient condition that Eq.8b is satisfied within the frozen orbital approximation turns out to be:

$$f(\vec{r}1)u_k(\vec{r}_1) +$$
$$\sum_j \left[\int dV_2 \dfrac{u^*_j(\vec{r}2)}{r_{12}} \left(\begin{array}{l} u_k(\vec{r}1)u_j(\vec{r}2) \\ -\delta(m_{s_k},m_{s_j})u_k(\vec{r}2)u_j(\vec{r}1) \end{array} \right) \right] \tag{9}$$
$$= -\sum_j \delta(m_{s_k},m_{s_j})\lambda_{kj}\,u_j(\vec{r}_1)$$

The above equation is called as the 'single-particle Hartree-Fock equation'. The N^2 Lagrange's variational parameters λ_{ij} for $i,j = 1,...,N$ can be written as elements of a self-adjoint matrix $\left[\lambda_{ij}\right]$ which can be diagonalized through a unitary transformation. The Eq.9 can then be re-written in terms of new orthonormal functions $u_i(q_j)$ which result from the same unitary transformation applied to the basis of the one-electron wavefunctions of Eq.9. In terms of the representation in which $\left[\lambda_{ij}\right]$ is diagonal, Eq.8 then takes the following form:

$$f(\vec{r}1)u_i(\vec{r}_1) +$$
$$\sum_j \left[\int dV_2 \dfrac{u^*_j(\vec{r}2)}{r_{12}} \left(\begin{array}{l} u_i(\vec{r}1)u_j(\vec{r}2) - \\ \delta(m_{s_i},m_{s_j})u_i(\vec{r}2)u_j(\vec{r}1) \end{array} \right) \right] \tag{10}$$
$$= \varepsilon_i\,u_i(\vec{r}_1)$$

where

$$\varepsilon_i = -\lambda_{ii}. \tag{11}$$

Eq.10 is referred to as the HF equation in the diagonal form. There are N such coupled integro-differential equations and these are amenable to numerical solutions.

Within the framework of the frozen orbital approximation, it can be shown that

$$\langle k \mid f \mid k \rangle + \sum_j \left[\langle kj \mid g \mid kj \rangle - \langle kj \mid g \mid jk \rangle \right] = \varepsilon_k \tag{12}$$

and

$$E\left(\psi^{(N)}\right) - E\left(\psi^{(N-1)}\right)_{(n_k = 0)} = \varepsilon_k = -\lambda_{kk}, \tag{13}$$

where, $E\left(\psi^{(N)}\right) = \left\langle \psi^{(N)} \left| H^{(N)} \right| \psi^{(N)} \right\rangle \tag{14}$

and

$$E\left(\psi^{(N-1)}\right)_{n_k = 0} = \left\langle \psi^{(N-1)}_{n_k = 0} \left| H^{(N-1)} \right| \psi^{(N-1)}_{n_k = 0} \right\rangle. \tag{15}$$

It needs to be emphasized that the $N-1$ orbitals referred to in Eq.15 for the $N-1$ electron system are exactly the same as the corresponding orbitals in Eq.14, except for the fact that the kth orbital u_k is now considered to be unoccupied. The result expressed in Eq.13 is a consequence of the frozen orbital approximation. It lends a direct physical interpretation to the Lagrange's variational multipliers $\lambda_{kk} = -\varepsilon_k$, since the left hand side of Eq.13 can be directly interpreted as the ionization energy required to produce a hole in the kth one-electron state, within the framework of the approximation that other orbitals, and their associated charge densities, are considered 'frozen'. The result expressed in Eq.13 is known as an extremely famous theorem, known as the Koopmans theorem [7]. It provides an immediate connection of the HF methodology with experimentally measurable ionization energies.

3. ALGORITHMS FOR HF SCF SCHEME

The atomic HF scheme is readily extended to determine the electronic structure of molecules, clusters, solids etc. For molecules, one develops the SCF procedure for symmetry-adapted Hartree-Fock equations [8-10]. The iterative procedure for a molecular system is indicated by the flowchart appended below. Often, molecular point group symmetry codes as used in conjunction with the HF SCF procedure, since the molecular wavefunctions must have the point group symmetry of the molecular Hamiltonian. The iterative process is continued to improve orbitals until the change in electronic energy falls below a certain threshold and a set of SCF one-electron orbitals are calculated.

As mentioned above, the HF SCF method provides the electronic structure of a many-electron system within the framework of the IPA since the Slater determinantal wavefunction it generates is expressible in terms of a product of single-electron wavefunctions, thereby implying that the many-electron problem is separable in single-particle coordinates. This separation is enabled by arriving at a static average of the electron-electron interaction through the iterative SCF procedure. Nevertheless, unlike the Hartree method which was the precursor to the Hartree-Fock method, the latter does take into account certain correlations that result from the fact that due to the identity of the electrons, it is impossible to separate the dynamics of an electron from another! There is thus a certain amount of 'correlation' that is built into the HF scheme.

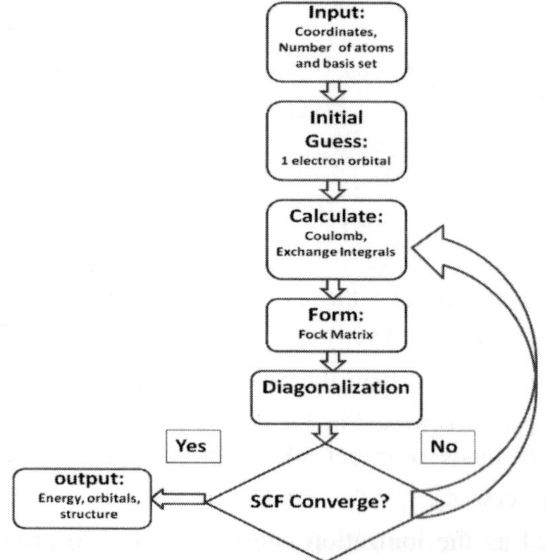

The correlation that is built into this framework is the one that stems from the identity of the electrons: an exchange of one electron with other results in a configuration that is indistinguishable from the former, but it must result in a wavefunction for the pair whose sign is -1 times that of the former. This statement is an expression of the anti-symmetry of the electron wavefunctions, since an electronic system needs to be described by the Fermi-Dirac statistics. The 'EXCHANGE CORRELATIONS' are thus incorporated in the HF-SCF method, and are also equivalently referred to as the Fermi-Dirac (or sometimes simply 'Fermi') correlations, or also as Pauli-correlations since the Pauli exclusion principle is also governed by essentially the same phenomenology. Often, these correlations are also known as 'SPIN CORRELATIONS', since they result from the electron's intrinsic spin angular momentum.

The only superposition of the product of one-electron wavefunctions that are included in the HF method are those that result from the $N!$ permutations of the N identical electrons, which is just what the Slater determinant stands for. It is for this reason that the HF method belongs to the family of the IPA, inspite of the fact that the HF scheme does include the 'spin correlations'.

The two-electron integrals (see the Appendix below) corresponding to the spin/exchange correlations are often very cumborsome to evaluate, and hence some approximations to this term are often employed. The exchange potential that results from the exchange integral is non-local, which makes the Hartree-Fock model gauge-dependent when it is employed to determine atomic properties such as the photoionization cross-section [9]. A number of approximations to the exchange terms have been developed, which trace their origins to the method introduced by John Slater, known as the local density approximation [10,11,12]. These methods do not ignore the Fermi-Dirac statistics, but account for the same only in an approximate manner, by making a 'local' density approximation to the 'non-local' exchange potential.

The many-electron (often called 'many-body') correlations that are not included in the HF scheme are called as the 'COULOMB CORRELATIONS'. The Coulomb correlations are important when the $N!$ permutations that result only from the identity of the N electrons are insufficient to describe the electronic configuration. These result from the fact that an alternative set of N single-particle states may be occupied by the N electrons resulting in a different electronic configuration which is nearly degenerate with the previous one. The net wavefunction of the N-electron system must then be written as a linear superposition of two (or more) Slater determinants. A complete description may well require a superposition of an infinite alterative Slater determinants, each corresponding to a different 'configuration' that spells out the occupancy number (which is 1 or 0 for fermions) of the possible one-electron spin-orbitals. An iterative self-consistent-field can then be generated as before. Such a scheme, that includes superposition of Slater determinants for different configurations is then called as the Multi-Configuration Hartree-Fock method (MCHF) and/or Configuration Interaction (CI) method [13-15]. The MCHF/CI methods take (partial) account of the Coulomb correlations that are left out of the HF formulation.

Unfortunately, there is no formalism that can be developed, even in principle, which can include the Coulomb correlations completely. This is because of the fact that a many-body problem is simply not amenable to exact solutions – "if one is looking for exact solutions, having no body at all is already too many"! The challenge before a many-body theorists therefore is not one of getting exact solutions to a many-body problem, but to procure the best approximations to the same. The MCHF/CI is one of the several approximate methods that have been developed to include the Coulomb correlations in one's anaysis. Other common many-body approximations are the random phase approximation (RPA) [16],

Feynman-Goldstone diagrammatic perturbation theory [17], coupled-cluster methods [18] etc.

The HF, MCHF, RPA and other methods mentioned above are primarily based on the Schrodinger equation. However, the Schrodinger equation is non-relativistic; it is not Lorentz covariant and does not therefore accommodate consequences of the fact that the speed of light is finite, and essentially the same in all inertial frames of references. Relativistic effects are sometimes included, if only partially, in methods based on the Schrodinger equation by using perturbative corrections that model some of the relativistic effects. However, such an approach is often trecherous as it is not easy to take care that all perturbative terms of equal importance are properly incorporated. A better and safer approach is to base the many-electron formalism on Dirac's relativistic equation rather than the Schrodinger equation. Yet again, a self-consisten-field many-electron formalism based on the Dirac equation can be built, anologous to the Hartree-Fock method, and the resulting relativistic scheme is then called as the Dirac-Hartree-Fock (DHF) method [19-21]. Unfortunately, this is often referred to only as just the 'Dirac-Fock' method, which does injustice [21,22] to Hartree, who developed the first SCF idea. Again, just as the HF method, the DHF method includes the spin correlations, but leaves out the Coulomb correlations. Many-body relativistic methods can then be developed to address the Coulomb correlations, such as the Relativistic Multi Configuration Dirac (Hartree) Fock (MCDHF/MCDF) methods [23,24], the relativistic random phase approximation (RRPA) [25], the relativistic coupled cluster (RCC) method [26], etc.

Conclusion

Hartree-Fock method has central importance in atomic and molecular physics, quantum chemistry, and in all studies of electron structure studies in condensed matter physics, including band structure calculations. It is an approximate method derived through variational approach; its solutions provide the basic frame work for systematic improvement of accuracy through enlarging the variational space by considering linear combination of determinants: Configuration interaction (CI) methods, Multi-configuration Self Consistent Field (MCSCF) methods. Through the Koopmans theorem, the Lagrange's variational parameters introduced in the scheme attain significant physical measurable attributes which connect the HF methodology to experimental observables. It is superfluous to add that all collision dynamics involving matter-probe interactions require an accurate description of the target, no matter what the probe is: electromagnetic radiation and/or elementary and/or composite particles. Most theoretical/computational studies of this kind therefore begin with the HF methodology, or some approximation to it, or some improvisation to it, which keep the HF method at the center stage.

Appendix A

A few useful mathematical expressions are provided in this Appendix which will be useful in filling in the steps that have been omitted in the formulation of the HF SCF method described in the main text of this article.

A.1   For every operator Ω which is symmetric with respect to identical electrons,

$$\langle \psi^{(N)} | \Omega | \psi^{(N)} \rangle =$$

$$= \frac{1}{\sqrt{N!}} P \sum_{P=1}^{N!} (-1)^P \int ... \int dq_1..dq_N \left[P^{-1} \psi^*(q_1,..,q_N) \right] \Omega \{ u_{\alpha_1}(q_1)...u_{\alpha_N}(q_N) \}$$

$$= \frac{1}{\sqrt{N!}} P \sum_{P=1}^{N!} (-1)^P \int ... \int dq_1..dq_N \left[(-1)^P \psi^*(q_1,..,q_N) \right] \Omega \{ u_{\alpha_1'}...u_{\alpha_N'}(q_N) \}$$

$$= \frac{1}{\sqrt{N!}} P \sum_{P=1}^{N!} (-1)^{2P} \int ... \int dq_1..dq_N \psi^*(q_1,..,q_N) \Omega \{ u_{\alpha_1}(q_1)...u_{\alpha_N}(q_N) \}$$

$$= \frac{1}{\sqrt{N!}} P \sum_{P=1}^{N!} \int ... \int dq_1..dq_N \psi^*(q_1,..,q_N) \Omega \{ u_{\alpha_1}(q_1)...u_{\alpha_N}(q_N) \}$$

A2. The average/expectation value of the one-electron part of the N-electron Hamiltonian in the N-electron Slater determinant of Eq.1 is given by:

$$\langle \psi^{(N)} | H_1 | \psi^{(N)} \rangle = \sum_{i=1}^{N} \int dq u^*_{\alpha_i}(q) f(q) u_{\alpha_i}(q)$$

$$= \sum_{i=1}^{N} \int dV u^*_{\alpha_i}(\vec{r}) f(\vec{r}) u_{\alpha_i}(\vec{r}) \sum_{\zeta} \langle \zeta | \alpha_i \rangle^* \langle \zeta | \alpha_i \rangle$$

$$= \sum_{i=1}^{N} \langle \alpha_i | f | \alpha_i \rangle = \sum_{i=1}^{N} \langle i | f | i \rangle,$$

since

$$\sum_{\zeta} \langle \zeta | \alpha_i \rangle^* \langle \zeta | \alpha_i \rangle = \sum_{\zeta} \langle \alpha_i | \zeta \rangle \langle \zeta | \alpha_i \rangle = 1.$$

A.3 The Coulomb integral is given by:

$$\int \int dq_1 dq_2 u^*_{\alpha_i}(q_1) u^*_{\alpha_j}(q_2) \frac{1}{r_{12}} u_{\alpha_i}(q_1) u_{\alpha_j}(q_2)$$

$$= \int \int dV_1 dV_2 u^*_{\alpha_i}(\vec{r}_1) u^*_{\alpha_j}(\vec{r}_2) \frac{1}{r_{12}} u_{\alpha_i}(\vec{r}_1) u_{\alpha_j}(\vec{r}_2) \times$$

$$\times \sum_{\zeta_1} \sum_{\zeta_2} \langle \zeta_1 | m_{s_i} \rangle^* \langle \zeta_2 | m_{s_j} \rangle^* \langle \zeta_1 | m_{s_i} \rangle \langle \zeta_2 | m_{s_j} \rangle$$

$$= \int \int dV_1 dV_2 u^*_{\alpha_i}(\vec{r}_1) u^*_{\alpha_j}(\vec{r}_2) \frac{1}{r_{12}} u_{\alpha_i}(\vec{r}_1) u_{\alpha_j}(\vec{r}_2)$$

A4. The Exchange integral is given by:

$$\int\int dq_i dq_j u^*_{\alpha_i}(q_j)u^*_{\alpha_j}(q_i)\frac{1}{r_{ij}}u_{\alpha_i}(q_i)u_{\alpha_j}(q_j) =$$

$$= \int\int dq_1 dq_2 u^*_{\alpha_i}(q_2)u^*_{\alpha_j}(q_1)\frac{1}{r_{12}}u_{\alpha_i}(q_1)u_{\alpha_j}(q_2)$$

$$= \int\int dV_1 dV_2 u^*_{\alpha_i}(\vec{r}_2)u^*_{\alpha_j}(\vec{r}_1)\frac{1}{r_{12}}u_{\alpha_i}(\vec{r}_1)u_{\alpha_j}(\vec{r}_2) \ \times$$

$$\times \sum_{\zeta_1}\sum_{\zeta_2}\langle\zeta_2|m_{s_i}\rangle^*\langle\zeta_1|m_{s_j}\rangle^*\langle\zeta_1|m_{s_i}\rangle\langle\zeta_2|m_{s_j}\rangle$$

$$= \int\int dV_1 dV_2 u^*_{\alpha_i}(\vec{r}_2)u^*_{\alpha_j}(\vec{r}_1)\frac{1}{r_{12}}u_{\alpha_i}(\vec{r}_1)u_{\alpha_j}(\vec{r}_2)\ \delta(m_{s_i},m_{s_j})$$

A.5 The average/expectation value of the two-electron part of the N-electron Hamiltonian in the N-electron Slater determinant of Eq.1 is given by:

$$\langle\psi^{(N)}|H_2|\psi^{(N)}\rangle =$$

$$\sqrt{N!}\int..\int dq_1.dq_N\psi^*(q_1,..,q_N)\left[\frac{1}{2}\sum_{j=1,j\neq i}^{N}\sum_{i=1}^{N}\frac{1}{r_{ij}}\right]\{u_{\alpha_1}(q_1)..u_{\alpha_N}(q_N)\}$$

$$= \frac{1}{2}\sum_{j=1,j\neq i}^{N}\sum_{i=1}^{N}\left[\begin{array}{c}\int\int dq_i dq_j u^*_{\alpha_i}(q_i)u^*_{\alpha_j}(q_j)\frac{1}{r_{ij}}u_{\alpha_i}(q_i)u_{\alpha_j}(q_j) \\[2ex] -\int\int dq_i dq_j u^*_{\alpha_i}(q_j)u^*_{\alpha_j}(q_i)\frac{1}{r_{ij}}u_{\alpha_i}(q_i)u_{\alpha_j}(q_j)\end{array}\right]$$

$$= \frac{1}{2}\sum_{j=1}^{N}\sum_{i=1}^{N}\left[\langle ij|v|ij\rangle - \langle ij|v|ji\rangle\right]$$

A6. The variation in the average/expectation value of the N-electron Hamiltonian in a Slater determinantal wavefunction is give by:

$$\delta\langle\psi^{(N)}|H|\psi^{(N)}\rangle =$$

$$\left[\begin{array}{l}\int dV_1\{\delta u^*_k(\vec{r}_1)\}f(\vec{r}_1)u_k(\vec{r}_1) + \\[3ex] +\sum_j\left[\begin{array}{l}\int\int dV_1 dV_2\{\delta u^*_k(\vec{r}_1)\}u^*_j(\vec{r}_2)\frac{1}{r_{12}}u_k(\vec{r}_1)u_j(\vec{r}_2) \\[2ex] -\delta(m_{s_k},m_{s_j})\int\int dV_1 dV_2\{\delta u^*_k(\vec{r}_2)\}u^*_j(\vec{r}_1)\frac{1}{r_{12}}u_k(\vec{r}_1)u_j(\vec{r}_2)\end{array}\right] \\[3ex] + \ \int dV_1 u^*_k(\vec{r}_1)f(\vec{r}_1)\{\delta u_k(\vec{r}_1)\} + \\[3ex] + \ \sum_j\left[\begin{array}{l}\int\int dV_1 dV_2 u^*_k(\vec{r}_1)u^*_j(\vec{r}_2)\frac{1}{r_{12}}\{\delta u_k(\vec{r}_1)\}u_j(\vec{r}_2) \\[2ex] -\delta(m_{s_k},m_{s_j})\int\int dV_1 dV_2 u^*_k(\vec{r}_2)u^*_j(\vec{r}_1)\frac{1}{r_{12}}\{\delta u_k(\vec{r}_1)\}u_j(\vec{r}_2)\end{array}\right]\end{array}\right]$$

$$\delta\langle\psi^{(N)}\,|\,H\,|\,\psi^{(N)}\rangle =$$

$$\left[\int dV_1\left\{\delta u^*_{k}(\vec{r_1})\right\}\left(f(\vec{r_1})u_k(\vec{r_1})+\sum_j\left[\int dV_2\,\frac{u^*_{j}(\vec{r_2})}{r_{12}}\left(u_k(\vec{r_1})u_j(\vec{r_2})-\delta(m_{s_k},m_{s_j})u_k(\vec{r_2})u_j(\vec{r_1})\right)\right]\right)+\int dV_1\left\{\delta u_k(\vec{r_1})\right\}(\;)^*\right]$$

A7. Since $f\left(\vec{r_1}\right)$ is a Hermitian operator,

$$\int dV_1\,u^*_k\left(\vec{r_1}\right)f\left(\vec{r_1}\right)\left\{\delta u_k\left(\vec{r_1}\right)\right\}=$$

$$=\int dV_1\left\{\delta u_k\left(\vec{r_1}\right)\right\}\left\{f\left(\vec{r_1}\right)u^*_k\left(\vec{r_1}\right)\right\}$$

References

[1]   B.H.Bransden,C.J.Joachain,"Physics of Atoms and Molecules", Prentice Hall 2003

[2]   Hans.A   Bethe,   Roman   W   Jackiew,"Intermediate   Quantum mechanics",Addison Wesley,1997

[3]   D.R Hartee,"Proc.R.Soc", A141-282, 1933 ibid A143-506, 1933;
      D.R.Hartee, "Calculation of Atomic Structure, Wiley, 1952

[4]   J. C. Slater, "Quantum Theory of Atomic Structure" McGraw-Hill, 1960

[5]   Feynman Lectures, Vol-3, page: 4-3

[6]   G. B. Arfken, H. J. Weber "Mathematical methods for Physicists", Elsevier Academic Press, sixth edition (2005)

[7]   T. A. Koopmans, Physica 1:104,(1933)

[8]   Carl Trindle, Donald Shillady, "Electronic Structure Modeling", CRC Press 2008

[9]   S. Chandrasekhar, 'On the Continuous Absorption Coefficient of the Negative Hydrogen Ion' in Astrophysics Journal, Vol. 102, Page 223 (1945); see also the article 'Length and Velocity Formulas in Approximate Oscillator Strength Calculations' by Anthony F Starace in Physical Review A, Vol. 3 No.4, Page 1242.

[10] J. C. Slater, Phys.Rev.34: 1293,(1929); Adv. In Quant. Chem. S 5, 411 (1972)

[11] I Lindgren and K. Schwartz, Phys. Rev. A 5, 542 (1972)

[12] W. Kohn and L. J. Sham, "Self-consistent Equation Including Exchange and Coulomb affects", Phys. Rev 140,A1133, (1965).

[13] R.McWeeny,B. T. Sutcliffe, "Fundamentals of self consistent filed(SCF), Hartee-Fock (HF), Multiconfiguration (MC) SCF and Configuration Interaction (CI) Schemes", Computer Physics Reports 2 (1984) 217-278.

[14] Ira N. Levine,"Quantum Chemistry", Fifth edition, Prentice Hall, New Jersey 2000

[15]   C.F Fischer, T. Brage, P Joensson "Computational Atomic Structure, An MCHF Approach", IOP Publishing.

[16]   David Bohm and David Pines," A collective Description of Electron Interaction: III, coulomb Interaction in a degenerate Electron Gas", Phys.Rev 92,609-625, (1953).

[17]   A.L.Felter and J.D Walecka," Quantum Theory of many-particle systems", McGraw-Hill Book Company, New York

[18]   I.Lindgren,J.Morrison, "Atomic many-body theory", Springer-Verlag, Second edition.

[19]   F.C Smith and W.R Johnson, "Relativistic Self-consistent Fields with Exchange", Phys.Rev 160, (1967).

[20]   J.P.Desclaux, Comput. Phys. Commun.9, 31; (1975)

[21]   I.P.Grant, Adv.Phys.19,747;1975

[22]   Vojilav Radojevic (2011, private communication)

[23]   I.P.Grant,B.J.Mckenzie,P.H Norrnigton.D.F Meyers and N.C.Pyper, "An atomic multi-configuration Dirac-Fock package, Comput. Phys. Commun,2i, 207;1980

[24]   I.P.Grant, "Relativistic Quantum Theory of Atoms and Molecules", Springer, ISBN-10: 0-387-34671-6,2007

[25]   W.R. Johnson and C.D Lin," Multichannel Relativistic Random-phase Approximation for the Photoionization of Atoms", Phys, Rev A 20,964 (1974)

[26]   I.Lindgren, "Relativistic Atomic many-body theory", Springer, ISBN 978-1-4419-8308-4

Symmetry in Electron-Atom Collisions and Photoionization Process

Pranawa C. Deshmukh[1*], Dilip Angom[2], and Alak Banik[3]

[1]*Indian Institute of Technology Madras, Chennai;*
[2]*Physical Research Laboratory, Ahmadabad;*
[3]*Space Applications Centre, Ahmadabad.*
email: pcd@physics.iitm.ac.in[*]

1. INTRODUCTION

Physical processes are governed by conservation laws, which in turn are connected intimately to symmetry processes. For example, many conservation laws are obtained from physical laws, but often the physical laws themselves are obtainable from symmetry considerations that govern them [1,2]. The connections between symmetry and conservation laws have far reaching implications in physics that impact our understanding of the laws of nature. This subject is both vast and deep, but the purport of the present article is limited; it is

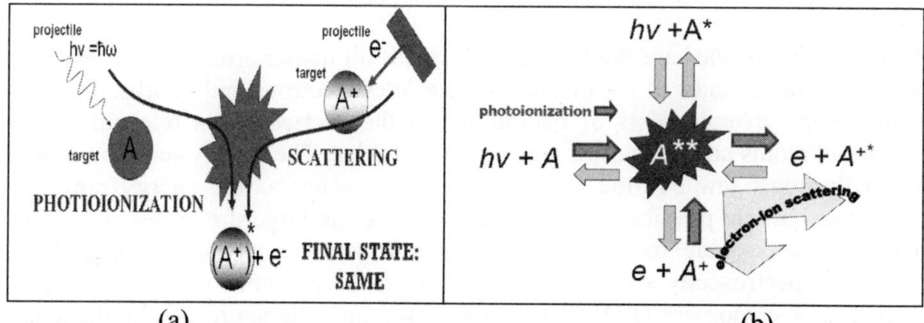

Fig.1. The final state of two reactions whose initial state ingredients are quite different is the same for the two processes: (a) atomic photoionization and (b) electron-ion collision.

only aimed at providing an introduction to mathematical connections based on the quantum mechanical time reversal symmetry which relate solutions of the quantum collision problem for an electron impinging on an atomic ion to those of atomic photoionization process in which an atom absorbs electromagnetic radiation resulting in an atomic electron get knocked out of the atom and escape as a 'free' electron. This is illustrated in Fig.1.

It is thus natural to expect that the quantum mechanical description of the two processes, (a) atomic photoionization and (b) electron-ion collision, must be related. Indeed it is, but the relationship is subtle. The difficulty in relating the two processes comes from the fact that even if the final state of the two processes is the same, the initial states are different: even the ingredients of the initial states are different for the two processes! By simply running the process backward in time, one cannot recover the original ingredients because of the multiplicity of channels in which the central complex can decay! The temporal evolution of the process does not regenerate the history of the process simply by letting the time t go to $-t$ in the equation of motion! In this respect, it is necessary to understand the difference in the role time-reversal symmetry plays in quantum mechanics as opposed to classical mechanics. In classical mechanics, the equation of motion contains either the second order differential operator with

respect to time, namely the operator $\dfrac{d^2}{dt^2}$, in Newton's equation $\vec{F} = m\dfrac{d^2\vec{r}}{dt^2}$, or

two first order differential operators in time, $\dfrac{d}{dt}$, as in Hamilton's equations

$\dot{q} = \dfrac{\partial H}{\partial p}$ and $\dot{p} = -\dfrac{\partial H}{\partial q}$. The classical equations of motion are symmetric with

respect to the transformation $t \rightarrow -t$. This result is of course independent of the formulation, whether Newtonian, Lagrangian, or Hamiltonian as long as there are no unspecified degrees of freedom that lead to dissipation.

Atoms and molecules are the building blocks of all matter around us. These are, however, bound states of the elementary particles electrons and quarks. There is a large gap, several orders of magnitude, in the energy scales relevant to the physics of atoms and physics of elementary particles. The energy scales in atoms are at the most few eV, the excitation energies, whereas the energy range of interest in particle physics is TeV (10^{12}). Despite the large difference in energy scales, it is possible to probe phenomena in particle physics through high precision spectroscopy of atoms. One remarkable example is the permanent electric dipole moment (EDM). It is the observable signature of simultaneous violations of parity and time-reversal symmetries. Among the two, a proper understanding of time-reversal violation is of paramount importance to resolve the preponderance of matter in the Universe. Another observable of equal importance arises from the parity violation, which modifies the selection rules of radiative transitions in atoms.

2. COLLISIONS: DESCRIPTION IN TERMS OF OUTGOING WAVE BOUNDARY CONDITION

In the context of the relationship between (i) the solution to the electron-ion collision process and (ii) photoionization of a neutral atom, shown in Fig.1, we shall first briefly review the well-known solution to the Schrodinger equation formulation of the scattering problem given by [3]:

$$\psi_{\vec{k}_i}(\vec{r}; r \to \infty) \to A\left[e^{i\vec{k}\Box\vec{r}} + \frac{f(\hat{\Omega})}{r} e^{ikr} \right], \tag{1}$$

in which $f(\hat{\Omega})$ is the well-known scattering amplitude.

The total wavefunction is then given by:

$$\psi_{Total}(\vec{r}) \xrightarrow{r \to \infty} \sum_l c_l\, i^l\, (2l+1)\, P_l(\cos\theta)\, \frac{\sin(kr - \frac{l\pi}{2} + \delta_l)}{kr} \tag{2}$$

in which δ_l is the phase-shift caused by the scattering potential, and the 'normalization' constant c_l must be chosen 'appropriately' i.e., as per the 'boundary conditions'. We restrict ourselves to central field potentials for which the current formulation is applicable.

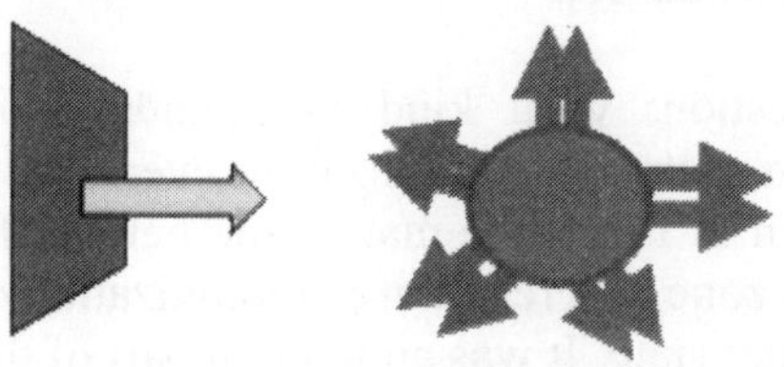

Fig.2: Pictorial depiction of an electron-ion collision process

In the collision problem described pictorially in Fig.2, a mono-energetic beam of electrons is incident from the left along the Z-axis of a Cartesian coordinate system on a scattering central field atomic potential. The electron flux scattered by the target cannot have any spherical ingoing wave, and this requirement fixes the coefficient $c_l = e^{+i\delta_l}$, which gives the following scattered wave solution:

$$\psi_{scattered}(\vec{r};\, r \to \infty) \to \frac{e^{ikr}}{r} \sum_l (2l+1)\, P_l(\cos\theta) \left[\frac{e^{i2\delta_l} - 1}{2ik} \right]. \tag{3}$$

The scattering amplitude of Eq.1 is then given by:

$$f(\hat{\Omega}) = \sum_l (2l+1)\, P_l(\cos\theta) \left[\frac{e^{2i\delta_l}-1}{2ik}\right], \tag{4}$$

and the differential scattering cross-section is given by:

$$\frac{d\sigma}{d\Omega} = \left|f(\hat{\Omega})\right|^2 \tag{5}$$

We note that the time-dependence of the wavefunction is given by:

$$e^{-i\frac{E}{\hbar}t} = e^{-i\omega t} \tag{6}$$

and accordingly the time-dependent solution to the scattering problem is given by:

$$\psi_T^+(\vec{r},t)\xrightarrow[r\to\infty]{} e^{i(kz-\omega t)} + \frac{e^{i(kr-\omega t)}}{r}\sum_l (2l+1)\, P_l(\cos\theta)\left[\frac{e^{2i\delta_l}-1}{2ik}\right] \tag{7}$$

We observe from the solution given in Eq.7 that the two terms on the right hand side provide the asymptotic description of the incident plane wave and a scattered outgoing spherical wave shown in Fig.2. To highlight the fact that the solution is based on the outgoing spherical wave in the final total solution, a superscript ' + ' is placed on the symbol for the wavefunction on the left hand side of Eq.7. The boundary condition that has been used is based on cancellation of all spherical ingoing waves in the scattered solution and is referred to as outgoing wave boundary condition.

We now raise the question: what kind of boundary conditions should be employed to describe an atomic photoionization event, as opposed to electron-ion scattering? As shown in Fig.1, the main issue here is that the photoelectron that escapes the reaction zone as a result of photoionization did not really exist as a free electron in the initial state. It was an integral part of the neutral atom in the 'nucleus + electron(s)' bound system. Following the description in Reference [4,5] which is both the inspiration and the primary source for this article, we shall first discuss the one-dimensional analogue that would relate the collision dynamics to photoionization.

We consider a collision process in which the electron is incident from the left and impinges on a one-dimensional scattering center as shown in Fig.3. The reflection and transmission coefficients can be determined readily by employing the equation of continuity for the conservation of electron charge density flux.
The experiment we envisage has an electron incident on the reaction zone from the left, along the X-axis. It is thus clear that the boundary condition is set by requiring $G = 0$ (see Fig.3).This boundary

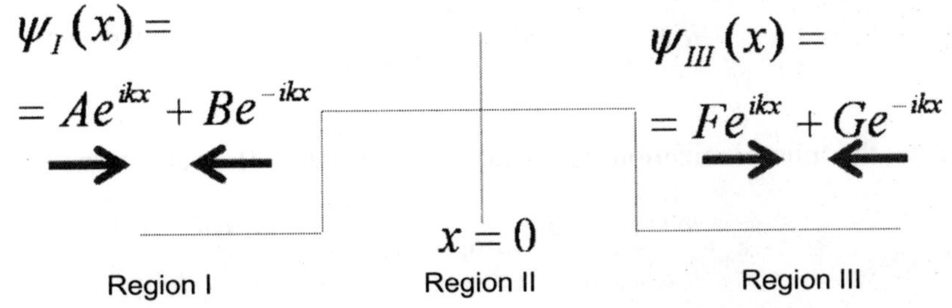

$$\psi_I(x) = $$
$$= Ae^{ikx} + Be^{-ikx}$$

$$\psi_{III}(x) = $$
$$= Fe^{ikx} + Ge^{-ikx}$$

Fig.3: Relationship between one-dimensional electron-ion scattering/collision process and photoionization. Collision of an electron incident from the left is described by the boundary condition G=0, while photoionization resulting in the photoelectron escaping to the left is described by F=0.

condition determines the collision experiment in which the entrance channel is unique; it has an electron incident from the left. In a (one-dimensional) photoionization experiment in which we envisage the photoelectron to escape to the left, the boundary condition is then represented by the choice $F = 0$ (Fig.3). It is the exit channel which is unique in this case, represented by the photoelectron flux escaping to the left as a result of photoionization. We observe that the relationship of collision to photoionization is thus in some sense one of 'motion reversal', except that this is not merely the time-reversal of classical mechanics in which the equations of motion are symmetric under the transformation $t \rightarrow -t$.

3. PHOTOIONIZATION: DESCRIPTION IN TERMS OF INGOING WAVE BOUNDARY CONDITION

To understand the relationship between solutions to the collision problem with those of photoionization, we re-write the traveling wave solutions of the one-dimensional problem in terms of a new base pair: $\{u_+(x), u_-(x)\}$, defined as follows:

$$x \leq -a \text{ (region I): } u_+(x) = N_E \cos(kx - \delta_+); \quad u_-(x) = -N_E \cos(kx - \delta_-) \qquad (8)$$

and

$$x \geq a \text{ (region III); } u_+(x) = N_E \cos(kx + \delta_+); \quad u_-(x) = N_E \cos(kx + \delta_-) \qquad (9)$$

In terms of the new base pair, the traveling wave solutions shown in Fig.3 become:

$$u^{(I)}(x) = \frac{1}{2} N_E \left[(c_+ e^{-i\delta_+} - c_- e^{-i\delta_-}) e^{ikx} + (c_+ e^{i\delta_+} - c_- e^{i\delta_-}) e^{-ikx} \right] \qquad (10)$$

$$u^{(III)}(x) = \frac{1}{2} N_E \left[(c_+ e^{i\delta_+} + c_- e^{i\delta_-}) e^{ikx} + (c_+ e^{-i\delta_+} + c_- e^{-i\delta_-}) e^{-ikx} \right] \qquad (11)$$

The 'collision' boundary condition $G = 0$ is then expressed as:

$$(c_+ e^{-i\delta+} + c_- e^{-i\delta-}) = 0; \quad i.e. \quad \frac{c_+}{c_-} = -e^{+i(\delta_+ -\delta_-)} \tag{12}$$

Likewise, the 'photoionization' boundary condition $F = 0$ is expressed as:

$$c_+ e^{i\delta+} + c_- e^{i\delta-} = 0; \quad i.e. \quad \frac{c_+}{c_-} = -e^{-i(\delta_+ -\delta_-)} \tag{13}$$

We observe the complex-conjugation of $\dfrac{c_+}{c_-}$ in the description of the photoionization boundary condition (Eq.13) in relation to the collision boundary condition (Eq.12). This complex conjugation is characteristic of 'motion reversal' in quantum mechanics, usually referred to as 'time reversal'. Using a quantum mechanical operator Θ for 'time/motion reversal', one can depict, as in Fig.4, the photoionization process as time/motion reversed electron-ion collision. Photoionization is referred to in the literature often as 'half-scattering' on account of this relation. The term 'motion reversal' was preferred by Wigner, since the relationship involves complex-conjugation of the wavefunction in addition to $t \rightarrow -t$ under the operator Θ which nevertheless is most often referred to as the 'Time-reversal Operator'. Time-reversal is a discrete symmetry, just like parity and charge-conjugation [1]. It is effected through an operator generally denoted by Θ which is an anti-unitary operator. The (anti-unitary) time-reversal operator is certainly not the inverse of the (unitary) time-evolution operator in quantum mechanics; it has the following commutation/anti-commutation properties with respect to the position, momentum and angular momentum operators:

$$\begin{aligned}
[\vec{r},\Theta]_- &= 0 \quad : \quad commute \\
[\vec{p},\Theta]_+ &= 0 \quad : \quad anticommute \\
[\vec{J},\Theta]_+ &= 0 \quad : \quad anticommute
\end{aligned} \tag{14}$$

It is important to understand the difference between the implications of time-reversal in quantum mechanics as opposed to what it is in classical mechanics [7]. As mentioned above, the classical equations of motion are symmetric under time-reversal. From the equation of motion, one can thus predict the future, and also determine the past of the mechanical state of the system. The meaning in quantum mechanics is however different. Suppose it is known that the system at time t is in the state $|\psi(t)\rangle$. Then, the system is said to be in the time-reversed state $|\psi_R(t)\rangle$ under the time-reversal transformation, if the transformation ensures that the probability of finding the system in a state $|\phi_R(t)\rangle$ is equal to finding it at time $-t$ in the state $|\phi(t)\rangle$

We recall that Eq.7 gave us the scattering/collision solution subject to the so-called 'OUTGOING WAVE BOUNDARY CONDITION'. We must now write the total wavefunction given in Eq.2 with a different set of boundary conditions as would be appropriate for the photoionization process.

Now,

$$\psi_{Tot}(\vec{r}) \xrightarrow[r \to \infty]{} \sum_l c_l \, i^l \, (2l+1) \, P_l(\cos\theta) \, \frac{\sin(kr - \frac{l\pi}{2} + \delta_l)}{kr}$$

$$\psi_{Tot}(\vec{r}) \xrightarrow[r \to \infty]{} \frac{1}{2ikr} \sum_l c_l \, (2l+1) \left[P_l(\cos\theta) e^{+i(\delta_l + kr)} - P_l(-\cos\theta) \, e^{-i(\delta_l + kr)} \right]$$

$$(15)$$

and the incident wavefunction is:

$$\psi_{inc}(\vec{r}) \to \frac{1}{2ikr} \sum_l (2l+1) \, P_l(\cos\theta) \left[e^{ikr} - (-1)^l e^{-ikr} \right].$$

$$(16)$$

Thus,

$$\psi_{Scattered}(\vec{r}) = \psi_{Tot}(\vec{r}) - \psi_{incident}(\vec{r})$$

$$\psi_{Scattered}(\vec{r}) \to$$

$$\to \frac{1}{2ikr} \sum_l (2l+1) \, P_l(\cos\theta) \left(c_l e^{i\delta_l} - 1 \right) e^{ikr}$$

$$- \frac{1}{2ikr} \sum_l (2l+1) \, P_l(-\cos\theta) \left(c_l e^{-i\delta_l} - 1 \right) e^{-ikr}$$

$$(17)$$

It is clear that the choice $c_l = e^{-i\delta_l}$ cancels the outgoing spherical waves in the scattered solution as must happen in the photoionization event whose exit channel is unique which represents the escaping photoelectron plane wave, receding away from the reaction zone. This choice of the coefficient is, not surprisingly, the complex conjugate of that employed to describe the collision process.

The total wavefunction, now written with a superscript '-', with the new choice $c_l = e^{-i\delta_l}$, and inclusive of the time-dependence then is:

$$\psi_T^-(\vec{r}, t; r \to \infty) \to e^{i(kz - \omega t)}$$

$$- \frac{e^{-i(kr - \omega t)}}{r} \sum_l (2l+1) \, P_l(-\cos\theta) \left[\frac{e^{-2i\delta_l} - 1}{2ik} \right]$$

$$(18a)$$

On complex-conjugation, it becomes:

$$\psi_T^{-}(\vec{r},t;r\to\infty)^{*} \to e^{-i(kz-\omega t)} + \frac{e^{+i(kr-\omega t)}}{r}\sum_{l}(2l+1)\,P_l(-\cos\theta)\left[\frac{e^{2i\delta_l}-1}{2ik}\right] \qquad (18b)$$

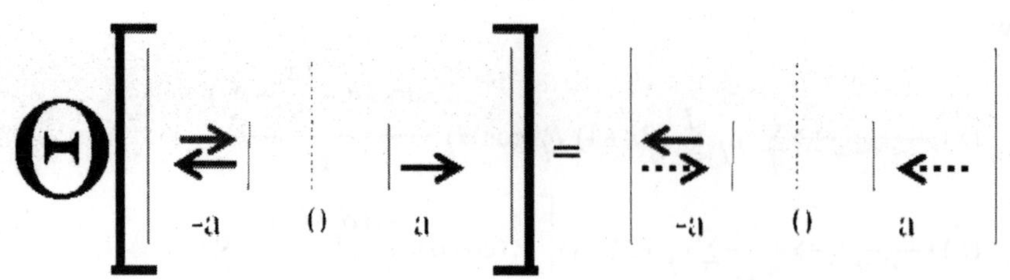

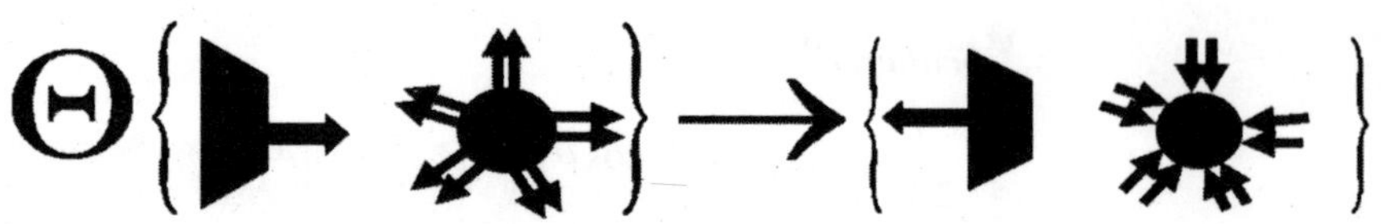

Fig.4a and Fig.4b: Shown in the above two panels respectively is the one-dimensional and three-dimensional pictorial depiction of the relationship between atomic photoionization and electron-ion collision process effected by the 'time/motion reversal operator.

Further, by letting $t \to -t$, we get:

$$\psi_T^{-}(\vec{r},-t;r\to\infty)^{*} \to e^{-i(kz+\omega t)} + \frac{e^{+i(kr+\omega t)}}{r}\sum_{l}(2l+1)\,P_l(-\cos\theta)\left[\frac{e^{2i\delta_l}-1}{2ik}\right].$$

$$(18c)$$

Note that the surface of constant phase of the wave represented by the first term must have:

$$d\left(kz+\omega t\right) = 0 \ \ i.e. \ \ kdz + \omega dt = 0,$$

which gives:

$$\frac{dz}{dt} = -\frac{\omega}{k}$$

The fact that $\dfrac{dz}{dt}$ is intrinsically negative implies that the surface of constant phase this term represents is a plane wave moving toward $z \to -\infty$ from the reaction zone. It represents the plane wave moving toward the left in Fig.4b showing the photoelectron's escape along the unit exit channel.

Likewise, the surface of constant phase of the wave represented by the second term must have:

$$d(kr + \omega t) = 0; \quad i.e.\ kdr + \omega dt = 0 \qquad \text{which gives}$$

$$\frac{dr}{dt} = -\frac{\omega}{k}$$

$\frac{dr}{dt}$ being intrinsically negative, it corresponds to a spherical wavefront of diminishing radius representing the spherical ingoing wave shown in Fig.4b.

We see that the choice $c_l = e^{-i\delta_l}$ has provided us the correct boundary condition on the total wavefunction appropriate for the description of the photoionization process.

The scattered solution for the photoionization process is thus given by Eq.17, with $c_l = e^{-i\delta_l}$, which is referred to as INGOING WAVE BOUNDARY CONDITION since this choice cancels the outgoing spherical waves in the scattered solution. Hence,

$$\psi_{Scattered}(\vec{r}) \to -\frac{e^{-ikr}}{r} \sum_l (2l+1)\, P_l(-\cos\theta) \left(\frac{e^{-i2\delta_l}-1}{2ik}\right), \qquad (19)$$

and the final state of the photoelectron then becomes:

$$|\psi_f^-\rangle \to e^{ikz} - \frac{e^{-ikr}}{r} \sum_l (2l+1)\, P_l(-\cos\theta) \left(\frac{e^{-i2\delta_l}-1}{2ik}\right). \qquad (20)$$

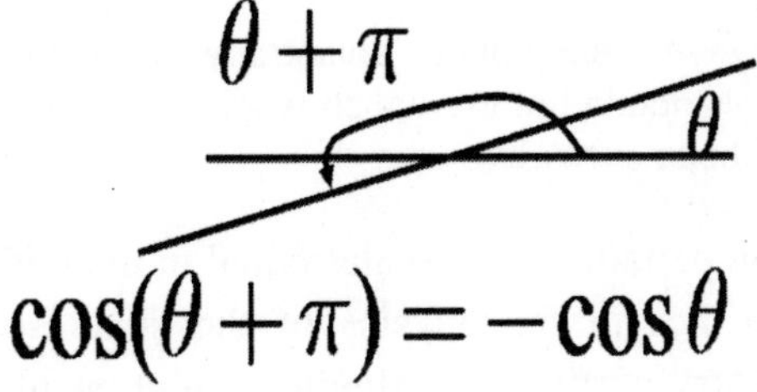

$$\cos(\theta + \pi) = -\cos\theta$$

Fig.5: Geometry showing the reversal of motion corresponding to Fig.4

To determine the photoionization transition matrix element $\langle \psi_f | T | \psi_i \rangle$ for transition from an initial state $|\psi_i\rangle$ to a final continuum state $|\psi_f\rangle$, one must therefore employ the final state with INGOING WAVE BOUNDARY CONDITION expressed in Eq.20. The form given in Eq.20, not the one given in Eq.3, must therefore be used to determine the angular distribution of the photoelectrons, as for example in the famous Cooper-Zare formula [8]. The two forms differ in respect of the outgoing/ingoing wave boundary conditions.

4. DISCRETE SYMMETRY VIOLATIONS

We have, so far, examined the importance of symmetry in the collision calculations and in particular, we have utilized the invariance of dynamics under time reversal. The other equally important facet of the same symmetry is the time reversal violation. The related phenomena and associated observables can probe fundamental problems in physics. One of the unresolved riddles of nature is the dearth of antimatter in the Universe. All observable signatures points to Big Bang as the event which created the Universe [9]. One important outcome is then the creation of matter and antimatter in equal amounts. However, till date the astronomical observations up to the edge of the Universe have detected only matter. This begs an explanation of how and where have all the antimatter vanished?

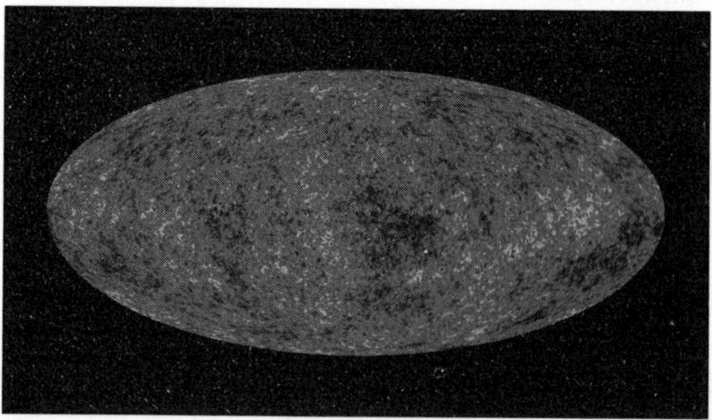

Fig.6: Cosmic microwave background anisotropy from the WMAP data, it is represents the matter distribution at the epoch when matter and radiation decoupled. (Image from WMAP website of NASA).

Although, the question pertains to the scales equal in magnitude to the size of the Universe, the answer lies in the physics of the smallest constituents of the Universe; the elementary particles. A simple resolution of the puzzle is, there must be physical process or processes which convert antimatter into matter. The necessary condition for this is the violation of time reversal symmetry. This, however, leads to another question: how to detect time reversal violation? The answer, it turns out, is to detect an observable which could arise from time reversal violation. Following symmetry conditions, the all familiar EDM is the observable. EDM, the ones we are familiar with, are induced by an external field or arise from degeneracy of opposite parity states. However, a permanent EDM in a non-degenerate quantum system is a signature of time reversal and parity violation. Of the two the former is of overriding importance as the latter, which we shall dwell upon subsequently, is well established in weak interactions.

To prove permanent EDM violates parity and time reversal; consider a particle or a composite non-degenerate quantum system has a permanent EDM d.

Under parity transformation, $PdP^{-1} = -d$.

that is, d is odd under parity transformation as it is a vector observable. Following projection theorem in quantum mechanics, the experimentally observable EDM of a system is the component along an internal vector quantity. It is the spin s for an elementary particle like electron or the total angular momentum J for a composite quantum system like atoms, then we can write $d = cJ$, where, c is a constant. Since the angular momentum is $r \times p$, there is time dependence through the momentum p, under time reversal transformation J is transformed to $-J$. The EDM transform under time reversal as

$$\Theta d\Theta^{-1} = \Theta cJ\Theta^{-1} = -cJ.$$

Again, like in parity transformation; the EDM is odd under time reversal transformation. These transformations are schematically represented in Fig.7.

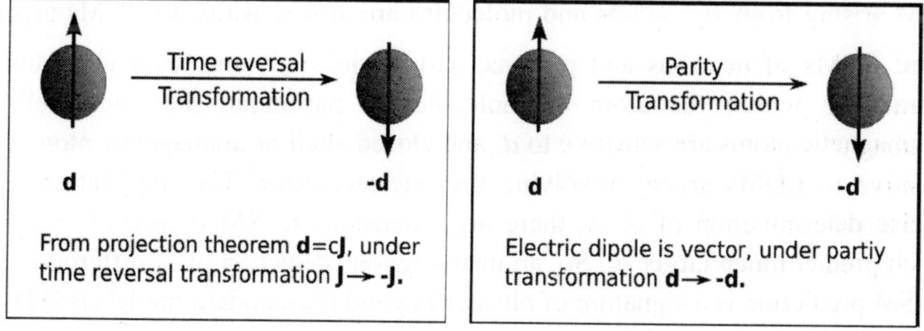

Fig.7: Schematic diagram of parity and time reversal transformation of EDM. The arrow represents the direction of the angular momentum and EDM.

It turns out that the standard model (SM) of particle physics, the most successful and well accepted theory of elementary particle physics, does predict a non-zero EDM of electron. And, the value is $|d_e| < 2.9 \times 10^{-38} e$ cm.

This is an extremely small value. Perhaps this is surprising as one tends to think of electrons as point particles and not associated with a charge distribution. This is not a precise description, in the proper quantum description of an electron; a cloud of virtual particles surrounds it (vacuum polarization). An asymmetry in the distribution of the virtual charges is the origin of electron EDM. To measure d_e one must apply an external electric field E and observe the energy shift arising from the interaction $-\mathbf{d}_e \bullet \mathbf{E}$. This interaction, like Larmor precession, causes precession about E. But, it is an impossible task as the electron accelerates away in presence of E. It is, however, possible to measure the EDM of neutral particle like neutron. EDM of neutron, surprising, isn't it? Not really, it is a bound state of quarks (two down quarks and one up quark) which are charged elementary particles. Experiments with neutrons are very challenging and the best bound is [10] $|d_n| < 2.9 \times 10^{-26} e$ cm.

Even better candidates are atoms and molecules. These are charge neutral and one can apply large external electric fields. An atom or molecule can have non-zero EDM due to the EDM of electron d_e [11]. From detailed theoretical studies, it is now well established that there is an enhancement of EDM in atoms and molecules due to relativistic corrections [12]. That is, for an atom, the EDM,

$$d_a = \eta d_e.$$

Where, $\eta \, \square \, 1$, is the enhancement factor. Determining or extracting the electron EDM from experimentally measured d_a requires accurate theoretical calculations to obtain η. This is where reliable atomic many-body theories like coupled-cluster are extensively used. Though, we have discussed about atomic EDM arising from d_e, atoms and molecules are also sensitive to EDMs arising from: EDMs of neutrons and protons, and; parity and time reversal violating interactions within the atom or molecule. In particular, the open-shell or paramagnetic atoms are sensitive to d_e and closed-shell or diamagnetic atoms are sensitive to EDMs arising involving the nuclear sector. The implication of a precise determination of d_e is, there are extensions to SM of particle physics which predict much larger d_e. So, an unambiguous detection of d_e different from the SM prediction is a signature of physics beyond the standard model (BSM).

Parity violation, often referred to as parity non-conservation is another discrete symmetry violation in atoms and molecules which has important implications to elementary particle physics. Within the SM of particle physics, parity is maximally violated in weak interactions. It is the fundamental interaction associated with phenomena like beta decay. However, there are BSM which predicts larger degree of parity violation. In atoms and molecules, one consequence of parity violation is modification to selection rules of radiative transitions. As example, consider the 1s and 2s states of Hydrogen atom. The states are of same parity and electric dipole (E1) transition between the states is forbidden.

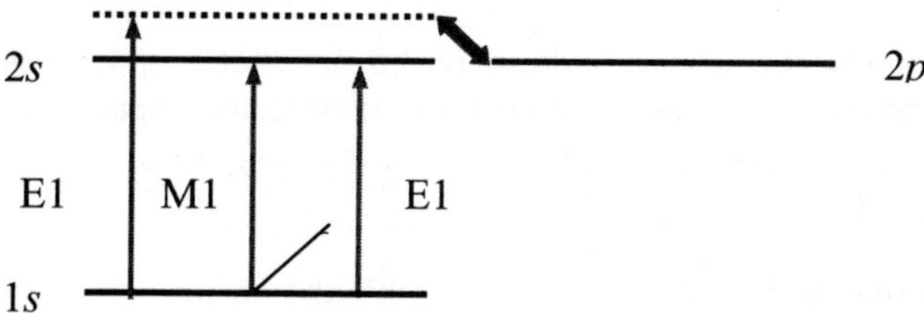

Fig 8: Electric dipole (E1) transition between the 1s and 2s states of Hydrogen is forbidden. However, in presence of parity violation 2s (solid line) acquires a mixture of opposite parity (dashed line). The odd parity state 2p dominates the opposite parity mixing and is denoted by the doubled sided arrow. The E1 transition between 1s and parity mixed 2s is then allowed.

This is schematically shown in Fig 8. However, in the presence of a parity violating interaction H_{PNC}, the 2s state acquires a mixture from the opposite parity states. From time-independent perturbation theory, the parity mixed 2s state is

$$|2\widetilde{s}\rangle = \sum_I |I\rangle \frac{\langle I|H_{PNC}|2s\rangle}{\varepsilon_{2s} - \varepsilon_I},$$

Where $|I\rangle$ are the odd parity intermediate states and ε_i are the energies of the states. Similarly, the 1s state also acquire an odd parity admixture. The E1 transition amplitude between the parity mixed states is then non-zero

$$E1_{PNC} = \langle 2|\widetilde{s}d|1\widetilde{s}\rangle = \sum_I \left[\frac{\langle 2s|d|I\rangle\langle I|H_{PNC}|1s\rangle}{\varepsilon_{1s} - \varepsilon_I} + \frac{\langle 2s|H_{PNC}|I\rangle\langle I|d|1s\rangle}{\varepsilon_{2s} - \varepsilon_I} \right] \neq 0,$$

where, $E1_{PNC}$ is the H_{PNC} induced electric dipole transition amplitude. In atomic experiments, $E1_{PNC}$ is measured using very sensitive interference techniques. Besides probing the physics of elementary particles, parity violation may be cause for handedness of organic molecules. That is, during any organic chemical reaction right and left handed molecules are produced in equal amounts. However, in nature, most of the organic molecules are right handed. To determine $E1_{PNC}$, like in EDM, one has to use accurate atomic theory calculations to extract the parameters related to particle physics.

In terms of the atomic theory calculations, the EDM and $E1_{PNC}$ calculations are very similar. The former is an expectation value and the later is transition amplitude. So, one may use the same method. In general, the $E1_{PNC}$ calculations are more complicated as it involves two different states and at least one is an excited state. Although, we have considered Hydrogen atom as a case study to show the role of parity violation in altering the selection rules, it is preferable to use heavier atoms. The reason is, the observables scales as Z^3. An important advantage of EDM or $E1_{PNC}$ as probes of particle physics is, the atomic experiments are table top experiments and far cheaper than accelerator based experiments. In addition, these probe parameter space complementary to the accelerators. So, it helps to constrain the parameter space.

Conclusion

The initial state ingredients, an electron and an ion, in an electron-ion scattering process are quite different from the ingredients (a photon and a neutral atom) of an atomic photoionization process. Nevertheless, their end-states both contain a 'free' electron and an ion. The quantum mechanical description of collision and photoionization is intimately related through the (discrete) time-reversal symmetry. This involves complex conjugation of the wavefunction in addition to $t \rightarrow -t$. While outgoing wave boundary conditions are employed to describe the quantum collision process, it is the ingoing wave boundary condition that must be employed to describe the photo-ionization process.

Atoms are suitable systems to probe the observable signatures of discrete symmetry violations. The results from the precision atomic and molecular experiments, when combined with theoretical results, provide stringent bounds on parameters in elementary particle physics.

References

[1]   P.C.Deshmukh and J.Libby 'Symmetry Principles and Conservation Laws in Atomic and Subatomic Physics', Resonance, September and   October issues, 2010.

[2]   P.C.Deshmukh and Shyamala Venkataraman 'Obtaining conservation principles from laws of   nature' Bull. Indian Assoc. of Physics Teachers (2011, in Press).

[3]   C.J.Joachain, 'Quantum Collision Theory' North-Holland Publishing Company, 1975.

[4]   G. Breit and H.A. Bethe in their highly cited famous paper: Phys. Rev. 93, 888(1954)   http://prola.aps.org/pdf/PR/v93/i4/p888_1

[5]   U.Fano and A.R.P. Rau, 'Theory of Atomic Collisions and Spectra' Academic Press, 1986.

[6]   J.M.Domingos 'Time Reversal in Classical and Quantum Mechanics' Int. J. Theor. Phys. Vol.18, No.3 p.213 (1979)

[7]   J.J.Sakurai, "Modern Quantum Mechanics", Pearson Education Inc,1994

[8]   J. Cooper and R.N. Zare Angular distribution of photoelectrons. J Chem Phys 48:942–943 (1968).

[9]   E. Kolb and M. Turner, 'The Early Universe', Westview Press, 1994.

[10]   C. A. Baker, *et al.*, Phys. Rev. Lett. **97**, 131801 (2006).

[11]   W. Bernreuther and M. Suzuki, Rev. Mod. Phys. **63**, 313 (1991).

[12]   J. S. M. Ginges and V. V. Flambaum, Phys. Rep. **397**, 63, (2004).

Atomic Photoionization in the Born Approximation and Angular Distribution of Photoelectrons

Pranawa C. Deshmukh[1*], Alak Banik[2] and Dilip Angom[3]

[1]*Indian Institute of Technology Madras, Chennai*
[2]*Space Applications Centre, Ahmadabad*
[3]*Physical Research Laboratory, Ahmadabad*
email: pcd@physics.iitm.ac.in[*]

1. INTRODUCTION

The atomic photo-effect has a fascinating history [1]. Other than the photoionization cross-sections which provide a measure of the transition probability, a physicist would of course wish to measure other physical properties related to the photoionization process. Any additional measurement is nevertheless constrained by the principle of uncertainty, which is the hallmark of quantum mechanics, since simultaneous measurements of certain physical properties, such as position and its conjugate momentum, are incompatible. A 'complete' experiment aims at carrying our all possible measurements that are compatible. In the case of atomic photo-effect, a 'complete' experiment consists of measurements of the integrated and differential cross sections, angular distribution of the photoelectrons, and spin-polarization parameters for the photoelectrons [2].

The angular distribution of the photoelectrons gives us information about the photoelectron yield in different directions, with respect to the direction of incidence of the electromagnetic radiation that is absorbed by the atom, and its direction of polarization.

In this article, we provide a pedagogical formulation of the interaction between electromagnetic radiation and an atomic electron in a bound state with the energy of the radiation above the ionization threshold. In Section II, we develop the basic expression for differential and total photoionization cross-section in the IPA (independent particle approximation) using the Born (high-energy) approximation. Essential elements are easily found in standard text-books [3-5], but are included in this article for completeness, and to provide the basis for the development of the expression for the angular distribution of the photoelectrons based on the famous exposition of this topic in the Cooper-Zare paper [6]. The aim of this article is to provide basic introduction for readers who are new to the field of photoionization spectroscopy so that they can appreciate the developments in this field wherein much more rigorous formulations have been developed well-beyond the approximations introduced in this article.

2. THE DIFFERENTIAL AND TOTAL PHOTOIONIZATION CROSS-SECTION

We shall consider the electromagnetic radiation to be polarized along the X-axis and incident along the Z-axis of a Cartesian frame of reference. As is found in standard text books on quantum mechanics and atomic physics, such as [3-5], we shall describe an electron in an atom interacting with the electromagnetic scalar and vector potentials $\left\{\phi(\vec{r},t), \vec{A}(\vec{r},t)\right\}$ by the Hamiltonian:

$$H = \left[\frac{1}{2m}\left(\vec{p} - \frac{q}{c}\vec{A}(\vec{r},t)\right)\bullet\left(\vec{p} - \frac{q}{c}\vec{A}(\vec{r},t)\right) + qf(\vec{r},t) - \frac{Zq^2}{r}\right] \tag{1}$$

Ignoring the quadratic term in the vector potential, an approximation that is valid for weak field, and employing the Coulomb gauge, we can express the electron-field interaction term in above Hamiltonian perturbatively as:

$$H = H_0 + \lambda H'' \tag{2a}$$

with

$$H' = -\frac{iq\hbar}{mc}\vec{A}(\vec{r},t)\bullet\vec{\nabla} = \frac{q}{mc}\vec{A}(\vec{r},t)\bullet\vec{p} \tag{2b}$$

where λ is the perturbation order parameter which we shall use to consider terms up to first order in perturbation theory, $q = -e$ is the electron's charge, m is the electron's mass, c is the speed of light, and $\vec{p} = -i\hbar\vec{\nabla}$ is the momentum operator. First order perturbation theory gives us, using the famous golden rule attributed to Fermi, the following expression for the transition rate (i.e. transition probability per unit time) for photoabsorption resulting in a transition from an initial bound state $\left|i\right\rangle$ to a final continuum state $\left|f\right\rangle$ effected by the transition perturbation operator:

$$W_{fi} = \left(\frac{qA_0(\omega)}{mc}\right)^2 \left|\left\langle f \left| e^{i\vec{k}\bullet\vec{r}}\hat{\varepsilon}\bullet\vec{\nabla}\right| i\right\rangle\right|^2 \times 2\pi\delta(\bar{\omega}) \tag{3}$$

The physical quantity of interest that gives us information about the photoionization cross-section is given by the ratio:

$$\frac{\text{Energy absorbed per unit time in } |i\rangle \to |f\rangle}{\textit{Energy flux of the EM radiation}} \tag{4}$$

$$=$$

$$\frac{\text{Energy absorbed per unit time in } |i\rangle \to |f\rangle}{\textit{Energy per unit area per unit time of the EM radiation}}$$

We shall use the geometry shown in Fig.1 to describe the photoionization process. The electromagnetic radiation that is absorbed is considered to be incident along the direction Z of a Cartesian frame of reference, and is polarized along $\hat{\varepsilon} = \hat{e}_x$. The ratio according to Eq.4 which gives us a measure of the differential cross-section for photoionization in an elemental solid angle $d\Omega$ resulting in a photoelectron's escape with momentum $\hbar\vec{k}$ along a direction given by the unit vector $\hat{k}_f$ is then given by:

$$\left[\frac{d\sigma}{d\Omega}\right]_{\hat{k}_f}^{\hat{\varepsilon}} = \frac{\hbar\omega \times \left[W_{fi}\right]_{\hat{k}_f}^{\hat{\varepsilon}}}{I(\omega)} , \tag{5}$$

Wherein $\tilde{\omega} = \omega - \omega_{fi}$, and $I(\omega)$ is the intensity of the electromagnetic radiation at the circular frequency ω.

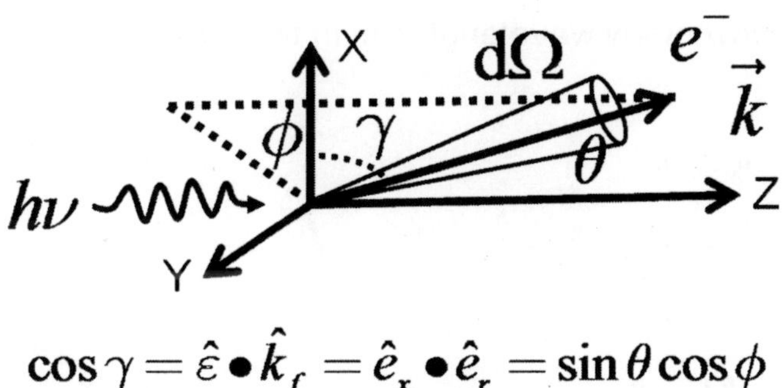

$$\cos\gamma = \hat{\varepsilon} \bullet \hat{k}_f = \hat{e}_x \bullet \hat{e}_r = \sin\theta\cos\phi$$

Fig.1: Geometry employed to describe the photoelectron angular distribution

From the magnetic vector potential given by:

$$\vec{A}(\vec{r},t) = A_0(\omega)\hat{\varepsilon}\left[e^{i(\vec{k}\bullet\vec{r} - \omega t)} + e^{-i(\vec{k}\bullet\vec{r} - \omega t)}\right] \tag{6a}$$

we obtain (in the Gaussian-CGS system of units):

$$\vec{E}(\vec{r},t) = -\vec{\nabla}\phi - \frac{1}{c}\frac{\partial\vec{A}}{\partial t} \qquad \text{in Coulomb Gauge } (\phi = 0; \vec{\nabla}\bullet\vec{A} = 0) \tag{6b}$$

and

$$\vec{H}(\vec{r},t) = \vec{\nabla}\times\vec{A}, \tag{6c}$$

from which we can find the intensity of the electromagnetic field as the average value of the Poynting vector as:

$$I(\omega)=\left\langle\left|\vec{S}\right|\right\rangle=\left\langle\left|\frac{c}{4\pi}\vec{E}\times\vec{H}\right|\right\rangle=\frac{\omega^2}{2\pi c}A_0^2(\omega)\,. \tag{7}$$

Form Eq.5 and 7 we get:

$$\left[\frac{d\sigma}{d\Omega}\right]_{\hat{k}_f}^{\hat{\varepsilon}}=\frac{\hbar\omega\times\left(\frac{qA_0(\omega)}{mc}\right)^2\left|\left\langle f\,|\,e^{i\vec{k}\bullet\vec{r}}\,\hat{\varepsilon}\bullet\vec{\nabla}\,|\,i\right\rangle\right|^2\times 2\pi\delta(\tilde{\omega})}{\frac{\omega^2}{2\pi c}A_0^2(\omega)} \tag{8a}$$

Hence,

$$\left[\frac{d\sigma}{d\Omega}\right]_{\hat{k}_f}^{\hat{\varepsilon}}=\frac{4\pi^2\alpha\hbar^3}{m^2\omega}\times\left|\left\langle f\,|\,e^{i\vec{k}\bullet\vec{r}}\,\hat{\varepsilon}\bullet\vec{\nabla}\,|\,i\right\rangle\right|^2\times\delta(E-E_{fi})\,. \tag{8b}$$

We urge the student-reader to verify, by considering the physical dimensions of each term on the right hand side that the dimension of the ratio described by Eq.8 is L^2, justifying the name 'cross-section'. It is usually measured in terms of the unit Mb (Mega-Barn) with $1\text{ Mb}=10^{-18}\text{cm}^2$.

Transitions to a number of degenerate states are possible, and we must therefore estimate the number of such states that must be considered. We shall use 'box normalization' over a cubical box of length L (Fig.2), but our final result will be independent of L.

How many wavelengths fit in the box?

$$n_x\lambda_x=L$$

$$n_x\frac{2\pi}{k_x}=L$$

$$k_x=\frac{2\pi n_x}{L}$$

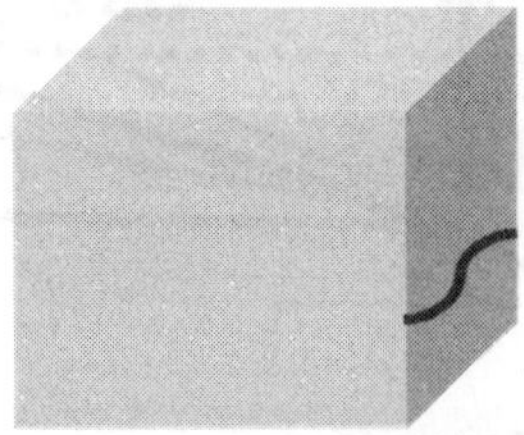

$$E=\frac{\hbar^2k^2}{2m}=\frac{\hbar^2}{2m}\left(k_x^2+k_y^2+k_z^2\right)=\frac{\hbar^2}{2m}\left(\frac{2\pi}{L}\right)^2\left(n_x^2+n_y^2+n_z^2\right)=\frac{2\pi^2\hbar^2}{mL^2}n^2$$

Fig. 2: Box normalization

The number of states in the volume element is then given by (Fig.3):

$$\begin{aligned}n^2dnd\Omega&=n^2\frac{dn}{dE}dEd\Omega\\&=n^2\frac{dk}{dE}\frac{L^2}{4\pi^2}\frac{2\pi}{L}dEd\Omega\,,\\&=\left(\frac{L}{2\pi}\right)^3\left(\frac{mk}{\hbar^2}\right)dEd\Omega\end{aligned} \tag{9}$$

Thus, integrating over the possible energy states (Eq.9), we get from Eq.8:

$$\left[\frac{d\sigma}{d\Omega}\right]_{\hat{k}_f}^{\hat{\varepsilon}} = \int \frac{4\pi^2\alpha\hbar^3}{m^2\omega} \times \left|\left\langle f\,|\,e^{i\vec{k}\bullet\vec{r}}\,\hat{\varepsilon}\bullet\vec{\nabla}\,|\,i\right\rangle\right|^2 \delta(E-E_{fi})\left(\frac{L}{2\pi}\right)^3\left(\frac{mk}{\hbar^2}\right)dE$$

(10a)

Carrying out the *Dirac* $-\delta$ function integration over energy, we get:

(10b)

$$\left[\frac{d\sigma}{d\Omega}\right]_{\hat{k}_f}^{\hat{\varepsilon}} = \frac{4\pi^2\alpha\hbar^3}{m^2\omega_{fi}} \times \left|\left\langle f\,|\,e^{i\vec{k}\bullet\vec{r}}\,\hat{\varepsilon}\bullet\vec{\nabla}\,|\,i\right\rangle\right|^2\left(\frac{L}{2\pi}\right)^3\left(\frac{mk}{\hbar^2}\right)^{\cdot}$$

The transition matrix element in Eq.10 is:

$$M = \left\langle f\,|\,e^{i\vec{k}\bullet\vec{r}}\,\hat{\varepsilon}\bullet\vec{\nabla}\,|\,i\right\rangle$$

$$= \int\left(\frac{1}{\sqrt{L^3}}e^{-i\vec{k}_f\bullet\vec{r}}\right)\left[e^{i\vec{k}\bullet\vec{r}}\,\hat{\varepsilon}\bullet\vec{\nabla}\right]\psi_i(\vec{r})dV$$

$$= \frac{-i}{\sqrt{L^3}}\left[\hat{\varepsilon}\bullet\vec{k}_f\right]\int\left[\psi_i(\vec{r})e^{i\left(\vec{k}-\vec{k}_f\right)\bullet\vec{r}}\right]dV$$

(11)

$$= \frac{-ik_f}{\sqrt{L^3}}\left[\cos\gamma\right]\int\left[\psi_i(\vec{r})e^{i\left(\vec{k}-\vec{k}_f\right)\bullet\vec{r}}\right]dV$$

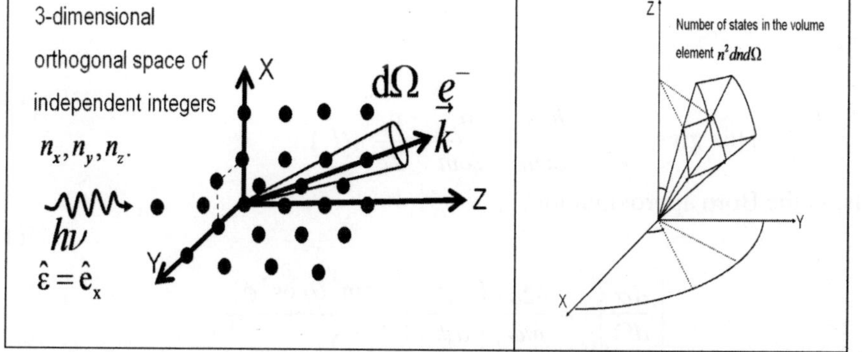

Fig.3: Estimation of the number of states to which the photoionization transition can result in. Each of these states can contribute to the transition from an initial bound state $|i\rangle$ to a final continuum state $|f\rangle$.

Since it is the absolute square of the matrix element that is involved in Eq.10, we see how the result is independent of the length L used in the 'box normalization'.

The integral in Eq.11 can be easily determined; it is proportional to the Fourier transform of the initial state wavefunction. If this state function is the 1s wavefunction $\left[\dfrac{1}{\sqrt{\pi}}\left(\dfrac{Z}{a_0}\right)^{\frac{3}{2}} e^{-\frac{Zr}{a_0}}\right]$ of the electron in atomic hydrogen, the integral becomes:

$$I = \int dV \psi_i(\vec{r}) e^{i\left(\vec{k} - \vec{k}_f\right)\cdot\vec{r}},$$

(12a)

i.e.

$$I = \frac{1}{\sqrt{\pi}}\left(\frac{Z}{a_0}\right)^{\frac{3}{2}} \frac{8\pi\left(\dfrac{Z}{a_0}\right)}{\left[\left\{Z^2 + a_0^2\left|\vec{k} - \vec{k}_f\right|^2\right\}\left\{\dfrac{1}{a_0^2}\right\}\right]^2}$$

(12b)

Using Eq.11 andEq.12 in Eq.10, we get:

(13)

$$\left[\frac{d\sigma}{d\Omega}\right]_{\hat{k}_f}^{\hat{\varepsilon}} = \frac{32\alpha\hbar k_f^3}{m\omega_{fi}} \frac{Z^5 a_0^3\left(\cos^2\gamma\right)}{\left[\left\{Z^2 + a_0^2\left|\vec{k} - \vec{k}_f\right|^2\right\}\right]^4}$$

We now make the high-energy (Born) approximation [4]:

$$\hbar\omega = \frac{\hbar^2 k_f^2}{2m} + \text{I.P.} \underset{\text{High Energy Approximation}}{\approx \;\rightarrow\; \approx} \frac{\hbar^2 k_f^2}{2m},$$

(14a)

whence

(14b)

$$kc = \frac{\hbar k_f^2}{2m}. \text{ This gives } \frac{k}{k_f} = \frac{\hbar k_f}{2cm} = \frac{p_f}{2cm} = \frac{v_f}{2c} \lll 1$$

Thus, in the Born approximation,

(15)

$$\left[\frac{d\sigma}{d\Omega}\right]_{\hat{k}_f}^{\hat{\varepsilon}} = \frac{32\alpha\hbar}{m\omega_{fi}}\left(\frac{Z}{a_0 k_f}\right)^5 \frac{\left(\sin^2\theta\cos^2\phi\right)}{\left(1 - \dfrac{v_f}{c}\cos\theta\right)^4}$$

For unpolarized light, we use the average value

$$\left\langle\cos^2\phi\right\rangle = \frac{1}{2\pi}\int_0^{2\pi}\cos^2\phi\, d\phi = \frac{1}{2},$$

which gives

$$\left[\frac{d\sigma}{d\Omega}\right]_{\hat{k}_f}^{unpolarized} \approx \frac{16\alpha\hbar}{m\omega_{fi}}\left(\frac{Z}{a_0 k_f}\right)^5 \left(\sin^2\theta\right)\left(1+4\frac{v_f}{c}\cos\theta\right) \tag{16a}$$

The total photoionization cross-section is then obtained by integrating the differential cross-section over all the angles θ:

$$\tag{16b}$$

$$\sigma_{Total}^{unpolarized} = \int\limits_{\theta=0}^{\pi}\int\limits_{\phi=0}^{2\pi}\left[\frac{d\sigma}{d\Omega}\right]_{\hat{k}_f}^{unpolarized}\sin\theta \, d\theta \, d\phi$$

$$= \frac{128\pi}{3m}\frac{\alpha\hbar}{\omega}\left(\frac{Z}{a_0}\right)^5\frac{1}{k_f^5}$$

$$Now, \text{ since } \hbar\omega \approx \frac{\hbar^2 k_f^2}{2m},$$

$$k_f^5 = \left(k_f^2\right)^{5/2} = \left(\frac{2m}{\hbar}\right)^{5/2}\omega^{5/2},$$

giving:

$$\tag{16c}$$

$$\sigma_{Total}^{unpolarized} = \frac{128\pi}{3m}\frac{\alpha\hbar}{\left(\dfrac{2m}{\hbar}\right)^{5/2}}\left(\frac{Z}{a_0}\right)^5\frac{1}{\omega^{7/2}}$$

We thus obtain the well-known result that in the Born approximation, the photoionization cross-section goes as $\sigma \rightarrow E^{-7/2}$, Z^5. If one carries out similar analysis with other hydrogenic wavefunctions for higher principal quantum numbers n, we find that the photoionization cross-section goes as n^{-3}. The Born approximation is very useful, but it does have limitations and strictly speaking it provides more of an exception than a rule [7,8].

3. THE ANGULAR DISTRIBUTION OF PHOTOELECTRONS

The presence of angle dependent terms in the differential cross-section (Eq.15, 16) tells us that the photoelectric effect would not generally generate an isotropic yield. To examine the angular dependence of the photoelectron yield, we examine the transition matrix element:

$$M = \left\langle f \mid e^{i\vec{k}\cdot\vec{r}}\hat{\varepsilon}\cdot\vec{\nabla} \mid i\right\rangle = \frac{i}{\hbar}\left\langle f \mid e^{i\vec{k}\cdot\vec{r}}\hat{\varepsilon}\cdot\vec{p} \mid i\right\rangle$$

$$= \frac{i}{\hbar}\left\langle f \mid p_x \mid i\right\rangle; \text{ (in the 'dipole approximation' } e^{i\vec{k}\cdot\vec{r}}\approx 1)$$

$$= \frac{i}{\hbar}\left\langle f \mid \frac{m}{i\hbar}[x,H_0] \mid i\right\rangle$$

$$\tag{17}$$

Since

$$\tag{18}$$

$$[r_k,H_0] = [r_k,\frac{p^2}{2m}] = \frac{i\hbar}{m}p_k$$

for $k=x,y,x$.

Using the fact that the states $|i\rangle$ and $|f\rangle$ are eigenstates of the unperturbed Hamiltonian H_0, we get, using Eq.8, Eq.17 and Eq.18:

$$\left[\frac{d\sigma}{d\Omega}\right]_{\hat{k}_f}^{\hat{\varepsilon}} = \frac{4\pi^2\alpha\hbar^2}{m^2\omega} \times \left|\frac{m}{\hbar^2}\left(E_i - E_f\right)\langle f\,|\,x\,|\,i\rangle\right|^2 \times \delta(\tilde{\omega}) \qquad (19a)$$

i.e.,

$$\left[\frac{d\sigma}{d\Omega}\right]_{\hat{k}_f}^{\hat{\varepsilon}} = \frac{4\pi^2\alpha\hbar^2}{m^2\omega} \times \left|\frac{m}{\hbar^2}\left(E_i - E_f\right)\langle\psi_f\,|\,T\,|\,\psi_i\rangle\right|^2 \times \delta(\tilde{\omega}) \qquad (19b)$$

From the form of the transition matrix element in Equations 17 and 19, we see that it can be obtained by taking the matrix element of either the momentum operator, as in Eq.17, or equivalently if the position/length operator, as long as the commutation property described in Eq.18 holds. Thus, alternative forms of the matrix elements can be used for their calculations [9]. Computationally, the matrix elements are then evaluated either or both forms, and accordingly known as the momentum or the length form of the matrix element. Eq.18, however, holds only as long as the potential term in the Hamiltonian is local. The exchange density term in the electron-electron interaction is however non-local, as in the Hartree-Fock Self Consistent Field potential [10], and when such a potential is employed the equality of the length and the momentum forms breaks down. Note that we have written $\langle f\,|\,x\,|\,i\rangle = \langle\psi_f\,|\,T\,|\,\psi_i\rangle$ so that we can follow the formulation in the Cooper-Zare paper [6] with the provision that Cooper-Zare paper [6] has the operator $T = z$ whereas we have $T = x$, the difference being only on account of the fact that in the Cooper-Zare paper, the incident electromagnetic radiation is considered to be polarized along the Cartesian Z-axis, whereas in the present article, it is considered to be polarized along the X-axis.

The matrix element in Eq.19 is an integral over the radial and angular variables and can be written as a product of the two:

$$\langle\psi_f\,|\,T\,|\,\psi_i\rangle = \sum_{l,m} a(l,m)\,\rho_{ll'}\left\{(-1)^{l-m}(-1)^g\sqrt{l_>}\begin{pmatrix} l & 1 & l' \\ -m & 0 & m' \end{pmatrix}\right\}, \qquad (20)$$

where $\rho_{ll'} = \int\limits_0^\infty r^2\,dr\,R_{kl}(r)\,r\,R_{n'l'}(r)$, $g = \dfrac{l'-l+1}{2}$, and $l_>$ is the greater of l and l'.

It is important to note that the final state wavefunction used in Eq.20 must have the correct boundary condition, known as the 'ingoing wave boundary condition' [11] which describe the photoionization   process as time-reversed electron-ion collision.

By taking the modulus square of Eq.20 and averaging over all states with azimuthal quantum numbers m' assumed to be equally populated, we thus see that the differential cross-section for photoionization,

$$
\left[\frac{d\sigma}{d\Omega}\right]_{\hat{k}_f}^{\hat{\varepsilon}} \text{ is thus proportional to:}
$$

(21)

$$
\frac{1}{2l'+1}\sum_{m'}\sum_{l_1} a(l_1,m')\,\rho_{l_1 l'}\left\{(-1)^{l_1}(-1)^{g_1}\sqrt{l_>}\begin{pmatrix} l_1 & 1 & l' \\ -m' & 0 & m' \end{pmatrix}\right\}\times
$$

$$
\sum_{l_2} a^*(l_2,m')\,\rho_{l_2 l'}\left\{(-1)^{l_2}(-1)^{g_2}\sqrt{l_>}\begin{pmatrix} l_2 & 1 & l' \\ -m' & 0 & m' \end{pmatrix}\right\}
$$

in which $a(l,m)=i^l 4\pi e^{-i\delta_l}Y_l^{m*}(\hat{k}_f)$,

(22)

using the notation in Ref.[6].

The summation over the orbital angular momentum quantum numbers is restricted to be over no more than 4 terms, since under the dipole selection rules ($\delta l=\pm 1$), transitions to two possible values of l_1 can be reached at from a given value of the orbital angular momentum l' of the initial bound state function, and similarly transitions to two possible values of l_2 are possible. The resulting four possible terms are explicitly treated below.

Now, the Wigner 3j symbols are well-known and can be obtained readily even from the internet and the following two relations can now be used:

$$
\begin{pmatrix} l+1 & 1 & l \\ -m & 0 & m \end{pmatrix}=(-1)^{l-m-1}\left[\frac{(l+1)^2-m^2}{(2l+1)(2l+3)(l+1)}\right]^{1/2}
$$

(23a)

and

$$
\begin{pmatrix} l-1 & 1 & l \\ -m & 0 & m \end{pmatrix}=(-1)^{l-m}\left[\frac{l^2-m^2}{(2l+1)(2l-1)(l)}\right]^{1/2}.
$$

(23b)

Using Eq.22 and 23, the proportionality in the relation 22 becomes:

$$\left[\frac{d\sigma}{d\Omega}\right]_{\hat{k}_f}^{\hat{\varepsilon}} \quad \alpha$$

$$\frac{16\pi^2}{2l+1}\sum_m \left[\begin{array}{c} \rho_{l-1,r}^2 \dfrac{l^2-m^2}{(2l+1)(2l-1)}\left|Y_{l-1,m}\right|^2 \\[2ex] +\rho_{l+1,r}^2 \dfrac{(l+1)^2-m^2}{(2l+1)(2l+3)}\left|Y_{l+1,m}\right|^2 \\[2ex] +\rho_{l+1}\rho_{l-1}\sqrt{\dfrac{l^2-m^2}{(2l+1)(2l-1)}}\sqrt{\dfrac{(l+1)^2-m^2}{(2l+1)(2l+3)}}\,Y_{l+1,m}^*Y_{l-1,m}e^{-i(\delta_{l+1}-\delta_{l-1})} \\[2ex] +\rho_{l+1}\rho_{l-1}\sqrt{\dfrac{l^2-m^2}{(2l+1)(2l-1)}}\sqrt{\dfrac{(l+1)^2-m^2}{(2l+1)(2l+3)}}\,Y_{l+1,m}Y_{l-1,m}^*e^{+i(\delta_{l+1}-\delta_{l-1})} \end{array}\right] \qquad (24)$$

The spherical harmonics are functions of the polar angles Θ (and the azimuthal angles Φ) defined with respect to the polar axis taken along the direction of polarization of the incident electromagnetic radiation. The form of the relation (24) is such that the Φ-dependence drops out. The angular dependence of the photoelectron yield is then determined by the polar angle Θ which is the angle between the direction of polarization of the incident electromagnetic radiation, and the direction along which the photoelectron escapes. The angular distribution of the photoelectrons is determined, as can be seen from the above expression, also by the difference in the phase-shifts $\left(\delta_{l+1}-\delta_{l-1}\right)$ in the dipole-dipole channels. The interference term between the dipole channels has important consequences on the photoelectron angular distributions. Using a few identities with regard to the spherical harmonics and angular momentum coupling coefficients which are explicitly spelled out in Ref.[6], the differential cross-section (Eq.19) can be put in the following, very compact, form using the relation 24:

$$(25)$$

$$\left[\frac{d\sigma}{d\Omega}\right]_{\hat{k}_f}^{\hat{\varepsilon}} = \frac{\sigma_{Total}}{4\pi}\left[1+\beta P_2\cos\Theta\right]$$

in which the parameter β that is introduced is the famously known dipole angular distribution asymmetry parameter:

$$\beta = \frac{l(l-1)\rho_{l-1}^2+(l+1)(l+2)\rho_{l+1}^2-6l(l+1)\rho_{l+1}\rho_{l-1}\cos\left(\delta_{l+1}-\delta_{l-1}\right)}{(2l+1)\left[l\rho_{l-1}^2+(l+1)\rho_{l+1}^2\right]} \qquad (26a)$$

In Eq.25

$$(26b)$$

$$P_2\cos\Theta = \frac{1}{2}\left(3\cos^2\Theta-1\right)$$

This result provides a quantitative relation that describes the angular distribution of the photoelectron yield resulting from absorption electromagnetic radiation incident along the Z-axis and polarized along the X-axis. Since the cross-section cannot go negative, the term in the Legendre polynomial in the Cooper-Zare formula (Eq.25) restricts the angular distribution asymmetry parameter to the range $-1 \leq \beta \leq 2$ which keeps $\left[1 + \beta P_2 \cos \Theta\right] \geq 0$..

4. BEYOND THE BORN APPROXIMATION, AND BEYOND THE COOPER-ZARE FORMULA

We have already mentioned above that the Born approximation used in Section II breaks down [7,8], and the dipole approximation also breaks down [12-15]. The 'dipole approximation' discussed above results from truncating the expansion of the exponent in $e^{i\vec{k} \bullet \vec{r}}$ powers of $\left(\dfrac{r}{\lambda}\right)$, and is a 'long-wavelength' or 'low-energy' approximation. It is considered to be an excellent approximation for photon energies up to ~5 keV above the ionization threshold. Terms beyond the dipole approximation are often needed, and there is considerable interest both in theory and experiment in examining non-dipole effects [12-15] in atomic photoionization.

Conclusion

The formulation of the photoionization cross-section and the angular distribution of the photoelectrons presented in this article is based on standard text-book [3-5] and/or well-known papers [6]. It is hoped that this formulation is useful to introduce a student-reader who is new to the literature on atomic photoionization. The two major approximations used in the present formulation, namely the Born approximation and the dipole approximation are both useful for very many applications in atomic physics, but most studies of current interests require one to go well beyond these approximations [3-5, 12-15]. An introduction to problems of current interest requires familiarity with the vocabulary and results described above, and it is hoped that it will be useful to some young readers. It was felt necessary to include a discussion on experimental aspects of atomic photoionization measurements of cross-sections, angular distribution of the photoelectrons etc., but we finally decided, considering the length of the article, to leave it out. Ref. [2, 12] can provide a useful introduction for the beginner.

References

[1]   P. C. Deshmukh and Shyamala Venkataraman   "100 years of Einstein's Photoelectric Effect" Bulletin of the Indian Association of Physics Teachers, Vol. 23, 320-327 (2006)

[2]   Volker Schmidt, "Photoionizatio of Atoms using Synchrotron Radiation" Rep. on Progress in Phys. 55 1483-1659 (1992).

[3]   H.A.Bethe and E.E. Salpeter "Quantum Mechanics of One and Two Electron Atoms"

[4]   Bransden and Joachain "Physics of Atoms and Molecules", Prentice Hall 2003.

[5]   J.J.Sakurai, "Modern Quantum Mechanics", Pearson Education Inc, 1994

[6]   J. Cooper and R.N. Zare, "Angular distribution of photoelectrons" J Chem Phys 48:942–943 (1968); and in Lectures in Theoretical Physics, edited by S. Geltman, K. T. Mahanthappa, and W. E. Brittin (Gordon and Breach, New York,1969), Vol. XI-C, pp. 317—337.

[7]   E.W.B.Dias, H. S. Chakraborty, P. C. Deshmukh, S. T. Manson, O.Hemmers, G.Fisher, P.Glans, D.L.Hansen, H.Wang, S.B.Whitfield, D.W.Lindle, R.Wehlitz, J.C.Levin, I.A.Sellin, R.C.C.Perera. "Breakdown of the Independent Particle Approximation in High-Energy Photoionization", Physical Review Letters 78:24 p.4553-4556 (1997)

[8]   D.L.Hansen, O.Hemmers, H.Wang, D.W.Lindle, P.Focke, I.A.Sellin, C.Heske, H. S. Chakraborty, P. C. Deshmukh and S. T. Manson "Validity of the independent particle approximation: The exception, not the rule", Phys.Rev. A 60, R2641-2644 (1999).

[9]   S.Chandrasekhar "On the Continuous Absorption Coefficient of the Negative Hydrogen Ion" in Astrophysics Journal, Vol. 102, Page 223 (1945). Also, see the article 'Length and Velocity Formulas in Approximate Oscillator Strength Calculations' by Anthony F Starace in Physical Review A, Vol. 3 No.4, Page 1242

[10]   P.C.Deshmukh, Alak Banik, Dilip Angom, "Hartree-Fock Self-Consistent Field Method for Many-Electron Systems", DST SERC SCHOOL, PILANI

[11]   P.C.Deshmukh, Dilip Angom, Alak Banik, "Symmetry in Electron-Atom Collision and Photoionization Process" DST SERC SCHOOL, PILANI

[12]   O.Hemmers, G.Fisher, P.Glans, D.L.Hansen, H.Wang, S.B.Whitfield, D.W.Lindle, R.Wehlitz, J.C.Levin, I.A.Sellin, R.C.C.Perera, E.W.B.Dias, H. S. Chakraborty, P. C. Deshmukh and S. T. Manson "Beyond the dipole approximation: angular distribution effects in valence photoemission" Journal of Physics B: Atomic, Molec. & Opt. Phys. 30 p.L727-L733 (1997)

[13]   P. C. Deshmukh, Tanima Banerjee , Hari R. Varma, O. Hemmers, R. Guilllemin, D. Rolles, A. Wolska, S. W. Yu, D. W. Lindle, W. R. Johnson and S. T. Manson "Theoretical and experimental demonstration of the existence of quadrupole Cooper minima" J. Phys. B: At. Mol. & Opt. Phys. (Fast Track) 41, 021002 (2008)

[14]   Tanima Banerjee, P. C. Deshmukh and S. T. Manson "Dipole and Quadrupole Cooper Minima and Their Effects on Dipole and Non-dipole Photoelectron Angular Distribution in Hg 6s" Phys. Rev. A 75, 042701 (2007)

[15]   G.B.Pradhan,J. Jose, P. C. Deshmukh, L. A LaJohn, R. H. Pratt, and S. T. Manson "Cooper Minima: A   Window on Non-dipole Photoionization at Low Energy" (To be submitted, 2011).

Non-relativistic and Relativistic Distorted Wave Theory for Electron-Atom Collisions

Rajesh Srivastava and Lalita Sharma

Indian Institute of Technology Roorkee, Roorkee-247667
e-mail: rajsrfph@iitr.ernet.in

1   INTRODUCTION

The study of electron impact excitation of atoms has been continuously growing field of Atomic and Molecular Collision Physics. Electron-atom collision processes have remained at the forefront of research due to their significant contributions to fundamentals of physics. They have practical applications in many fields of science which include production of atomic data for the modeling of fusion plasma, astrophysics, laser physics and the physics of planetary atmosphere. The constant development in the quality of experimental technology and the increasing power of modern computers have moved the level of comparison between experiment and theory to a very high degree of sophistication.

In an electron-atom collision process, the experiments measure the differential cross section (DCS) or the total cross section (TCS) when the incident beam of electrons interacts with the target atoms and is scattered after the collision. The number of electrons scattered per unit time into an element of solid angle yields DCS. The angle integration of DCS can provide TCS. It is found that in comparison to TCS, the DCS is better scattering parameter to report as the effect of the choice of target wave functions and the merit of the theoretical method employed for the calculation are clearly reflected.

In spite of the sophisticated experimental techniques, the growing demand of electron-atom collision data can not be fulfilled solely through the experimental measurements. Therefore, to meet the requirement of atomic data the theoretical studies have dominated the research in this area. In a theoretical study of electron - atom collision process, all the scattering amplitudes are to be determined primarily to further obtain various scattering parameters. However, the exact evaluation of the scattering amplitudes can not be accomplished practically and thus the theoretical calculations of scattering parameters depend on various approximation methods. The approximation methods available for electron-atom collisions can be categorized as perturbative and non-perturbative approaches. Among the perturbative methods are the Born series, distorted wave series and their variants. The second category contains the close-coupling, R-matrix, variational methods etc.

The simple first Born approximation theory (with only first term of the Born series) has been extensively used in the early days to calculate the total cross sections at high energies. However, it failed to explain the experiments at low (near threshold) and intermediate energies. It also could not describe properly differential cross section results at large scattering angles. Consequently, efforts were made to include the second term of the Born series and also to explore other methods like distorted wave Born series so that the reliable results at intermediate energies can be obtained.

The methods in the second category, *i.e.*, the close-coupling expansion methods have been found successful and extensively used in the low energy region. These methods are based on the eigen function expansion technique and work well when only few channels are open. However, as the energy increases more channels open up and it becomes almost impractical to include all of them. There have been efforts to develop techniques to extend the low energy methods to intermediate energy region. Some such approaches are the R-matrix, coupled-channel optical (CCO) and convergent close-coupling (CCC) methods [1]. However, among the perturbative methods, distorted wave approximation has been well known to be quite successful in explaining the experiments especially, at intermediate and high energies (to which the article addresses) [2,3]. The main advantage of using the distorted wave approximation is its flexibility which allows inclusion of more of the physical effects in the leading terms of the per-turbation series expansion. This leads distorted wave series to converge faster than the Born series and therefore, even the first order term of distorted wave series has been found to produces reasonably accurate results [3–8]. There have also been some attempts to include up to second order term in the distorted wave series [9–11]. The second order term involves an infinite summation over all the target states making its exact evaluation intractable and does not nec-essarily improves the results [2, 3, 8]. Though various simplifications have also been introduced which give results of varying accuracy [3]. As a result, the wide and successful applications of the first order distorted wave theory can be found in the literature [2]. Earlier applications of the distorted wave theory in its non-relativistic forms [2, 3, 8] have been very successful for electron impact excitation of lighter atoms. However, it has been observed that for inert gases and heavier atoms the relativistic effects *viz.* electron exchange and spin-orbit interaction are important and must be considered systematically. The feature which makes relativistic distorted wave (RDW) method distinct from others is its capability of dealing the fine-structure transitions of the atoms in a direct manner. In this method, the target atom is represented by multi-configuration Dirac-Fock wavefunctions while the wavefunction for the scattered electron is calculated via Dirac equations, thus the relativistic effects are included to all orders. Such a RDW theory has been successfully applied to various closed shell atoms for electron impact excitations from their ground state as well excited states [12–17].

In this article, theory of both the non–relativistic as well as the relativistic distorted wave method has been introduced and described in detail. The direct and exchange T-matrices have been evaluated for the atoms having one– and two– electrons in their valance shell.

2 DISTORTED WAVE APPROXIMATION THEORY

Consider the scattering of electrons from an atom having nuclear charge Z and N electrons. The total Hamiltonian of the electron–atom $i.e.$, the combined $(N+1)$ electron system can be written as

$$H = H_0 + V, \tag{1}$$

with

$$H_0 = H_{atom} + K_{N+1}, \tag{2}$$

where, H_0 is the unperturbed Hamiltonian for the non-interacting incident electron and the atom $i.e.$, when the projectile and target are far away, K_{N+1} is the kinetic energy operator of the projectile electron and V is the total interaction potential of the projectile–target atom and can be expressed as (atomic units will be used through out)

$$V = -\frac{Z}{r_{N+1}} + \sum_{j=1}^{N} \frac{1}{|\mathbf{r}_j - \mathbf{r}_{N+1}|}. \tag{3}$$

Here $\mathbf{r}_j$ and $\mathbf{r}_{N+1}$ refer respectively to the position vectors of the jth electron of the atom and projectile electron with respect to the target nucleus.

Further, we consider transition of the electron–atom system from the initial channel with the atom in its ground state, denoted by 'a' to the final channel with the atom in its excited state, denoted by 'b'. Following elementary theory of scattering we can write T-matrix for the transition $a \to b$ of the system as [6]

$$T_{a \to b} = \left\langle \psi_b(\mathbf{1,2,...,N+1}) \left| V \right| \Psi_a^+(\mathbf{1,2,...,N+1}) \right\rangle. \tag{4}$$

Here each of the numbers **1,2,...,N+1** refers the combined spin and position coordinates of the electron. ψ_b is the total unperturbed wavefunction of the system in the final channel and Ψ_a^+ is the wavefunction for the total system in the initial channel. These satisfy the following equations

$$(H_0 - E)\psi_b = 0, \tag{5}$$

and

$$(H - E)\Psi_a^+ = 0, \tag{6}$$

where, the '+' superscript refers to usual outgoing asymptotic wave boundary condition. E is the total energy of the system given by

$$E = E_a + \epsilon_a = E_b + \epsilon_b, \tag{7}$$

with $\epsilon_{a(b)}$ the energy of the atom in initial(final) channel with the corresponding projectile electron energy $E_{a(b)}$.

We see from equation (4) that to exactly evaluate this T-matrix for an excitation process we require exact solution of Ψ, satisfying equation (6) which as well known to us is a formidable task to obtain. Because of this reason several approximation methods are adopted to evaluate Ψ approximately thereby giving or defining the T-matrix in that particular approximation. Here we introduce one of such approximation methods known as *"distorted wave approximation"*.

Introducing the distortion potential, U and splitting the interaction potential into two parts such that

$$V = U_b + W_b, \tag{8}$$

and the scattering solution of the Hamiltonian

$$\bar{H} = H_0 + U_b, \tag{9}$$

is already known or can be found. Adopting the two potential formulation [6] we can write the exact T-matrix (equation (4)) alternatively as (electron exchange has been ignored for the moment)

$$\begin{aligned} T_{a \to b} &= \left\langle \chi_b^-(\mathbf{1,2,...,N+1}) \left| U_b \right| \psi_a(\mathbf{1,2,...,N+1}) \right\rangle \\ &+ \left\langle \chi_b^-(\mathbf{1,2,...,N+1}) \left| V - U_b \right| \Psi_a^+(\mathbf{1,2,...,N+1}) \right\rangle . \end{aligned} \tag{10}$$

Here the wavefunction χ_b, in the final channel satisfies the following equation

$$\bar{H}\chi = E\chi, \tag{11}$$

or

$$(H_0 + U)\chi = E\chi, \tag{12}$$

with $U = U_b$ and follows asymptotic boundary conditions. The division of V expressed by equation (8) is such that the scattering problem with U_b is aimed to be solved exactly while effects of the other part (*i.e.*, $W = V - U_b$) can be treated in some approximation. It should be noted that equation (10) for the T-matrix is still exact for any choice of distortion potential U_b and involves the exact solution Ψ_a^+ of equation (6) for which we adopt here "distorted wave approximation". We first expand Ψ_a^+ in terms of the usual distorted wave Green function

$$G_{U_a}^+ = \lim_{\eta \to 0^+} (E - H_0 - U_a + i\eta)^{-1}, \tag{13}$$

in the following form

$$\Psi_a^+ = \chi_a^+ + \sum_{j=0}^{\infty} \left[G_{U_a}^+ V \right]^j G_{U_a}^+ (V - U_a)\chi_a^+, \tag{14}$$

and substitute it in the T-matrix (equation(10)) as well as consider its first term only, which gives "first order distorted wave approximation" to the T-matrix expressed as $T_{a \to b}^{DW}$ *i.e.*,

$$\begin{aligned} T_{a \to b} &= \left\langle \chi_b^-(\mathbf{1,2,...,N+1}) \left| U_b \right| \psi_a(\mathbf{1,2,...,N+1}) \right\rangle \\ &+ \left\langle \chi_b^-(\mathbf{1,2,...,N+1}) \left| V - U_b \right| \chi_a^+(\mathbf{1,2,...,N+1}) \right\rangle . \end{aligned} \tag{15}$$

If further terms are taken from the series of equation (14) T-matrix in higher order distorted wave approximation can be obtained. Further, the distortion potentials $U = U_a$ or U_b can be chosen such that they depend only on the coordinate of the projectile electron only, *i.e.* $U = U_{a(b)}(N+1)$, then for inelastic scattering of electron the first term of the T-matrix becomes zero due to orthogonality of the atomic wavefunctions and thus the T-matrix takes the following form

$$T_{a\to b} = \langle \chi_b^-(\mathbf{1,2,...,N+1})|V - U_b|\chi_a^+(\mathbf{1,2,...,N+1})\rangle. \tag{16}$$

Representing $\chi_{a(b)}^{+(-)}$ in the above equation as the product of target atomic wavefunction $\Phi_{a(b)}(\mathbf{1,2,...,N})$ and the projectile electron distorted wavefunction $F_{a(b)}^{DW+(-)}(\mathbf{k}_{a(b)},\mathbf{N+1})$ *i.e.*,

$$\chi_{a(b)}^{+(-)}(\mathbf{1,2,...,N+1}) = \mathcal{A}\Phi_{a(b)}(\mathbf{1,2,...,N})\, F_{a(b)}^{DW+(-)}(\mathbf{k}_{a(b)},\mathbf{N+1}). \tag{17}$$

Here $\mathcal{A}$ is antisymmetrization operator that takes into account the exchange of projectile electron with the target electrons. $\mathbf{k}_a$ is the incident projectile electron wave vector and $\mathbf{k}_b$ is scattered electron wave vector after the excitation process. The target atom wavefunction in the initial (final) state satisfies the following equation

$$H_{atom}\Phi_{a(b)} = \epsilon_{a(b)}\Phi_{a(b)}. \tag{18}$$

The total energy of the system can be expressed in terms of these wave vectors as

$$E = \frac{1}{2}k_a^2 + \epsilon_a = \frac{1}{2}k_b^2 + \epsilon_b. \tag{19}$$

Consequently, the T-matrix becomes,

$$T_{a\to b} = \langle \Phi_b(\mathbf{1,2,...,N})F_b^{DW-}(\mathbf{k}_b,\mathbf{N+1})\,|V - U_b(N+1)| \tag{20}$$
$$\mathcal{A}\,\Phi_a(\mathbf{1,2,...,N})F_a^{DW+}(\mathbf{k}_a,\mathbf{N+1})\rangle.$$

Further the T-matrix can be decomposed into its direct and exchange components

$$T_{a\to b} = T_{a\to b}^d + T_{a\to b}^{ex}, \tag{21}$$

with

$$T_{a\to b}^d = \langle \Phi_b(\mathbf{1,2,...,N})F_b^{DW-}(\mathbf{k}_b,\mathbf{N+1})\,|V - U_b(N+1)| \tag{22}$$
$$\times \quad \Phi_a(\mathbf{1,2,...,N})F_a^{DW+}(\mathbf{k}_a,\mathbf{N+1})\rangle,$$

and

$$T_{a\to b}^{ex} = \langle \Phi_b(\mathbf{1,2,...,N})F_b^{DW-}(\mathbf{k}_b,\mathbf{N+1})\,|V - U_b(N+1)| \tag{23}$$
$$\times \quad \sum_{j=1}^{N}(-1)^{N+1-j}\Phi_a(\mathbf{-j})F_a^{DW+}(\mathbf{k}_a,\mathbf{j})\rangle,$$

where $\Phi_a(\mathbf{-j})$ denotes atomic wavefunction with coordinate $\mathbf{j}$ absent and coordinate $\mathbf{N+1}$ included. The exchange T-matrix (23) can be written as the sum of three exchange terms;

$$T^{ex}_{a\to b} = T^{ex1}_{a\to b} + T^{ex2}_{a\to b} + T^{ex3}_{a\to b}, \tag{24}$$

where

$$T^{ex1}_{a\to b} = -N \left\langle \Phi_b(\mathbf{1,2,...,N})F^{DW-}_b(\mathbf{k}_b, \mathbf{N+1}) \left| -\frac{Z}{r_{N+1}} - U_b(N+1) \right| \right. \tag{25}$$
$$\left. \times \quad \Phi_a(\mathbf{1,2,..,N-1,N+1})F^{DW+}_a(\mathbf{k}_a, \mathbf{N}) \right\rangle,$$

$$T^{ex2}_{a\to b} = -N \left\langle \Phi_b(\mathbf{1,2,...,N})F^{DW-}_b(\mathbf{k}_b, \mathbf{N+1}) \left| \frac{1}{|\mathbf{r}_{N+1} - \mathbf{r}_N|} \right| \right. \tag{26}$$
$$\left. \times \quad \Phi_a(\mathbf{1,2,..,N-1,N+1})F^{DW+}_a(\mathbf{k}_a, \mathbf{N}) \right\rangle,$$

$$T^{ex3}_{a\to b} = -N \left\langle \Phi_b(\mathbf{1,2,...,N})F^{DW-}_b(\mathbf{k}_b, \mathbf{N+1}) \left| \sum_{j=1}^{N} \frac{1}{|\mathbf{r}_{N+1} - \mathbf{r}_j|} \right| \right. \tag{27}$$
$$\left. \times \quad \Phi_a(\mathbf{1,2,..,N-1,N+1})F^{DW+}_a(\mathbf{k}_a, \mathbf{N}) \right\rangle.$$

One way of the simplification of the exchange T-matrix is that the distorted waves are forced to be orthogonalized to the bound orbitals. This retains only the two-electron exchange term (equations (26) and (27)). Since exchange of electrons from core electrons is expected to be very small, the third term (equation (27)) can be neglected. Therefore, only second term (26) is left to be evaluated in the exchange T-matrix (23). In the following sections non-relativistic and relativistic distorted wave theory are developed separately for evaluation of the T-matrices.

3   NON-RELATIVISTIC DISTORTED WAVE THEORY

In order to derive the T-matrices in their non-relativistic form we need to describe the atomic structure of the atom and express the distorted waves, non-relativistically. For this purpose Schrödinger equation is solved to obtain both the bound and continuum orbitals. In the following subsections we describe all necessary ingredients to evaluate the non-relativistic T-matrices. Thereafter, we derive the T-matrices for one– and two– valance electron systems.

3.1 The bound state wavefunctions

In its non-relativistic form the Hamiltonian for the atom can be expressed as

$$H_{atom} = \sum_{i=1}^{N} \left[-\frac{1}{2}\nabla_i^2 - \frac{Z}{r_i} + \frac{1}{2}\sum_{j\neq 1}^{N} \frac{1}{|\mathbf{r}_i - \mathbf{r}_j|} \right]. \tag{28}$$

Consequently, the equation (18) can be solved using Hartree–Fock method and the atomic wave function for N electrons representing the atoms in initial or final state can be expressed in terms of Slater determinants from orthogonal orbitals as

$$\Phi_{a(b)}(\mathbf{1,2,...,N}) = \frac{1}{\sqrt{N}} \begin{bmatrix} \phi_{n_1 l_1 m_1}(\mathbf{1}) & \phi_{n_1 l_1 m_1}(\mathbf{2}) & \cdots & \phi_{n_1 l_1 m_1}(\mathbf{N}) \\ \phi_{n_2 l_2 m_2}(\mathbf{1}) & \phi_{n_2 l_2 m_2}(\mathbf{2}) & \cdots & \phi_{n_2 l_2 m_2}(\mathbf{N}) \\ \cdot & \cdot & \cdots & \cdot \\ \cdot & \cdot & \cdots & \cdot \\ \phi_{n_N l_N m_N}(\mathbf{1}) & \phi_{n_N l_N m_N}(\mathbf{2}) & \cdots & \phi_{n_N l_N m_N}(\mathbf{N}) \end{bmatrix},$$

$$\tag{29}$$

where, each Slater orbital $\phi_{nlm}(\mathbf{j})$ of the **j**th electron can be expressed by

$$\phi_{nlm}(\mathbf{j}) = \frac{1}{r_j} P_{nl}(r_j)\, Y_{lm}(\hat{\mathbf{r}}_j)\, S(\boldsymbol{\sigma}_j). \tag{30}$$

Here P_{nl} represents the radial part of the wavefunction, angular part is represented by $Y_{lm}(\hat{\mathbf{r}})$, the spherical harmonics with l and m respectively, the total orbital angular momentum of the atom and the associated magnetic number and S is the spin function. The radial part of the wavefunction can be obtained in numerical form by using the Hartree–Fock computer code of Froese Fischer [18] or in suitable analytic form using table of Clementi [19] or CIV3 code of Hibbert [20].

The radial wavefunction satisfies the orthogonality condition

$$\int_0^\infty dr P_{n'l}(r) P_{nl}(r) = \delta_{n'n}. \tag{31}$$

3.2 The continuum distorted wavefunctions

Further, the distorted waves $F_{a(b)}^{DW+(-)}(\mathbf{k}_{a(b)}, \mathbf{N+1})$ are expressed using the following form of the partial wave expansion,

$$F_{a(b)}^{DW+(-)}(\mathbf{k}_{a(b)}, \mathbf{N+1}) = \frac{4\pi}{k_{a(b)} r_{N+1}} \sum_{l_{a(b)} m_{a(b)}} i^{l_{a(b)}} e^{\pm i\delta_{l_{a(b)}}} \tag{32}$$

$$\times\ u_{l_{a(b)}}(k_{a(b)}, r_{N+1}) Y^{*}_{l_{a(b)} m_{a(b)}}(\hat{\mathbf{k}}_{a(b)}) Y_{l_{a(b)} m_{a(b)}}(\hat{\mathbf{r}}_{N+1})$$

$$\times\ S(\boldsymbol{\sigma}_{N+1}),$$

where, δ_l is the phase shift of the lth partial wave and $u_{l_{a(b)}}(k_{a(b)}, r_{N+1})$ is the radial part of the distorted wave. S_{N+1} is the spin function. On substituting $F_{a(b)}^{DW+(-)}$ in the following equation

$$\left[K_{N+1} + U_{a(b)}(N+1) \right] F_{a(b)}^{DW}(\mathbf{k}_{a(b)}, \mathbf{N+1}) = (E - \epsilon_{a(b)}) F_{a(b)}^{DW}(\mathbf{k}_{a(b)}, \mathbf{N+1}), \tag{33}$$

we find that $u_{l_{a(b)}}(k_{a(b)}, r_{N+1})$ satisfies the equation

$$\left[\frac{d^2}{dr_{N+1}^2} + k_{a(b)}^2 - \frac{l_{a(b)}(l_{a(b)}+1)}{r_{N+1}^2} - 2U_{a(b)}(r_{N+1})\right]u_{l_{a(b)}} = 0. \tag{34}$$

The above equation is solved to obtain the radial wavefunction and phase shift by subjecting to the following usual boundary conditions

$$u_{l_{a(b)}}(k_{a(b)}, r_{N+1})_{r_{N+1}\to 0} = 0, \tag{35}$$

and

$$u_{l_{a(b)}}(k_{a(b)}, r_{N+1})_{r_{N+1}\to\infty} = \frac{1}{\sqrt{k_{a(b)}}}\sin\left(k_{a(b)}r_{N+1} - \frac{\pi l_{a(b)}}{2} + \delta_{l_{a(b)}}k_{a(b)}^2\right). \tag{36}$$

3.3   Choice of the distortion potential

In order to evaluate the T-matrix and to obtain continuum distorted waves we need to choose the distortion potential $U_{a(b)}$ in the initial(final) channel. This has been taken as the sum of the static potential V_{ch}^{st} and the exchange potential V_{ch}^{ex}, i.e.,

$$U_{ch} = V_{ch}^{st} + V_{ch}^{ex} \qquad (\text{where, } ch = \text{a or b}). \tag{37}$$

The static potential V_{ch}^{st} is the spherical average of the static interaction $\langle\phi_{ch}|V|\phi_{ch}\rangle$ between the incident or scattered electron and the target atom in the initial or final state depending on whether ch is $'a'$ or $'b'$. Consequently,

$$\begin{aligned}
V_{ch}^{st}(r_{N+1}) &= -\frac{Z}{r_{N+1}} + \sum_{j=1}^{n}\bar{n}_j\int_0^\infty\frac{|R_j(r)|^2}{r_>}r^2dr \tag{38}\\
&= -\frac{Z}{r_{N+1}} + \sum_{j=1}^{n}\bar{n}_j\left(\frac{1}{r_{N+1}}\int_0^{r_{N+1}}|P_{n_jl_j}(r)|^2dr\right.\\
&\quad + \left.\int_{r_{N+1}}^\infty\frac{1}{r}|P_{n_jl_j}(r)|^2dr\right),
\end{aligned}$$

where $\bar{n}_j$ is the occupation number of the electrons in the jth orbital and $R_j(r)$ is the corresponding radial part of the orbital wavefunction $\phi_{ch}(r)$ of the atomic target state. The summation is over all the core and valance orbitals i.e., for all the number of orbitals for an N-electron atom. The wavefunction of the outer orbitals, as well as the core orbitals are taken to be the Hartree–Fock wavefunctions obtained using Froese Fischer code [18].

The dynamic effects in the atomic potential such as electron exchange distortion V_{ch}^{ex} can be included in the distortion potential U_{ch}. Since the exchange potential is non local, the following form is commonly used equivalent local energy dependent exchange potential [21],

$$\begin{aligned}
V_{ch}^{ex} &= \frac{1}{2}\left[\left(\frac{1}{2}k^2 - V_{st}(r) + \frac{3}{10}(3\pi^2\rho(r))^{\frac{2}{3}}\right)\right.\\
&\quad - \left.\left\{\left(\frac{1}{2}k^2 - V_{st}(r) + \frac{3}{10}(3\pi^2\rho(r))^{\frac{2}{3}}\right)^2 + 4\pi\rho(r)\right\}^{\frac{1}{2}}\right], \tag{39}
\end{aligned}$$

where $\rho(r)$ is the spherical charge density of the atom given by

$$\rho(r) = \sum_{j=1}^{n} \bar{n}_j \left| \frac{P_{n_j l_j}(r)}{r} \right|^2. \tag{40}$$

3.4 The T-matrix for one-valance electron system

The direct T-matrix as given by equation (22) can be reduced for target atom having one electron (or a filled core with one valance electron) in the following form after substituting the target wavefunction of N electron system in terms of one electron Slater orbitals (29) and (30). It can be shown that in the expression (22) for the T- matrix only $1/|\mathbf{r}_{N+1} - \mathbf{r}_N|$ contributes [22] resulting in

$$
\begin{aligned}
T_{a \to b}^d &= \left\langle \phi_b(\mathbf{N}) F_b^{DW-}(\mathbf{k}_b, \mathbf{r}_{N+1}) \left| \frac{1}{|\mathbf{r}_{N+1} - \mathbf{r}_N|} \right. \right. \\
&\times \left. \phi_a(\mathbf{N}) F_a^{DW+}(\mathbf{k}_a, \mathbf{r}_{N+1}) \right\rangle \\
&\times \langle S_b(\boldsymbol{\sigma}_N; \boldsymbol{\sigma}_{N+1}) | S_a(\boldsymbol{\sigma}_N; \boldsymbol{\sigma}_{N+1}) \rangle,
\end{aligned}
\tag{41}
$$

where, the symmetry property of the potential operator and the Slater determinants have been used. Here the coordinate N corresponds to the valance electron.

Further, expressing in terms of space (r) and spin $(\boldsymbol{\sigma})$ coordinates we can re-write the T-matrix (equation(41)) as

$$
\begin{aligned}
T_{a \to b}^d &= \langle F_b^{DW-}(\mathbf{k}_b, \mathbf{r}_{N+1}) | V_{a \to b}(\mathbf{r}_{N+1}) | F_a^{DW+}(\mathbf{k}_a, \mathbf{r}_{N+1}) \rangle \\
&\times \langle S_b(\boldsymbol{\sigma}_N; \boldsymbol{\sigma}_{N+1}) | S_a(\boldsymbol{\sigma}_N; \boldsymbol{\sigma}_{N+1}) \rangle \\
&= \hat{T}_{a \to b}^d \times \langle S_b(\boldsymbol{\sigma}_N; \boldsymbol{\sigma}_{N+1}) | S_a(\boldsymbol{\sigma}_N; \boldsymbol{\sigma}_{N+1}) \rangle,
\end{aligned}
\tag{42}
$$

where,

$$\hat{T}_{a \to b}^d = \langle F_b^{DW-}(\mathbf{k}_b, \mathbf{r}_{N+1}) | V_{a \to b}(\mathbf{r}_{N+1}) | F_a^{DW+}(\mathbf{k}_a, \mathbf{r}_{N+1}) \rangle, \tag{43}$$

with

$$V_{a \to b}(\mathbf{r}_{N+1}) = \left\langle \phi_b(\mathbf{N}) \left| \frac{1}{|\mathbf{r}_{N+1} - \mathbf{r}_N|} \right| \phi_a(\mathbf{N}) \right\rangle. \tag{44}$$

In the above equation (44) putting the value of initial and final target wavefunctions as expressed in (30) and using the following form of expression for the interaction term

$$\frac{1}{|\mathbf{r}_{N+1} - \mathbf{r}_N|} = \sum_{p=0}^{\infty} \frac{4\pi}{2p+1} \frac{r_<^p}{r_>^{p+1}} \sum_{q=-p}^{p} Y_{pq}^*(\hat{\mathbf{r}}_{N+1}) Y_{pq}(\hat{\mathbf{r}}_N), \tag{45}$$

we can write

$$V_{a \to b}(\mathbf{r}_{N+1}) = \sum_{pq} U_{a \to b}^p(r_{N+1}) Y_{pq}^*(\hat{\mathbf{r}}_{N+1}), \tag{46}$$

where,

$$U^p_{a\to b}(r_{N+1}) = \frac{4\pi}{2p+1}\left[\frac{(2L_a+1)(2p+1)}{4\pi(2L_b+1)}\right]^{1/2}\begin{pmatrix} L_a & p & L_b \\ 0 & 0 & 0 \end{pmatrix} \tag{47}$$

$$\times \begin{pmatrix} L_a & p & L_b \\ M_a & q & M_b \end{pmatrix} I^p_{L_a L_b}(r_{N+1}).$$

Here

$$I^p_{L_a L_b}(r_{N+1}) = \int_0^\infty P_{n_b L_b}(r_N)\,\gamma_p\,P_{n_a L_a}(r_N)dr_N, \tag{48}$$

and,

$$\gamma_p(r_N, r_{N+1}) = \frac{r_<^p}{r_>^{p+1}}, \tag{49}$$

with $r_< = \min(r_N, r_{N+1})$ and $r_> = \max(r_N, r_{N+1})$.

Now using the equation (46) and the partial wave expansion for the distorted wave as given by (32) the expression for direct T- matrix given by (41) takes the following simplified form after carrying out the involved angular momentum algebra [23]

$$\hat{T}^d_{a\to b} = \frac{8\pi^{3/2}}{k_a k_b}\sum_{\substack{p\,l_a\,l_b \\ q\,m_a\,m_b}} i^{l_a-l_b}e^{i(\delta_{l_a}+\delta_{l_b})}Y^*_{l_a m_a}(\hat{\mathbf{k}}_a)Y^*_{l_b m_b}(\hat{\mathbf{k}}_b)D^{l_a l_b}_{p L_a L_b} \tag{50}$$

$$\times \frac{[l_b, l_a]^{1/2}}{[p]^{1/2}}\begin{pmatrix} l_a & l_b & p \\ 0 & 0 & 0 \end{pmatrix}\begin{pmatrix} l_a & l_b & p \\ m_a & m_b & q \end{pmatrix}$$

where,

$$D^{l_a l_b}_{p L_a L_b} = \int_0^\infty u_{l_b}(k_b, r_{N+1})\,U^p_{a\to b}(r_{N+1})\,u_{l_a}(k_a, r_{N+1})\,dr_{N+1}, \tag{51}$$

and $[a, b, ...] = (2a+1)(2b+1)....$

Similarly, the exchange T-matrix as given by equation (26) can be further expressed in the following form for one-electron system (or a system having one valence electron)

$$T^{ex}_{a\to b} = -\left\langle \phi_b(\mathbf{N})F^{DW-}_b(\mathbf{k}_b, \mathbf{r}_{N+1})\left|\frac{1}{|\mathbf{r}_{N+1}-\mathbf{r}_N|}\right|\right. \tag{52}$$

$$\times \left. \phi_a(\mathbf{N+1})F^{DW+}_a(\mathbf{k}_a, \mathbf{r}_N)\right\rangle \langle S_b(\boldsymbol{\sigma}_{N+1};\boldsymbol{\sigma}_N)|\,S_a(\boldsymbol{\sigma}_N;\boldsymbol{\sigma}_{N+1})\rangle$$

$$= -\hat{T}^{ex}_{a\to b}\times\langle S_b(\boldsymbol{\sigma}_{N+1};\boldsymbol{\sigma}_N)|\,S_a(\boldsymbol{\sigma}_N;\boldsymbol{\sigma}_{N+1})\rangle,$$

where,

$$\hat{T}^{ex}_{a\to b} = \left\langle \phi_b(\mathbf{N})F^{DW-}_b(\mathbf{k}_b, \mathbf{r}_{N+1})\left|\frac{1}{|\mathbf{r}_{N+1}-\mathbf{r}_N|}\right|\phi_a(\mathbf{N+1})F^{DW+}_a(\mathbf{k}_a, \mathbf{r}_N)\right\rangle. \tag{53}$$

Using the expansion for the interaction term (45), the initial and final orbital wavefunctions (30) and the partial wave expansion of the distorted waves from equation (32) and after carrying out the angular momentum algebra we obtain

$$\hat{T}^{ex}_{a\to b} = -\frac{16\pi^2}{k_a k_b} \sum_{\substack{p\,l_a l_b \\ q\,m_a m_b}} i^{l_a-l_b} e^{i(\delta_{l_a}+\delta_{l_b})} Y^*_{l_a m_a}(\hat{\mathbf{k}}_a) Y^*_{l_b m_b}(\hat{\mathbf{k}}_b) E^{l_a l_b}_{p\,L_a L_b} \quad (54)$$

$$\times \frac{[L_a, l_b, l_a]^{1/2}}{[p^2 L_b]^{1/2}} \begin{pmatrix} l_b & L_a & p \\ 0 & 0 & 0 \end{pmatrix} \begin{pmatrix} l_b & L_a & p \\ m_b & M_a & q \end{pmatrix}$$

$$\times \begin{pmatrix} p & l_a & L_b \\ 0 & 0 & 0 \end{pmatrix} \begin{pmatrix} p & l_a & L_b \\ q & m_a & M_b \end{pmatrix},$$

where,

$$E^{l_a l_b}_{p\,L_a L_b} = \int_0^\infty \int_0^\infty P_{n_b l_b}(r_{N+1}) u_{l_b}(k_b, r_N) \frac{r^p_<}{r^{p+1}_>} \quad (55)$$

$$\times \; P_{n_a l_a}(r_N) u_{l_a}(k_a, r_{N+1}) \, dr_N \, dr_{N+1}.$$

Since we are considering the "collision frame of reference" *i.e.*, the z-axis is chosen to the incident beam direction $\mathbf{k}_a$, we can write

$$Y^*_{l_a m_a}(\hat{\mathbf{k}}_a) = \frac{[l_a]}{\sqrt{4\pi}} \delta_{m_a,0}. \quad (56)$$

Using this in equations (50) and (54) we obtain for the direct and exchange T-matrices

$$\hat{T}^d_{a\to b} = \frac{4\pi}{k_a k_b} \sum_{\substack{p\,l_a l_b \\ q\,m_b}} i^{l_a-l_b} e^{i(\delta_{l_a}+\delta_{l_b})} Y^*_{l_b m_b}(\hat{\mathbf{k}}_b) D^{l_a l_b}_{p\,L_a L_b} \quad (57)$$

$$\times \frac{[l_b, l_a^2]^{1/2}}{[p]^{1/2}} \begin{pmatrix} l_a & l_b & p \\ 0 & 0 & 0 \end{pmatrix} \begin{pmatrix} l_a & l_b & p \\ 0 & m_b & q \end{pmatrix},$$

and

$$\hat{T}^{ex}_{a\to b} = -\frac{8\pi^{3/2}}{k_a k_b} \sum_{\substack{p\,l_a l_b \\ q\,m_b}} i^{l_a-l_b} e^{i(\delta_{l_a}+\delta_{l_b})} Y^*_{l_b m_b}(\hat{\mathbf{k}}_b) E^{l_a l_b}_{p\,L_a L_b} \frac{[L_a, l_b, l_a^2]^{1/2}}{[p^2 L_b]^{1/2}} \quad (58)$$

$$\times \begin{pmatrix} l_b & L_a & p \\ 0 & 0 & 0 \end{pmatrix} \begin{pmatrix} l_b & L_a & p \\ m_b & M_a & q \end{pmatrix} \begin{pmatrix} p & l_a & L_b \\ 0 & 0 & 0 \end{pmatrix} \begin{pmatrix} p & l_a & L_b \\ q & 0 & M_b \end{pmatrix}.$$

Thus the combined T-matrix following equations (50) and (54) as

$$T_{a\to b} = \hat{T}^d_{a\to b} \langle S_b(\boldsymbol{\sigma}_N; \boldsymbol{\sigma}_{N+1}) | S_a(\boldsymbol{\sigma}_N; \boldsymbol{\sigma}_{N+1}) \rangle \quad (59)$$

$$- \hat{T}^{ex}_{a\to b} \langle S_b(\boldsymbol{\sigma}_{N+1}; \boldsymbol{\sigma}_N) | S_a(\boldsymbol{\sigma}_N; \boldsymbol{\sigma}_{N+1}) \rangle,$$

where the first term in the bracket comes from the direct part of the expression and the second term from the exchange contribution. This T-matrix expressed here is for transition of an atom for $L_a M_a \to L_b M_b$ in their doublet states.

The above T-matrix (59) can be further simplified for any excitation process from an initial doublet state of the target to a final doublet state by carrying out integration over the spin coordinates. In the doublet–doublet excitation of the target the scattering can take place either in the singlet mode (s) or triplet mode (t). The spin function in these modes are referred to as $S_{s,t}(\sigma_N; \sigma_{N+1})$. Scattering in the singlet mode occurs in one fourth of all collisions and in the triplet mode occurs in three fourths of all collisions. The singlet normalized spin function $S_s(\sigma_N; \sigma_{N+1})$ in terms of usual Dirac matrices α and β for the two electron system with total magnetic component $M_a = 0$ is given by

$$S_s(\sigma_N; \sigma_{N+1}) = \frac{1}{\sqrt{2}} \left[\alpha_N \beta_{N+1} - \alpha_{N+1} \beta_N \right]. \tag{60}$$

Substituting $S_a = S_b = S_s$ into the T-matrix (59) and carrying out the spin integration using usual orthogonal property of individual spin function $i.e.$,

$$\langle \alpha_i | \alpha_j \rangle = \delta_{ij}, \tag{61}$$

$$\langle \beta_i | \beta_j \rangle = \delta_{ij}, \tag{62}$$

$$\langle \alpha_i | \beta_j \rangle = 0, \tag{63}$$

we get,

$$T^s_{a \to b} = \hat{T}^d_{a \to b} + \hat{T}^{ex}_{a \to b}, \tag{64}$$

where, $\hat{T}^d_{a \to b}$ and $\hat{T}^{ex}_{a \to b}$ can be said as spin averaged direct and exchange T-matrices given by equations (57) and (58).

Similarly, the triplet mode spin function $S_t(\sigma_N; \sigma_{N+1})$ for the total magnetic components $M_s = -1, 0$ and 1 are given by

$$\begin{aligned}
S_t(\sigma_N; \sigma_{N+1}) &= \alpha_N \alpha_{N+1}, & M_a &= +1, & (65)\\
&= \beta_N \beta_{N+1}, & M_a &= -1, & (66)\\
&= \frac{1}{\sqrt{2}} \left[\alpha_N \beta_{N+1} + \alpha_{N+1} \beta_N \right], & M_a &= 0. & (67)
\end{aligned}$$

Substituting $S_a = S_b = S_t$ into the T-matrix (59) and carrying out spin integration we get

$$T^t_{a \to b} = \hat{T}^d_{a \to b} - \hat{T}^{ex}_{a \to b}. \tag{68}$$

Thus the differential cross section for the doublet–doublet excitation can be written as

$$\frac{d\sigma}{d\Omega} = \frac{1}{4\pi^2} \frac{k_b}{k_a} \sum_{M_b} \left[\frac{1}{4} |T^s_{a \to b}|^2 + \frac{3}{4} |T^t_{a \to b}|^2 \right]. \tag{69}$$

3.5   The T-matrix for two-valance electron system

The general T-matrix (equation(21)) can be simplified for two electron system (or system with two valance electrons) similar to as done for the one electron system in the previous subsection. Following Lindgren and Morrison [22] the direct T-matrix (equation(22)) for target atom having two valence electrons is then given by

$$
\begin{aligned}
T^d_{a\to b} &= \Big\langle \phi_b(\mathbf{N\text{-}1,N})F^{DW-}_b(\mathbf{k}_b,\mathbf{r}_{N+1}) \Big| \frac{2}{|\mathbf{r}_{N+1}-\mathbf{r}_N|} \Big| \\
&\times \quad \phi_a(\mathbf{N\text{-}1,N})F^{DW+}_a(\mathbf{k}_a,\mathbf{r}_{N+1}) \Big\rangle \\
&\times \quad \langle S_b(\boldsymbol{\sigma}_{N-1},\boldsymbol{\sigma}_N;\boldsymbol{\sigma}_{N+1})| S_a(\boldsymbol{\sigma}_{N-1},\boldsymbol{\sigma}_N;\boldsymbol{\sigma}_{N+1})\rangle .
\end{aligned}
\tag{70}
$$

Here $\mathbf{N}$ and $\mathbf{N\text{-}1}$ are taken to be the coordinates of the outer two valence electrons in the N-electron atomic target.

Further writing in terms of space ($\mathbf{r}$) and spin ($\boldsymbol{\sigma}$) coordinates we can re-express the T-matrix (70)

$$
\begin{aligned}
T^d_{a\to b} &= \big\langle F^{DW-}_b(\mathbf{k}_b,\mathbf{r}_{N+1}) \,|V_{a\to b}(\mathbf{r}_{N+1})|\, F^{DW+}_a(\mathbf{k}_a,\mathbf{r}_{N+1})\big\rangle \\
&\times \quad \langle S_b(\boldsymbol{\sigma}_{N-1},\boldsymbol{\sigma}_N;\boldsymbol{\sigma}_{N+1})|\, S_a(\boldsymbol{\sigma}_{N-1},\boldsymbol{\sigma}_N;\boldsymbol{\sigma}_{N+1})\rangle \\
&= \hat{T}^d_{a\to b} \times \langle S_b(\boldsymbol{\sigma}_{N-1},\boldsymbol{\sigma}_N;\boldsymbol{\sigma}_{N+1})|\, S_a(\boldsymbol{\sigma}_{N-1},\boldsymbol{\sigma}_N;\boldsymbol{\sigma}_{N+1})\rangle ,
\end{aligned}
\tag{71}
$$

where, $\hat{T}^d_{a\to b}$ is again spin–avaraged direct T-matrix and can be given by equation (43) with

$$
V_{a\to b}(\mathbf{r}_{N+1}) = \Big\langle \phi_b(\mathbf{N\text{-}1,N}) \Big| \frac{2}{|\mathbf{r}_{N+1}-\mathbf{r}_N|} \Big| \phi_a(\mathbf{N\text{-}1,N}) \Big\rangle .
\tag{72}
$$

To evaluate the above expression the wavefunctions of the atom with two valence electrons in its initial and final states are required. For this purpose the Hartree–Fock wavefunctions [18] can be used for the bound states of the target atom.

Let us take the initial state to be singlet ground state having L_a and M_a as its orbital and magnetic quantum numbers and both the electrons are in the same orbital represented by $n_A l_A$ then we can write

$$
\begin{aligned}
\phi_a(\mathbf{r}_{N-1},\mathbf{r}_N) &= \sum_{\nu_A} \begin{pmatrix} l_A & l_A & L_a \\ \nu_A & M_a-\nu_A & M_a \end{pmatrix} \phi_{n_A l_A \nu_A}(\mathbf{r}_{N-1}) \\
&\times \quad \phi_{n_A l_A M_a - \nu_A}(\mathbf{r}_N).
\end{aligned}
\tag{73}
$$

Similarly, we can write the excited state in its triplet or singlet states having L_b and M_b as its orbital and magnetic quantum numbers and both the electrons are in orbitals represented by $n_A l_A$ and $n_B l_B$ respectively, by the following form

$$
\begin{aligned}
\phi_b(\mathbf{r}_{N-1},\mathbf{r}_N) &= \frac{1}{\sqrt{2}} \sum_{\nu_B} \begin{pmatrix} l_A & l_B & L_b \\ \nu_B & M_b-\nu_B & M_b \end{pmatrix} \\
&\times \quad (\phi_{n_A l_A \nu_B}(\mathbf{r}_{N-1})\phi_{n_B l_B M_b-\nu_B}(\mathbf{r}_N) \\
&\pm \quad \phi_{n_A l_A \nu_B}(\mathbf{r}_N)\phi_{n_B l_B M_b-\nu_B}(\mathbf{r}_{N-1})),
\end{aligned}
\tag{74}
$$

where $+(-)$ is associated with singlet (triplet) state.

Substituting the expressions of the wavefunctions given by equations (73) and (74) in equation (72) as well as using the expansion of the interaction term as given by equation (45) we can write

$$
V_{a\to b}(\mathbf{r}_{N+1}) = \sum_{\nu_A \nu_B pq} \frac{8\pi}{\sqrt{2}(2p+1)} \begin{pmatrix} l_A & l_A & L_a \\ \nu_A & M_a - \nu_A & M_a \end{pmatrix} \tag{75}
$$

$$
\times \begin{pmatrix} l_A & l_B & L_b \\ \nu_B & M_b - \nu_B & M_b \end{pmatrix}
$$

$$
\times \Bigg\langle (\phi_{n_A l_A \nu_B}(\mathbf{r}_{N-1}) \phi_{n_B l_B M_b - \nu_B}(\mathbf{r}_N)
$$

$$
\pm \ \phi_{n_A l_A \nu_B}(\mathbf{r}_N) \phi_{n_B l_B M_b - \nu_B}(\mathbf{r}_{N-1})) \left| \frac{r_<^p}{r_>^{p+1}} Y_{pq}^*(\hat{\mathbf{r}}_{N+1}) Y_{pq}(\hat{\mathbf{r}}_N) \right|
$$

$$
\times \ \phi_{n_A l_A \nu_A}(\mathbf{r}_{N-1}) \phi_{n_A l_A M_a - \nu_A}(\mathbf{r}_N) \Bigg\rangle .
$$

Considering initial state to be ^1S ground state and forcing orthonormalization between the orbitals of ground and excited states we have

$$
V_{a\to b}(\mathbf{r}_{N+1}) = \sum_{\nu_B pq} \frac{8}{\sqrt{2}(2p+1)} \begin{pmatrix} l_A & l_B & L_b \\ \nu_B & M_b - \nu_B & M_b \end{pmatrix} \tag{76}
$$

$$
\times \left\langle \phi_{n_B l_B M_b - \nu_B}(\mathbf{r}_N) \left| \frac{r_<^p}{r_>^{p+1}} Y_{pq}^*(\hat{\mathbf{r}}_{N+1}) Y_{pq}(\hat{\mathbf{r}}_N) \right| \phi_{n_A l_A M_a}(\mathbf{r}_N) \right\rangle .
$$

Here M_a takes the value zero. Now writing $V_{a\to b}(\mathbf{r}_{N+1})$ in a similar form as in one-electron system case (46), $U_{a\to b}^p(r_{N+1})$ can be written as,

$$
U_{a\to b}^p(r_{N+1}) = 8\pi \frac{[l_A]}{[p][l_B]} \sum_{\nu_B} I_{l_A l_B}^p \begin{pmatrix} l_A & p & l_B \\ 0 & 0 & 0 \end{pmatrix} \tag{77}
$$

$$
\times \begin{pmatrix} l_A & l_B & L_b \\ \nu_B & M_b - \nu_B & M_b \end{pmatrix} \begin{pmatrix} l_A & p & l_B \\ -\nu_A & q & M_b - \nu_B \end{pmatrix} ,
$$

where

$$
I_{l_A l_B}^p(r_{N+1}) = \int_0^\infty P_{n_B l_B}(r_N) \frac{r_<^p}{r_>^{p+1}} P_{n_A l_A}(r_N) \, dr_N. \tag{78}
$$

Using this form of the $V_{a\to b}(\mathbf{r}_{N+1})$ in equation (72) for the direct T-matrix and also the partial wave expansion for the distorted waves as given by (32) then the direct T-matrix in simplified form can be obtained after carrying out the angular momentum algebra in the form similar to equation (57) for excitation of one electron case. However, here the $D_{p L_b L_a}^{l_b l_a}$ in the equation is replaced by $D_{p l_B l_A}^{l_b l_a}$ which is given by

$$
D_{p l_B l_A}^{l_b l_a} = \int_0^\infty u_{l_b}(k_b, r_{N+1}) \, U_{a\to b}^p(r_{N+1}) \, u_{l_a}(k_a, r_{N+1}) \, dr_{N+1}. \tag{79}
$$

Further, following similar discussion as for excitation of one-electron system in previous subsection we can write the exchange T-matrix for two-electron system as

$$T_{a\to b}^{ex} = -2\left\langle \phi_b(\mathbf{N\text{-}1,N})F_b^{DW-}(\mathbf{k}_b,\mathbf{r}_{N+1})\left|\frac{1}{|\mathbf{r}_{N+1}-\mathbf{r}_N|}\right|\right. \tag{80}$$

$$\times\ \left.\phi_a(\mathbf{N\text{-}1,N+1})F_a^{DW+}(\mathbf{k}_a,\mathbf{r}_N)\right\rangle$$

$$\times\ \langle S_b(\boldsymbol{\sigma}_{N-1},\boldsymbol{\sigma}_N;\boldsymbol{\sigma}_{N+1})|\,S_a(\boldsymbol{\sigma}_{N-1},\boldsymbol{\sigma}_{N+1};\boldsymbol{\sigma}_N)\rangle$$

$$=\ -2\hat{T}_{a\to b}^{ex}\times\langle S_b(\boldsymbol{\sigma}_{N-1},\boldsymbol{\sigma}_N;\boldsymbol{\sigma}_{N+1})|\,S_a(\boldsymbol{\sigma}_{N-1},\boldsymbol{\sigma}_{N+1};\boldsymbol{\sigma}_N)\rangle,$$

where

$$\hat{T}_{a\to b}^{ex} = \left\langle \phi_b(\mathbf{N\text{-}1,N})F_b^{DW-}(\mathbf{k}_b,\mathbf{r}_{N+1})\left|\frac{1}{|\mathbf{r}_{N+1}-\mathbf{r}_N|}\right|\right. \tag{81}$$

$$\times\ \left.\phi_a(\mathbf{N\text{-}1,N+1})F_a^{DW+}(\mathbf{k}_a,\mathbf{r}_N)\right\rangle.$$

In the above equation using the atomic wavefunctions given by equations (73) and (74) and treating initial state to be ^1S ground state we have,

$$\hat{T}_{a\to b}^{ex} = \frac{1}{\sqrt{2}}\sum_{\nu_B}\begin{pmatrix} l_A & l_B & L_b \\ \nu_B & M_b-\nu_B & M_b \end{pmatrix} \tag{82}$$

$$\times\ \left\langle \left(\phi_{n_Al_A\nu_B}(\mathbf{r}_{N-1})\phi_{n_Bl_BM_b-\nu_B}(\mathbf{r}_N)\right.\right.$$

$$\pm\ \phi_{n_Al_A\nu_B}(\mathbf{r}_N)\phi_{n_Bl_BM_b-\nu_B}(\mathbf{r}_{N-1}))\,F_b^{DW-}(\mathbf{k}_b,\mathbf{r}_{N+1})$$

$$\times\ \left|\frac{1}{|\mathbf{r}_{N+1}-\mathbf{r}_N|}\right|\phi_{n_Al_AM_a}(\mathbf{r}_{N-1})\phi_{n_Al_AM_a}(\mathbf{r}_{N+1})F_a^{DW+}(\mathbf{k}_a,\mathbf{r}_N)\Big\rangle,$$

which on forcing orthogonalization leads to

$$\hat{T}_{a\to b}^{ex} = \frac{1}{\sqrt{2}}\sum_{\nu_B}\begin{pmatrix} l_A & l_B & L_b \\ \nu_B & M_b-\nu_B & M_b \end{pmatrix} \tag{83}$$

$$\times\ \left\langle \phi_{n_Bl_BM_b-\nu_B}(\mathbf{r}_N)F_b^{DW-}(\mathbf{k}_b,\mathbf{r}_{N+1})\left|\frac{1}{|\mathbf{r}_{N+1}-\mathbf{r}_N|}\right|\right.$$

$$\times\ \phi_{n_Al_AM_a}(\mathbf{r}_{N+1})F_a^{DW+}(\mathbf{k}_a,\mathbf{r}_N)\Big\rangle.$$

Following the substitution of the bound state wavefunctions (30) and partial wave expansion for distorted waves equation (32) and interaction term (45) we obtain

$$\hat{T}^{ex}_{a\to b} = \frac{8\sqrt{2}\pi^2}{k_a k_b} \sum_{\substack{\nu_B p\, l_a l_b \\ q\, m_a m_b}} i^{l_a - l_b} e^{i(\delta_{l_a}+\delta_{l_b})} Y^*_{l_a m_a}(\hat{\mathbf{k}}_a) Y^*_{l_b m_b}(\hat{\mathbf{k}}_b) E^{l_b l_a}_{p\, l_B l_A} \qquad (84)$$

$$\times \frac{[l_A, l^2_a, l_b]^{1/2}}{[p^2 l_B]^{1/2}} \begin{pmatrix} l_A & l_B & L_b \\ \nu_B & M_b - \nu_B & M_b \end{pmatrix} \begin{pmatrix} l_a & l_A & p \\ 0 & 0 & 0 \end{pmatrix}$$

$$\times \begin{pmatrix} l_a & l_A & p \\ m_a & M_a & q \end{pmatrix} \begin{pmatrix} l_b & p & l_B \\ 0 & 0 & 0 \end{pmatrix} \begin{pmatrix} l_b & p & l_B \\ m_b & q & M_b - \nu_B \end{pmatrix},$$

where,

$$E^{l_b l_a}_{p\, l_B l_A} = \int_0^\infty \int_0^\infty P_{n_B l_B}(r_N) u_{l_a}(k_a, r_N) \frac{r^p_<}{r^{p+1}_>} P_{n_A l_A}(r_{N+1}) \qquad (85)$$

$$\times\, u_{l_b}(k_a, r_{N+1})\, dr_N\, dr_{N+1}.$$

Again considering the "collision frame of reference" *i.e.*, the z-axis is chosen parallel to the incident beam direction $\mathbf{k}_a$, the exchange T-matrix $\hat{T}^{ex}_{a\to b}$ is given by

$$\hat{T}^{ex}_{a\to b} = \frac{4\sqrt{2}\pi^{3/2}}{k_a k_b} \sum_{\substack{\nu_B p\, l_a l_b \\ q\, m_a m_b}} i^{l_a - l_b} e^{i(\delta_{l_a}+\delta_{l_b})} Y^*_{l_b m_b}(\hat{\mathbf{k}}_b) E^{l_b l_a}_{p\, l_B l_A} \frac{[l_A, l^2_a, l_b]^{1/2}}{[p^2 l_B]^{1/2}} \qquad (86)$$

$$\times \begin{pmatrix} l_A & l_B & L_b \\ \nu_B & M_b - \nu_B & M_b \end{pmatrix} \begin{pmatrix} l_a & l_A & p \\ 0 & 0 & 0 \end{pmatrix} \begin{pmatrix} l_a & l_A & p \\ 0 & M_a & q \end{pmatrix}$$

$$\times \begin{pmatrix} l_b & p & l_B \\ 0 & 0 & 0 \end{pmatrix} \begin{pmatrix} l_b & p & l_B \\ m_b & q & M_b - \nu_B \end{pmatrix}.$$

Again the combined T-matrix can be expressed as

$$T_{a\to b} = \hat{T}^d_{a\to b} \times \langle S_b(\boldsymbol{\sigma}_{N-1}, \boldsymbol{\sigma}_N; \boldsymbol{\sigma}_{N+1}) | S_a(\boldsymbol{\sigma}_{N-1}, \boldsymbol{\sigma}_N; \boldsymbol{\sigma}_{N+1}) \rangle \qquad (87)$$

$$- 2\, \hat{T}^{ex}_{a\to b} \times \langle S_b(\boldsymbol{\sigma}_{N-1}, \boldsymbol{\sigma}_N; \boldsymbol{\sigma}_{N+1}) | S_a(\boldsymbol{\sigma}_{N-1}, \boldsymbol{\sigma}_{N+1}; \boldsymbol{\sigma}_N) \rangle.$$

Let us consider two types of transitions for the two electron system *viz.*, the excitation from the ground n^1S state to excited n^1P or n^1D state *i.e.*, singlet to singlet (SS) transition and the other excitation from the ground n^1S state to n^3P state or n^3D state *i.e.*, singlet to triplet (ST) excitations. In the SS excitaion the scattering takes palce in the doublet mode *i.e.*, the total spin of the system $S = 1/2$. The doublet mode spin function $S_a(S_b)$ in terms of the usual Dirac matrices α and β for the composite system is given by

$$S_{a(b)}(\boldsymbol{\sigma}_{N-1}, \boldsymbol{\sigma}_N; \boldsymbol{\sigma}_{N+1}) = \frac{1}{\sqrt{2}} \alpha_{N+1} \left[\alpha_{N-1}\beta_N - \alpha_N\beta_{N-1} \right], \qquad (88)$$

where $N-1, N$ and $N+1$ denote the two valence electrons and the projectile electron, respectively.

Substituting $S_{a(b)}$ into the T-matrix (87) and carrying out the spin integration using orthogonal properties of individual spin function given by equations

(61), (62) and (63), we get for $^1S \to ^1 L_b M_b$ excitation, the combined T-matrix (87) as

$$T_{a \to b} = \hat{T}^d_{a \to b} - \hat{T}^{ex}_{a \to b}.$$ (89)

The differential cross section for the singlet-singlet excitations therefore, is given by

$$\frac{d\sigma}{d\Omega} = \frac{1}{4\pi^2} \frac{k_b}{k_a} \sum_{M_b} \left[|T^d_{a \to b} - T^{ex}_{a \to b}|^2 \right].$$ (90)

In the singlet-triplet excitations, the scattering takes place in the doublet mode, *i.e.*, in this mode also the total spin of the composite system $S = 1/2$. This is due to conservation of the total spin of the system during the excitation process. The expression for S_a and S_b in the ST excitation are given by

$$S_a(\boldsymbol{\sigma}_{N-1}, \boldsymbol{\sigma}_N; \boldsymbol{\sigma}_{N+1}) = \frac{1}{\sqrt{2}} \alpha_{N+1} \left[\alpha_{N-1}\beta_N - \alpha_N\beta_{N-1} \right],$$ (91)

$$S_b(\boldsymbol{\sigma}_{N-1}, \boldsymbol{\sigma}_N; \boldsymbol{\sigma}_{N+1}) = \frac{1}{\sqrt{6}} \left[2\alpha_{N-1}\alpha_N\beta_N - \alpha_{N+1}(\alpha_{N-1}\beta_N + \alpha_N\beta_{N-1}) \right].$$ (92)

On substituting these in T-matrix equation (87) and performing spin integration we get the combined T-matrix for $^1S \to ^3 L_b M_b$ excitation

$$T_{a \to b} = \sqrt{3}\, \hat{T}^{ex}_{a \to b}.$$ (93)

It should be noted here that the direct T-matrix $\hat{T}^d_{a \to b}$ does not appear because the ST excitation can not take place without an interchange of the projectile and target electron. Consequently, the differential cross section for the ST excitation *i.e.*, $^1S \to ^3 L_b$ is

$$\frac{d\sigma}{d\Omega} = \frac{1}{4\pi^2} \frac{k_b}{k_a} \sum_{M_b} \left[\sqrt{3}\, |T^{ex}_{a \to b}| \right]^2.$$ (94)

Finally, electron impact excitation of various one– and two– electron systems have been investigated using the non-relativistic distorted wave method as described above. The results obtained for various parameters of lighter atoms are found in good agreement with experiments as well as other theoretical calculations [24–26]. However, non-relativistic methods have limitations when spin-orbit interaction is significant and fine–structure levels of atoms are clearly resolved, *e.g.*, noble gases and heavier atoms. Therefore, below we describe the relativistic distorted wave method which is most suitable for fine–structure transitions.

4   RELATIVISTIC DISTORTED WAVE THEORY

In order to calculate the T-matrices equations (22) and (26) in relativistic distorted wave approximation we need to represent the atomic wavefunctions $\Phi_{a(b)}$ and the distorted waves $F^{DW+(-)}_{a(b)}$ in their relativistic forms which are obtained by solving Dirac equations.

4.1 The atomic wavefunctions

The relativistic description of atomic structure has been discussed by Grant [27] and Grant *et al.* [28]. The Dirac form of the Hamiltonian for an atom with N electrons can be written as

$$H_{atom} = \sum_{j=1}^{N} h_j + \sum_{ij,\, i<j}^{N} \frac{1}{|\mathbf{r}_i - \mathbf{r}_j|},$$ (95)

where

$$h_j = -ic\,\boldsymbol{\alpha} \cdot \boldsymbol{\nabla}_j + \beta c^2 + V_{nuc}(r_j).$$ (96)

$\boldsymbol{\alpha}$ and β are usual Dirac matrices and $V_{nuc}(r) = -Z/r$ is the potential due to the nucleus. In order to avoid the complexity of the problem, here, we do not take into account the Breit operator *i.e.*, the spin-spin interaction. This enables us to use the multi-configuration Dirac-Fock (MCDF) algorithm to calculate the atomic orbitals and is also consistent with the approximation of Grant [27] and Grant *et al.* [28] who use the same Hamiltonian.

The N-electron bound state wavefunction for the target is taken as a Slater determinant of single electron Dirac-Fock orbitals. Here we represent the N-electron bound state wavefunction by a compact notation $\frac{1}{\sqrt{N}}\left\{1s\frac{1}{2}\frac{1}{2}...n\kappa m\right\}$, given by

$$\Phi_{a(b)}(\mathbf{1,2,...,N}) = \frac{1}{\sqrt{N}}\left\{1s\frac{1}{2}\frac{1}{2}...n\kappa m\right\},$$ (97)

$$= \frac{1}{\sqrt{N}}\begin{vmatrix} \phi_{1s\frac{1}{2}\frac{1}{2}}(\mathbf{1}) & \phi_{1s\frac{1}{2}-\frac{1}{2}}(\mathbf{1}) & ... & \phi_{n\kappa m}(\mathbf{N}) \\ \phi_{1s\frac{1}{2}\frac{1}{2}}(\mathbf{2}) & \phi_{1s\frac{1}{2}-\frac{1}{2}}(\mathbf{2}) & ... & \phi_{n\kappa m}(\mathbf{2}) \\ \cdot & \cdot & ... & \cdot \\ \cdot & \cdot & ... & \cdot \\ \phi_{1s\frac{1}{2}\frac{1}{2}}(\mathbf{N}) & \phi_{1s\frac{1}{2}-\frac{1}{2}}(\mathbf{N}) & ... & \phi_{n\kappa m}(\mathbf{N}) \end{vmatrix},$$

where $n\kappa m$ are the quantum numbers corresponding to the outermost electron. The relativistic quantum number κ is defined as,

$$\kappa = \begin{cases} l & \text{if } j = l - \frac{1}{2} \\ -l-1 & \text{if } j = l + \frac{1}{2} \end{cases}.$$ (98)

The single electron central field Dirac orbitals are given by,

$$\phi_{n\kappa m}(\mathbf{r}, \boldsymbol{\sigma}) = \frac{1}{r}\begin{pmatrix} P_{n\kappa}(r)\xi_{\kappa m}(\hat{\mathbf{r}}, \boldsymbol{\sigma}) \\ iQ_{n\kappa}(r)\xi_{-\kappa m}(\hat{\mathbf{r}}, \boldsymbol{\sigma}) \end{pmatrix},$$ (99)

where $P_{n\kappa}$ and $Q_{n\kappa}$ are the large and small components of the radial wavefunctions and the spin–angular function ξ is given by

$$\xi_{\kappa m}(\hat{\mathbf{r}}, \boldsymbol{\sigma}) = \sum_{\mu\nu}\left(l\mu\,\frac{1}{2}\nu\Big|jm\right)Y_{l\mu}(\hat{\mathbf{r}})\psi_{\frac{1}{2}\nu}(\boldsymbol{\sigma}),$$ (100)

$$\xi_{-\kappa m}(\hat{\mathbf{r}}, \boldsymbol{\sigma}) = \sum_{\mu\nu}\left(\tilde{l}\mu\,\frac{1}{2}\nu\Big|jm\right)Y_{\tilde{l}\mu}(\hat{\mathbf{r}})\psi_{\frac{1}{2}\nu}(\boldsymbol{\sigma}).$$ (101)

Here j is the total angular momentum of the electron, m is the z-component of j, $\bar{l} = 2j - l$, $(l_1 m_1 l_2 m_2 | l_3 m_3)$ is a ClebschGordan coefficient, the $Y_{lm}(\hat{\mathbf{r}})$ is spherical harmonic and $\psi_{\frac{1}{2}\nu}(\boldsymbol{\sigma})$ is a spinor basis function.

The bound state orbitals satisfy the orthogonal conditions

$$\int_0^\infty \left(P_{n\kappa}(r) P_{n'\kappa}(r) + Q_{n\kappa}(r) Q_{n'\kappa}(r) \right) dr = \delta_{nn'}, \tag{102}$$

and

$$\langle \xi_{\kappa m}(\hat{\mathbf{r}}, \boldsymbol{\sigma}) | \xi_{\kappa' m'}(\hat{\mathbf{r}}, \boldsymbol{\sigma}) \rangle = \delta_{\kappa\kappa'} \delta_{mm'}. \tag{103}$$

4.2 Relativistic distorted wavefunction

The relativistic expansion of projectile electron distorted-wave functions $F^{DW\pm}_{a(b),\mu_{a(b)}}$ can be written as [29]

$$F^{DW\pm}_{ch,\mu_{ch}}(\mathbf{k}_{ch}, \mathbf{r}) = \frac{1}{(2\pi)^{3/2}} \sum_{\kappa m} e^{\pm i\eta_\kappa}\, a^{\mu_{ch}}_{ch,\kappa m}(\hat{\mathbf{k}}_{ch}) \tag{104}$$
$$\times \;\; \frac{1}{r} \left(\begin{array}{c} f_\kappa(r)\, \xi_{\kappa m}(\hat{\mathbf{r}}) \\ i g_\kappa(r)\, \xi_{-\kappa m}(\hat{\mathbf{r}}) \end{array} \right),$$

with

$$a^{\mu_{ch}}_{ch,\kappa m}(\hat{\mathbf{k}}_{ch}) = 4\pi i^l \left[\frac{E_{ch} + c^2}{2E_{ch}} \right]^{\frac{1}{2}} \tag{105}$$
$$\times \;\; \sum_{m_l} (l\, m_l\, 1/2\, \mu_{ch} | j\, m) Y^*_{l m_l}(\hat{\mathbf{k}}_{ch}).$$

We use 'ch' to denote either the initial channel 'a' or the final channel'b'. η_κ is the phase shift of the partial wave. f_κ and g_κ are the large and small components of the radial wavefunctions and the $\xi_{\pm\kappa m}$ are the spinor spherical harmonics as given by equations (100) and (101). E_{ch} is the relativistic energy of the projectile electron with linear momenta k_{ch} such that $E_{ch} = \sqrt{k_{ch}^2 c^2 + c^4}$ and μ_{ch} is the spin projections of the projectile electron.

The large and small components f_κ and g_κ of the continuum wave functions satisfy the coupled Dirac equations,

$$\left(\frac{d}{dr} + \frac{k}{r} \right) f_\kappa(r) - \frac{1}{c}(c^2 - U_{ch} + E_{ch}) g_\kappa(r) = 0, \tag{106}$$

$$\left(\frac{d}{dr} - \frac{k}{r} \right) g_\kappa(r) + \frac{1}{c}(-c^2 - U_{ch} + E_{ch}) f_\kappa(r) = 0, \tag{107}$$

for the distorted-wave potential U_{ch} as described below in next subsection. These coupled equations can be solved numerically and phase shift can be evaluated by accounting for the boundary conditions;

$$f_\kappa(r) \xrightarrow{r \to \infty} \frac{1}{k_{ch}} \sin\left(k_{ch} r - \frac{l\pi}{2} + \eta_\kappa \right), \tag{108}$$

$$g_\kappa(r) \xrightarrow{r\to\infty} \frac{c}{c^2 + E_{ch}} \cos\left(k_{ch}r - \frac{l\pi}{2} + \eta_\kappa\right).$$

$$(109)$$

4.3   Distortion potential

In order to obtain distorted waves (104), we decompose the distortion potential as the sum of a direct part $V_{st}(r)$ (or static potential) and an exchange part $V_{ex}(r)$,

$$U_b(r) = V_{st}(r) + V_{ex}(r).$$

$$(110)$$

The first term $V_{st}(r)$ is a spherically symmetric distorting potential which is obtained from the charge distribution of the atomic wavefunctions. This is a static atomic potential since the atomic wavefunctions are assumed not to be distorted by the projectile electron. The static potential of the atom with nuclear charge Z and number of electrons N is given by,

$$V_{st}(r) = -\frac{Z}{r} + \sum_{j\in all\ subshells} \mathcal{N}_j \int_0^\infty \left[P^2_{n_j\kappa_j}(r_j) + Q^2_{n_j\kappa_j}(r_j)\right]\frac{1}{r_>}\,dr.$$

$$(111)$$

Here $\mathcal{N}_j$ is the occupation number of the jth subshells and the electron in it is represented by quantum number $n_j\kappa_j$. P and Q are the larger and smaller components of the radial wavefunctions of atomin orbitals. These bound orbital wavefunctions can be calculated by using GRASP92 code of Parpia $et\ al$ [30] or GRASP2K code [31].

The exchange potential can also be taken into account by using equation (39) where the charge density $\rho(r)$ for relativistic orbitals is given by

$$\rho(r) = \frac{1}{4\pi r^2} \sum_{j\in all\ subshells} \mathcal{N}_j\left[P^2_{n_j\kappa_j}(r_j) + Q^2_{n_j\kappa_j}(r_j)\right].$$

$$(112)$$

4.4   The T-matrix for one-valance electron system

As we discussed earlier that the T-matrix can be decomposed into its direct and exchange components. The direct term of the T-matrix in relativistic representation is given by

$$\begin{aligned}
T^d_{a\to b} &= \langle\phi_b(\mathbf{1,2,...,N})F^{DW-}_{b,\mu_b}(\mathbf{k}_b,\mathbf{r}_{N+1})|V - U_b(N+1)| \\
&\quad \phi_a(\mathbf{1,2,...,N})F^{DW+}_{a,\mu_a}(\mathbf{k}_a,\mathbf{r}_{N+1})\rangle.
\end{aligned}$$

$$(113)$$

Due to the orthogonality of the initial and final atomic wavefunctions, the only components of $V - U_b(N+1)$ which contribute to the matrix element $T^d_{a\to b}$ are those which are functions of the atomic electrons [22]. Therefore, we are left with

$$\begin{aligned}
T^d_{a\to b} &= \left\langle\phi_b(\mathbf{1,2,...,N})F^{DW-}_{b,\mu_b}(\mathbf{k}_b,\mathbf{r}_{N+1})\middle|\sum_{j=1}^{N}\frac{1}{|\mathbf{r}_{N+1} - \mathbf{r}_j|}\middle| \right. \\
&\quad \left. \phi_a(\mathbf{1,2,...,N})F^{DW+}_{a,\mu_a}(\mathbf{k}_a,\mathbf{r}_{N+1})\right\rangle.
\end{aligned}$$

$$(114)$$

For one valence electron atomic system we can write the wavefunction in the following Slater determinant form in its ground and excited state as

$$\Phi_a(\mathbf{1,2,...,N}) \;=\; \frac{1}{\sqrt{N}}\left\{1s\frac{1}{2}\frac{1}{2}...nxj_s\bar{m}'\right\}, \tag{115}$$

and

$$\Phi_b(\mathbf{1,2,...,N}) \;=\; \frac{1}{\sqrt{N}}\left\{1s\frac{1}{2}\frac{1}{2}...nxjm'\right\}. \tag{116}$$

In this form we start from $1s$ orbital, each orbital is written with its j and m angular momenta values. Thus nx represents the valance electron with its initial state quantum numbers $j_s\bar{m}'$ and the final excited state quantum numbers jm'.

Using the above wavefunctions and following Lindgren and Morrison [22] the direct T-matrix can be written as

$$T^d_{a\to b} = \left\langle jm'(\mathbf{N})F^{DW-}_{b,\mu_b}(\mathbf{k}_b,\mathbf{r}_{N+1})\left|\frac{1}{|\mathbf{r}_{N+1}-\mathbf{r}_N|}\right|j_s\bar{m}'(\mathbf{N})F^{DW+}_{a,\mu_a}(\mathbf{k}_a,\mathbf{r}_{N+1})\right\rangle \tag{117}$$

This direct T-matrix can further be expressed in the form

$$T^d_{a\to b} = \left\langle F^{DW-}_{b,\mu_b}(\mathbf{k}_b,\mathbf{r}_{N+1})\left|V_{a\to b(r_{N+1})}\right|F^{DW+}_{a,\mu_a}(\mathbf{k}_a,\mathbf{r}_{N+1})\right\rangle, \tag{118}$$

where,

$$V_{a\to b(r_{N+1})} = \left\langle jm'(\mathbf{N})\left|\frac{1}{|\mathbf{r}_{N+1}-\mathbf{r}_N|}\right|j_s\bar{m}'(\mathbf{N})\right\rangle, \tag{119}$$

is the matrix element of the interaction potential. This matrix element can also be expressed in general as a product of the radial and angular parts as

$$V_{a\to b(r_{N+1})} \;=\; \sum_{pq}\sqrt{\frac{4\pi}{2p+1}}(-1)^{m'+1/2}[j_s,j]^{1/2}\begin{pmatrix} j & p & j_s \\ \frac{1}{2} & 0 & -\frac{1}{2} \end{pmatrix} \tag{120}$$

$$\times \begin{pmatrix} j & p & j_s \\ -m' & q & \bar{m}' \end{pmatrix}\langle jn'\kappa'|\gamma_p|j_s\bar{n}'\bar{\kappa}'\rangle Y^*_{pq}(\hat{\mathbf{r}_{N+1}}),$$

where $n'\kappa'$ and $\bar{n}'\bar{\kappa}'$ corresponds to the quantum numbers of the valence electron before and after scattering, respectively and γ_p is given by equation (49). The radial integral in the above equation is expressed as

$$\langle jn'\kappa'|\gamma_p|j_s\bar{n}'\bar{\kappa}'\rangle = \int_0^\infty [P_{n'\kappa'}(r)P_{\bar{n}'\bar{\kappa}'}(r)+Q_{n'\kappa'}(r)Q_{\bar{n}'\bar{\kappa}'}(r)]\,\gamma_p(r,r_{N+1})dr,$$

$$\tag{121}$$

subject to the parity condition $l+l_s+p=$ even.

Further, using the obtained form (equation (118)) for the matrix element of the interaction potential, and the relativistic partial wave expansion form of the distorted waves (equation (104)) we obtain the direct T-matrix as

$$T_{a\to b}^d = \frac{2}{\pi} \sum_{\substack{\kappa_b m_b p \\ \kappa_a m_a q}} i^{l_a-l_b} e^{i(\eta_{\kappa_b}+\eta_{\kappa_a})} \sqrt{\frac{E_a+c^2}{2E_a}\cdot\frac{E_b+c^2}{2E_b}} \tag{122}$$

$$\times \int_0^\infty (f_{\kappa_a}(r)f_{\kappa_b}(r) + g_{\kappa_a}(r)g_{\kappa_b}(r))V_{a\to b}(r_{N+1})dr$$

$$\times \langle \kappa_b m_b |Y_{pq}^*(\hat{\mathbf{r}}_{N+1})| \kappa_a m_a \rangle \sum_{m_{l_a} m_{l_b}} (l_b m_{l_b} \tfrac{1}{2}\mu_b | j_b m_b)$$

$$\times (l_a m_{l_a} \tfrac{1}{2}\mu_a | j_a m_a) Y_{l_b m_{l_b}}(\hat{\mathbf{k}}_b) Y_{l_a m_{l_a}}^*(\hat{\mathbf{k}}_a).$$

Following Grant [27] we can express

$$\langle \kappa_b m_b |Y_{pq}^*(\hat{\mathbf{r}}_{N+1})| \kappa_a m_a \rangle = (-1)^q \sqrt{\frac{2p+1}{4\pi}} (-1)^{m_b+1/2}[j_a,j_b]^{1/2} \tag{123}$$

$$\times \begin{pmatrix} j_b & p & j_a \\ \tfrac{1}{2} & 0 & \tfrac{1}{2} \end{pmatrix} \begin{pmatrix} j_b & p & j_a \\ -m_b & -q & m_a \end{pmatrix}.$$

Further, taking the direction $\mathbf{k}_a$ along the z-axis and converting all Clebsch–Gordan coefficients to $3-j$ symbols as well as eliminating the summation over many of the variables employing the fact that the three m values of the $3-j$ symbols must sum to zero we obtain

$$T_{a\to b}^d = \frac{1}{2\pi^2} \sum_{\kappa_a,\kappa_b p\, q} i^{l_a-l_b} e^{i(\eta_{\kappa_b}+\eta_{\kappa_a})} \sqrt{\frac{E_a+c^2}{2E_a}\cdot\frac{E_b+c^2}{2E_b}} Y_{l_b \mu_a-\mu_b-q}(\hat{\mathbf{k}}_b) \tag{124}$$

$$\times \int_0^\infty (f_{\kappa_a}(r)f_{\kappa_b}(r) + g_{\kappa_a}(r)g_{\kappa_b}(r))V_{a\to b}(r_{N+1})dr$$

$$\times (-1)^{-l_a-l_b+\frac{3}{2}+q-\mu_b}[j_a,j_b,l_a,p]^{1/2} \begin{pmatrix} l_b & \tfrac{1}{2} & j_b \\ \mu_a-\mu_b-q & \mu_b & q-\mu_a \end{pmatrix}$$

$$\times \begin{pmatrix} j_b & p & j_a \\ \tfrac{1}{2} & 0 & \tfrac{1}{2} \end{pmatrix} \begin{pmatrix} j_b & p & j_a \\ q-\mu_a & -q & \mu_a \end{pmatrix} \begin{pmatrix} l_a & \tfrac{1}{2} & j_a \\ 0 & -\mu_a & \mu_a \end{pmatrix},$$

with the parity condition $l_a + l_b + p = \text{even}$.

The exchange term of T-matrix is given by

$$T_{a\to b}^{ex} = -\Big\langle jm'(\mathbf{N})F_{b,\mu_b}^{DW-}(\mathbf{k}_b,\,\mathbf{N+1}) \Big| \frac{1}{|\mathbf{r}_{N+1}-\mathbf{r}_N|} \Big| \tag{125}$$

$$\times j_s\bar{m}'(\mathbf{N+1})F_{a,\mu_a}^{DW+}(\mathbf{k}_a,\mathbf{N}) \Big\rangle.$$

$$\tag{126}$$

From the formula for the relativistic Coulomb interaction which is,

$$\Big\langle A(r_i)B(r_j) \Big| \frac{1}{|\mathbf{r}_i-\mathbf{r}_j|} \Big| C(r_i)D(r_j) \Big\rangle = \sum_{\lambda\rho} d_\rho^\lambda(\kappa_C m_C, \kappa_A m_A) \tag{127}$$

$$\times d_\rho^\lambda(\kappa_B m_B, \kappa_D m_D)$$

$$\times \Gamma_\lambda(n_A\kappa_A, n_C\kappa_C; n_B\kappa_B, n_D\kappa_D),$$

where

$$d_\rho^\lambda(\kappa'm',\kappa m) = (-1)^{m+1/2}[j,j']^{1/2} \tag{128}$$

$$\times \begin{pmatrix} j & \lambda & j' \\ \frac{1}{2} & 0 & -\frac{1}{2} \end{pmatrix} \begin{pmatrix} j & \lambda & j' \\ -m & \rho & m' \end{pmatrix},$$

with the parity condition $l + l' + \lambda =$ even, and

$$\Gamma_\lambda(n_A\kappa_A, n_C\kappa_C; n_B\kappa_B, n_D\kappa_D) = \int_0^\infty \int_0^\infty [P_A(r_i)P_C(r_j) \tag{129}$$

$$+ \quad Q_A(r_i)Q_C(r_j)]\frac{r_<^\lambda}{r_>^{\lambda+1}}$$

$$\times \quad [P_B(r_i)P_D(r_j) + Q_B(r_i)Q_D(r_j)]dr_i dr_j.$$

Using equations (127) and (129) and taking $\mathbf{k}_a$ along the z-axis as well as converting all Clebsch–Gordan coefficients into $3 - j$ symbols, the exchange T-matrix reduces to

$$T_{a\to b}^{ex} = -\frac{1}{2\pi^{3/2}} \sum_{\substack{\kappa_a,\kappa_b,\lambda \\ m_b,m_{l_b}}} i^{l_b-l_a} e^{i(\eta_{\kappa_b}+\eta_{\kappa_a})} \sqrt{\frac{E_a+c^2}{2E_a}\cdot\frac{E_b+c^2}{2E_b}} \tag{130}$$

$$\times \quad Y_{l_b m_{l_b}}(\hat{\mathbf{k}}_b)[j_a,j_b,l_a,j,j_s]^{1/2} \begin{pmatrix} l_a & \frac{1}{2} & j_a \\ 0 & -\mu_a & \mu_a \end{pmatrix} \begin{pmatrix} l_b & \frac{1}{2} & j_b \\ m_{l_b} & \mu_b & -m_b \end{pmatrix}$$

$$\times \quad \begin{pmatrix} j & \lambda & j_a \\ \frac{1}{2} & 0 & -\frac{1}{2} \end{pmatrix} \begin{pmatrix} j & \lambda & j_a \\ -m' & m'-\mu_a & \mu_a \end{pmatrix} \begin{pmatrix} j_s & \lambda & j_b \\ \frac{1}{2} & 0 & -\frac{1}{2} \end{pmatrix}$$

$$\times \quad \begin{pmatrix} j_s & \lambda & j_b \\ -\bar{m}' & m'-\mu_a & m_b \end{pmatrix} \Gamma_\lambda(n\kappa, \epsilon_a\kappa_a; \epsilon_b\kappa_b, n_s\kappa_s),$$

with parity conditions

$$l_a + \lambda + l = \text{even} \tag{131}$$

$$l_s + \lambda + l_b = \text{even.}$$

4.5 The T-matrix for two-valance electron system

After evaluating the direct T-matrix for excitation of atom from initial state 'a' to final state 'b' for one –valance electron system, we are now interested in obtaining the same for two-valance electron system. For this purpose the atomic wavefunction for the initial and final states can be written in the form of Slater determinant as follows

$$\Phi_a(\mathbf{1,2,...,N}) = \frac{1}{\sqrt{N}} \sum_{\bar{m}m} (\bar{j}\bar{m}\ jm|JM)\{...\bar{n}\bar{\kappa}n\kappa\}, \tag{132}$$

and

$$\Phi_b(\mathbf{1,2,...,N}) = \frac{1}{\sqrt{N}} \sum_{\bar{m}m} (\bar{j}\bar{m}'\ j'm'|JM)\{...\bar{n}\bar{\kappa}n'\kappa'\}. \tag{133}$$

where $\bar{n}\bar{\kappa}$ represents the hole state with quantum numbers $\bar{n}\bar{j}\bar{l}\bar{m}$ and $n\kappa$ and $n'\kappa'$ represent the excited electron with quantum numbers $njlm$ and $n'j'l'm'$, respectively.

The atomic orbitals obtained from GRASP92 code [30] or GRASP2K code [31] are already orthogonal and we force the scattered wave to be orthogonal to the bound orbitals. Then only those transitions between states with same hole state are possible in the first order theory. Thus following Lindgren and Morrison [22] we have

$$T^d_{a\to b} = \sum_{\substack{\bar{m}m \\ \bar{m}'m'}} (\bar{j}\bar{m}' \, j'm'|J'M') \, (JM|\bar{j}\bar{m} \, jm) \, \delta_{\bar{m}\bar{m}'} \tag{134}$$

$$\times \; \left\langle j'm'(\mathbf{N})F^{DW-}_{b,\mu_b}(\mathbf{k}_b,\mathbf{r}_{N+1}) \left| \frac{1}{|\mathbf{r}_{N+1}-\mathbf{r}_N|} \right| jm(\mathbf{N})F^{DW+}_{a,\mu_a}(\mathbf{k}_a,\mathbf{r}_{N+1}) \right\rangle .$$

Following the similar steps as described in the previous subsection and carrying out the angular momentum algebra as well as assuming the incident electrons to be along z-axis, the direct T-matrix takes the form

$$T^d_{a\to b} = \frac{1}{\pi^{3/2}} \sqrt{\frac{E_a + c^2}{2E_a} \cdot \frac{E_b + c^2}{2E_b}} \sum_{\substack{p\kappa_a\kappa_b \\ qm_b,m_{l_b}}} [J',J,j',j,j_a,j_b,l_a]^{1/2} \tag{135}$$

$$\times \; i^{l_a-l_b} e^{i(\eta_{\kappa_b}+\eta_{\kappa_a})} (-)^{j'+j+j_a+j_b+\bar{j}-\mu_a-M'+1} \begin{pmatrix} j' & p & j \\ \frac{1}{2} & 0 & -\frac{1}{2} \end{pmatrix}$$

$$\times \; \begin{pmatrix} J' & p & J \\ -M' & q & M \end{pmatrix} \begin{Bmatrix} J' & p & J \\ j & \bar{j} & j' \end{Bmatrix} \begin{pmatrix} j_b & p & j_a \\ m_b & q & -\mu_a \end{pmatrix}$$

$$\times \; \begin{pmatrix} j_b & p & j_a \\ \frac{1}{2} & 0 & -\frac{1}{2} \end{pmatrix} Y_{l_b m_{l_b}}(\hat{\mathbf{k}}_b)(l_b m_{l_b} \, 1/2\mu_b|j_b m_b) \, (l_a 0 \, 1/2\mu_a|j_a \mu_a)$$

$$\times \; \int_0^\infty \int_0^\infty (f_{\kappa_a}(r_{N+1})f_{\kappa_b}(r_{N+1}) + g_{\kappa_a}(r_{N+1})g_{\kappa_b}(r_{N+1}))$$

$$\times \; [P_{n'\kappa'}(r_N)P_{n\kappa}(r_N) + Q_{n'\kappa'}(r_N)Q_{n\kappa}(r_N)] \gamma_p(r_N,r_{N+1}) dr_N \, dr_{N+1}.$$

Similarly, the exchange T-matrix can be expressed as

$$T^{ex}_{a\to b} = -\frac{1}{2\pi^{3/2}} \sqrt{\frac{E_a + c^2}{2E_a} \cdot \frac{E_b + c^2}{2E_b}} \sum_{\substack{p\kappa_a\kappa_b k \\ qm_b,m_{l_b}}} i^{l_a-l_b} e^{i(\eta_{\kappa_b}+\eta_{\kappa_a})} \tag{136}$$

$$\times \; [J',J,j',j,j_a,j_b,l_a,k^2]^{\frac{1}{2}} (-1)^{J+\bar{j}+p+j-j'+J'-M'-m_b+k}$$

$$\times \; \begin{pmatrix} j' & p & j_a \\ \frac{1}{2} & 0 & -\frac{1}{2} \end{pmatrix} \begin{pmatrix} j & p & j_b \\ \frac{1}{2} & 0 & -\frac{1}{2} \end{pmatrix} \begin{pmatrix} J' & k & J \\ -M' & q & M \end{pmatrix}$$

$$\times \; \begin{pmatrix} j_b & p & j_a \\ -m_b & q & m_a \end{pmatrix} \begin{Bmatrix} J & J' & k \\ j' & j & \bar{j} \end{Bmatrix} \begin{Bmatrix} j_b & j_a & k \\ j' & j & p \end{Bmatrix}$$

$$\times \; Y_{l_b m_{l_b}}(\hat{\mathbf{k}}_b)(l_b m_{l_b} \, 1/2\mu_b|j_b m_b) \, (l_a 0 \, 1/2\mu_a|j_a \mu_a)$$

$$\times \; \Gamma_\lambda(n'\kappa',\epsilon_a\kappa_a;\epsilon_b\kappa_b,n\kappa).$$

After evaluating the T-matrix, the parameters required to describe the electron–ion collision process can be calculated. For instance, with our normalization the differential cross section (DCS) is given by

$$\frac{d\sigma(\theta)}{d\omega} = \frac{(2\pi)^4}{2(2J_a + 1)} \frac{k_b}{k_a} \sum_{\substack{M_a, \mu_a \\ M_b, \mu_b}} |T_{a \to b}|^2,$$

(137)

where we have summed over the spins of the incident and scattered electron as well as magnetic quantum numbers corresponding to the total angular momenta of the atom in initial and final states. The integrated cross section (ICS) can be obtained by integrating the DCS (137) over the all scattering angles.

The relativistic distorted wave method has been successfully applied to various one– and two– electron systems and the results obtained agree very well with the experimental data and the other reliable CCC and R-matrix calculations [4, 7, 12–17].

Conclusions

A detailed derivation of the distorted wave theory in its non-relativistic and relativistic forms has been given. The T-matrices in both the forms are simplified to the electron impact excitation of one– and two– valance electron systems. It should be noted that relativistic distorted wave version presented here is most suitable to study excitation of fine–structure resolved transitions.

Acknowledgements

We thank Prof. A. D. Stauffer and our other co-workers for their collaborations in using the distorted wave theory presented here which resulted in the form of several joint publications.

References

[1] I. Bray, D. V. Fursa, A. S. Kheifets and A. T. Stelbovics, J. Phys. B **35**, R117 (2002).

[2] Y. Itikawa, Phys. Rep. **143**, 69 (1986).

[3] D. H. Madision and K. Bartschat, "The Distorted-Wave Method for Elatic Scattering and Atomic Excitation, Computational Atomic Physics: Electron and Positron Collision with Atoms and Ions", ed. Bartschat K. Springer-Verlag Berlin Heidelberg, p.65 (1996).

[4] S. Kaur, R. Srivastava, R. P. McEachran and A. D. Stauffer, J. Phys. B **33**, 2539 (2000).

[5] D. H. Madison, K. Bartschat and R. Srivastava, J. Phys. B **24**, 1839 (1991).

[6] C. J. Joachain, "Quantum Collision Theory", North Holland Publishing Co. (1975).

[7] S. Kaur, R. Srivastava, R. P. McEachran and A. D. Stauffer, J. Phys. B **31**, 157 (1998).

[8] H. R. J. Walters, Phys. Rep. **116**, 1 (1984).

[9] D. H. Madison, I. Bray and I. E. McCarthy, Phys. Rev. Lett. **64**, 2265 (1990).

[10] D. H. Madison, I. Bray and I. E. McCarthy,, J. Phys. B **24**, 3861 (1991).

[11] D. H. Madison, J. A. Hughes and D. S. McGinness, J. Phys. B **18**, 2737 (1985).

[12] G. Auendorf, F. Jüttemann, K. Muktavat, L. Sharma, R. Srivastava, A. D. Stauffer, K. Bartschat, D. V. Fursa, I. Bray, and G. F. Hanne, J. Phys. B **39**, 2403 (2006).

[13] L. Sharma, R. Srivastava, and, A. D. Stauffer, J. Phys. B **40**, 3025 (2007).

[14] L. Sharma, R. Srivastava, and, A. D. Stauffer, Phys. Rev. A **79**, 024701 (2007).

[15] L. Sharma, R. Srivastava, and A. D. Stauffer, Phys. Rev. A **78**, 014701 (2008).

[16] F. Jüttemann, G. F. Hanne, O. Zatsarinny, K. Bartschat, R. Srivastava, R. K. Gangwar, and A. D. Stauffer Phys. Rev. A **81**, 012705 (2010).

[17] R. K. Gangwar, L. Sharma, R. Srivastava, and A. D. Stauffer, Phys. Rev. A **81**, 052707 (2010).

[18] F.C. Fischer, Comput. Phys. Comm. **1**, 151 (1969).

[19] E. Clementi and C. Roetti, At. Data Nucl. Data Tables **14**, 177 (1974).

[20] A. Hibbert, Comput. Phys. Commun. **9**, 141 (1975).

[21] F. A. Gianturco and S. Scialla, J. Phys. B **20**, 3171 (1987).

[22] I. Lindgren and J. Morrison, "Atomic Many-Body Theory", 2nd ed. Berlin: Springer-Verlag (1986).

[23] M. E. Rose, "Elementary Theory of Angular Momentum", New York: London: Sydney (1967).

[24] Rajesh Srivastava, A. K. Katiyar, and D. K. Rai, Phys. Rev. A **40**, 2749 (1989).

[25] S. Verma and R. Srivastava, Phys. Rev. A **50**, 2269 (1994).

[26] S. Verma and R. Srivastava, J. Phys. B **29**, 3215 (1996).

[27] I. P. Grant, Adv. Phys. **19**, 747 (1970).

[28] I. P. Grant, B. J. McKenzie, P. H. Norrington, D. F. Mayers and N. C. Piper, Comput. Phys. Commun. **21**, 207 (1980).

[29] M. E. Rose, " Relativistic Electron Theory", New York: Wiley (1961).

[30] F. A. Parpia, C. F. Fischer, and I. P. Grant, Comput. Phys. Commun. **94**, 249 (1996).

[31] P. Jönsson, X. He, C. Froese Fischer and I.P. Grant, Comput. Phys. Commun. **177**, 597 (2007).

Theory of (e, 3e) Process on Atomic System

Rakesh Choubisa[1*] and K.K. Sud[2]

[1] *Department of Physics*
Birla Institute of Technology and Science, Pilani:333031 (India)
[2] *Department of Physics, School of Engineering*
Sir Padampat Singhania University, Bhatewar (Udaipur): 313601 (India)
*e-mail: rjchoubisa@yahoo.co.in**

Abstract: In this Chapter we introduce (e, 3e) process on atom with modest historical background. The Chapter describes the theoretical formalism for the calculation of fivefold differential cross section (FDCS) on He and He like ions in first and second Born approximation. This formalism can also be used for K-shell double ionization of many-electrons atoms, albeit some compromise like overlooking the interaction of other atomic electrons in pre and post collision interaction as well as on the K-shell electrons etc. This Chapter concludes with the calculation of FDCS in first and second Born approximation and its comparison with the available data. The second Born calculation clearly shows its signature by revealing symmetry breaking of FDCS about momentum transfer axis.

1. INTRODUCTION

The atomic and subatomic systems are governed by the rules of quantum mechanics. Much of the knowledge about this micro world has been obtained by the study of scattering process on given target. Particularly, scattering experiments provide one of the most important means of gathering information in the realm of atomic and subatomic particles. This type of study was initiated by the pioneering experiment of Rutherford, involving the scattering of alpha particles from gold atom which consequently led to the nuclear-atom model. Early experiment on electron scattering played a central role in the development of the quantum mechanics. The demonstration of diffraction of electron beam by gases confirmed the quantum mechanical duality of waves and particles and measurements of the energy losses in electron scattering in gases established the discrete nature of the energy level structure of atoms and molecules. Further, many of the features of the nuclear force such as its range, strength and spin dependence have been deduced from the data gathered from nucleon –nucleon scattering. With the development of multi-coincident measurement techniques in the collision process, in which momenta of the outgoing particles are fully resolved, it is possible to investigate the finer details of reaction mechanism involved in the collision process. The early coincident investigations of (p, 2p) and $(e,e'p)$ reactions in the Nuclear Physics [1-4] have made remarkable contributions in measuring momentum profile of proton in the nucleus.

This idea has led to the coincidence study of the electron impact single ionization (i.e., (e, 2e) process) on atoms during late sixties (see, [5-6]). Electron impact ionization of atoms, ions, molecules etc. is one of the basic processes of atomic and molecular physics and the knowledge of various types of collision cross sections of ionization processes are of great importance for understanding of the basic phenomena in the fields such as astrophysics, plasma physics, laser physics and fusion physics etc. Further, the (e, 2e) process on target yields a three-body Coulomb problem which is governed by infinite long range Coulomb potential making the exact treatment of the three-body problem extremely difficult. The study of three-body Coulomb problem has led to remarkable theoretical and experimental investigations in the last forty two years (e.g., see [7-10]). Similar type of three-body problem is encountered in the photo-double ionization process (i.e., $(\gamma, 2e)$ process) in which two electrons get ejected from the target due to photon impact (for details, see [11]). In addition to this, a remarkable advancement towards more complex many-body problem, i.e. four-body problem due to electron impact double ionization, has been achieved during last twenty two years since the first measurement of (e, 3e) process on Ar atom by Lahmam-Bennani et al (1989) [12]. Before, we present details of the (e, 3e) processes on atoms; we would like to present a brief introduction about the collision and the possible processes which occur during a collision.

A collision is basically an interaction between two or more systems, each containing one or more particles. In a collision process, two particles or systems approach each other from a large distance, interact (collide) for a short time and then recede from the collision zone and are detected at a very long distance from the collision zone. During the collision a large force acts between the colliding partners and the post- collision state is different from pre-collision state and thus from a study of these two states it should be possible to learn about the nature of the interaction between the colliding partners.

Considering a well collimated, mono-energetic electron beam incident on a thin scattering chamber of target Z in its ground state, various processes which may take place and some of these are given as follows;

- Elastic scattering $\quad\quad e^- + Z \rightarrow Z + e^-$
- Inelastic scattering $\quad\quad e^- + Z \rightarrow Z^* + e^-$
- Single ionization $\quad\quad e^- + Z \rightarrow Z^+ + e^- + e^-$

- double ionization $\qquad\qquad e^- + Z \rightarrow Z^{++} + e^- + e^- + e^-$

If there is no change in the internal states of the colliding particles, scattering is termed as elastic scattering while if one or both colliding particles suffer a change in their internal states, the process is inelastic. The process is called to be a reaction process (single ionization, double ionization) if there is rearrangement or the projectile / target gets split into two or more particles. The scattering theory is well documented in literature (see [13]).

1.1 Introduction to (e, 3e) processes

In an (e, 3e) process, an incident electron with its definite energy and momentum $(E_i, \mathbf{k}_i)$ is incident on a target (atom, ion, molecules etc.) which results in the ejection of two bound electrons into continuum state (see, Fig. 1) and these two ejected electrons as well as the scattered electron are detected coincidently with their momenta resolved (i.e., energies and angles are fully resolved). Such types of experiments are kinematically complete (except spin resolution of electrons) so these experiments produce a plenty of kinematical domains wherein we can learn a lot about the mechanism and physics of the electron impact double ionization process. A five-fold differential cross section (FDCS) (i.e., $d^5\sigma/dE_1\, dE_2\, d\Omega_s\, d\Omega_s d\Omega_2$) is required to fully characterized the (e, 3e) process which is differential in solid angles of all outgoing electrons and energies of two electrons (energy of other electron is determined by the energy conservation law). The minimum number of kinematical variables defining the final state of (e, 3e) process is eight (E_1, E_2, θ_s, θ_1, θ_2, ϕ_s, $\phi_1\, and\, \phi_2$) much larger than the (e, 2e) process. The variables are the energies E_1 and E_2 of the ejected electrons 1 & 2 (energy of the scattered electron can be determined by the energy conservation law), the polar angles θ_s, θ_1 and θ_2 of the scattered and ejected electrons 1 and 2 and the corresponding azimuthal angles ϕ_s, $\phi_1\, and\, \phi_2$ of the outgoing electrons relative to the incident electron beam axis (see, Fig. 1). In the (e, 3e) experimental measurement, we keep seven kinematical variables constant and vary only the remaining one and investigate FDCS as a function of this variable. So the (e, 3e) process gives much finer details compared to the (e, 2e) and $(\gamma, 2e)$ processes but unfortunately it requires tedious multiple coincidence technique which makes these experiments far more difficult than the (e, 2e) and $(\gamma, 2e)$ experiments.

The (e, 3e) experiments require an extra solid angle of detection for the third electron, $\Delta\Omega$, resulting in a reduction of the overall collection efficiency by a factor $\Delta\Omega/4\pi$. In addition to this, the double ionization cross section is typically two orders of magnitude smaller than the (e, 2e) process [14]. These two major constraints make the design of (e, 3e) experiments very difficult. As far as the theoretical view of the (e, 3e) process is concerned, it is more complicated problem as it has four particles in the continuum state and are highly correlated even at asymptotic state due to long range nature of the Coulomb interaction and demands appropriate theoretical model to describe the four-body problem exactly. Further, in the (e, 3e) processes, three mechanisms are believed to be dominated in the direct ejection of two target electrons: the shake-off (SO), the two-step1 (TS1) and the two-step2 (TS2) mechanisms so one has to incorporate these mechanisms in the calculation of FDCS which is a very hard task.

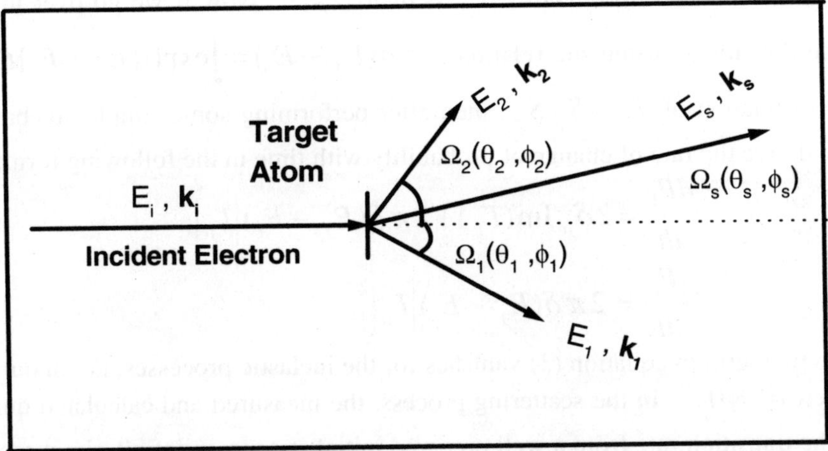

Figure 1: Schematic diagram showing an (e, 3e) process. An incident electron of energy E_i and momentum k_i produces three outgoing electrons having energy E_s, E_1, E_2 and momenta k_s, k_1 and k_2 after collision.

In the shake off mechanism the incident electron interacts with one of the bound electrons and ejects it, the second electron is no longer an eigen state of the target as there is a sudden change in the potential. This will induce the second electron to get ejected after a relaxation process. The other two mechanisms (TS1 and TS2) in (e, 3e) process involve interaction in two steps. The first step is common for the two mechanisms, in which the incident electron ejects one of the target electrons. The difference between the two mechanisms is in the second step. In the TS1 mechanism it is the ejected electron which ejects the other target electron whereas in the TS2 mechanism it is the scattered electron which ejects the other target electron [15].

2. THEORY

We define following S matrix element as (see [14]);

$$S_{fi} = \delta_{fi} - 2\pi \delta(E_f - E_i)\langle \psi^- | V_i | k_i, \phi_a \rangle \ . \tag{1}$$

Here, $V_i = V_{pe1} + V_{pe2} + V_{pc}$. We define $V_{pe1,e2}$ is the Coulomb potential between the projectile and the electrons 1 and 2, while, V_{pc} is the Coulomb potential between the projectile and nucleus of the target. The other symbols have their usual meanings.

The dynamics of the scattered flux is governed by the matrix elements,

$$T_{fi} = \left\langle \psi^- \left| V_i \right| k_i , \phi_a \right\rangle \tag{2}$$

The S matrix elements S_{fi} are the measure of transition probability amplitude.

Thus the transition probability P_{fi} is then $\left| S_{fi} \right|^2$. Now if we go over into the time domain by using the relation $2\pi \delta(E_f - E_i) = \int \exp(i[E_f - E_i]t)\, dt$ in the evaluation of $P_{fi} = S_{fi}^* S_{fi}$, then after performing some simple algebra, one can derive the rate of change of probability with time in the following form;

$$\frac{dP_{fi}}{dt} = 2\delta_{fi}\, \mathrm{Im}(T_{fi}) + 2\pi \delta(E_f - E_i)\left| T_{fi} \right|^2$$

$$\frac{dP_{fi}}{dt} = 2\pi \delta(E_f - E_i)\left| T_{fi} \right|^2 \tag{3}$$

The first term in equation (3) vanishes for the inelastic processes, i.e. in our case, where $\delta_{fi} = 0$.   In the scattering process, the measured and calculated quantity is the transition rate from a well prepared initial state to an infinitesimal group of final states which is characterized by a certain density of states. In the present case, the density of states in the momentum space is given by $d^3 k_s\, d^3 k_1\, d^3 k_2$. We further normalize this transition rate to the asymptotic probability flux density j_i of the incoming projectile and is termed as the multiple differential cross section $\sigma(k_s, k_1, k_2 ; k_i, \phi_a)$.

In mathematical language one can express it in the following form;

$$\sigma(k_s, k_1, k_2 ; k_i, \phi_a) = \frac{dP_{fi}}{dt} \cdot \frac{1}{j_i} \cdot d^3 k_s\, d^3 k_1\, d^3 k_2$$

$$= 2\pi \cdot \frac{1}{v_i \big/ (2\pi)^3} \cdot \delta(E_f - E_i) \cdot \left| T_{fi} \right|^2 d^3 k_s\, d^3 k_1\, d^3 k_2 \tag{4}$$

The velocity of the incoming projectile is v_i relative to the centre of mass of the target. Now, using $d^3 k_j = k_j^2\, dk_j\, d\Omega_j = k_j (k_j\, dk_j)\, d\Omega_j = k_j\, dE_j\, d\Omega_j$, one can write;

$$d^3 k_s\, d^3 k_1\, d^3 k_2 = (k_s\, dE_s\, d\Omega_s)(k_1\, dE_1\, d\Omega_1)(k_2\, dE_2\, d\Omega_2)$$

and, for incident electron as a projectile $v_i = k_i$ (in a.u.).

Once the energies of the two outgoing electrons are resolved in the (e, 3e) process, the energy of the third electron can easily be found using the energy conservation law so the cross section is differential of the two energy band passes $dE_1 \, and \, dE_2$ as well as the solid angles $d\Omega_s^*, d\Omega_1 \, and \, d\Omega_2$ of the outgoing electrons in the continuum state. So finally, the coincidence cross section of the (e, 3e) process can be expressed by the following form;

$$\sigma(E_1, E_2, \Omega_s, \Omega_1, \Omega_2) = (2\pi)^4 \cdot \frac{k_s \, k_1 \, k_2}{k_i} \cdot \left| T_{fi} \right|^2 \cdot dE_1 \, dE_2 \, d\Omega_s \, d\Omega_1 \, d\Omega_2$$

(5)

The scattering amplitude is given by the following expressions;

$$T_{fi} = -\left\langle \phi_f \left| V \right| \psi_i^+ \right\rangle$$

(6)

or,

$$T_{fi} = -\left\langle \psi_f^- \left| V \right| \phi_i^+ \right\rangle,$$

(7)

where, $V = -\dfrac{Z}{r_i} + \dfrac{1}{r_{i1}} + \dfrac{1}{r_{i2}}$ is the interaction potential between the incident electron and the target. The wave functions $\phi_i \, and \, \phi_f$ are the eigenstates of H_i, where $H_i = h_a + h_s$, the operators $h_a \, and \, h_c$ are respectively the Hamiltonians of the undisturbed atom in the initial state and that of the residual ion. Here $\psi_i^+ \, and \, \psi_f^-$ denotes the full scattering wave functions satisfying the Lippmann-Schwinger equations;

$$\psi_i^+ = \phi_i + G_0^+ \, V \, \psi_i^+$$

(8)

$$\psi_f^- = \phi_f + G_0^- \, V \, \psi_f^-$$

(9)

In these equations, the Green's operators $G_0^\pm$ are given by;

$$G_0^\pm = \frac{1}{(E - H_i \pm i\eta)}$$

(10)

Here, we have neglected the spin of the projectile as we are considering only non-relativistic collision. Let us return to the Lippmann-Schwinger equation (8) and (9) and solve them by iteration, starting from the free waves $\phi_i \, and \, \phi_f$, respectively, we obtain;

$$\psi_i^+ = \phi_i + G_0^+ V \phi_i + G_0^+ V G_0^+ V \phi_i + \dots\dots\dots\dots$$

(11)

and

$$\psi_f^- = \phi_f + G_0^- V \phi_f + G_0^- V G_0^- V \phi_f + \dots\dots\dots\dots$$

(12)

Substituting expressions (11) & (12) in the equations (8) and (9) respectively and considering only n leading terms, we have a Born series for the scattering amplitude;

$$T_{fi} = \sum_{j=1}^{j=n} T_{fi}^{B_j} \text{ , where} \tag{13}$$

$$\dot{T}_{fi}^{B_j} = \left\langle \phi_f \left| VG_0^+ V \ldots\ldots G_0^+ V \right| \phi_i \right\rangle. \tag{14}$$

In the equation (14), the interaction potential occurs n times while the Green's operator G_0^+ appears (n-1) times. So the matrix elements T_{fi} can be expressed as sum of first n Born series. In the present investigation, we consider only the first two leading terms of the Born series, so the FDCS in the (e, 3e) process in second Born approximation can be expressed by the following form

$$\frac{d^5\sigma(E_1,E_2,\Omega_s,\Omega_1,\Omega_2)}{dE_1\,dE_2\,d\Omega_s\,d\Omega_1\,d\Omega_2} = (2\pi)^4 \cdot \frac{k_s\,k_1\,k_2}{k_i} \cdot \left| T_{fi}^{B1} + T_{fi}^{B2} \right|^2 , \tag{15}$$

where, $$T_{fi}^{B1} = -\left\langle \phi_f \left| V \right| \phi_i \right\rangle, \tag{16}$$

and, $$T_{fi}^{B2} = -\left\langle \phi_f \left| VG_0 V \right| \phi_i \right\rangle. \tag{17}$$

Now consider the second Born term T_{fi}^{B2} ;

$$T_{fi}^{B2} = -\left\langle \phi_f \left| VG_0 V \right| \phi_i \right\rangle = -\int \phi_f\, V(r)\, G_0^+(r,r')\, V(r')\, \phi_i\, dr'\, dr . \tag{18}$$

Now defining,

$$\left\langle q \left| V \right| q' \right\rangle = (2\pi)^3 \lim_{\eta\to 0^+} \int \exp\{i(q'-q).r\} V(r)\, dr , \tag{19}$$

and using the integral representation of the Green's function $G_0^+(r,r')$ of the following form;

$$G_0^+(r,r') = -(2\pi)^3 \lim_{\eta\to 0^+} \int \frac{\exp\{ik'.(r-r')\}}{k'^2 - k^2 - i\eta}\, dk' , \tag{20}$$

we get the second Born matrix element in the following form in the momentum space ;

$$T_{fi}^{B2} = -2\pi^2 \int \frac{\left\langle k_f \left| V \right| k' \right\rangle \left\langle k' \left| V \right| k_i \right\rangle}{k^2 - k'^2 + i\eta}\, dk' . \tag{21}$$

In the above equation (21), we see the second Born matrix element T_{fi}^{B2} includes two step processes between the incident projectile and the target which is reflected in the mathematical form of it in which the two first order matrix elements $\left\langle k_f \left| V \right| k' \right\rangle$ and $\left\langle k' \left| V \right| k_i \right\rangle$ gets multiplied. Once the terms $\left\langle k_f \left| V \right| k' \right\rangle$ and $\left\langle k' \left| V \right| k_i \right\rangle$ are calculated after integrating over the radial space, it will be further integrated to the intermediate three dimensional momentum space on k'.

The calculation has been performed in first and second Born approximation using approximated 3 Coulomb wave approach (also known as 2 Coulomb Gamow (2CG) approach). We describe the present formalism with the following assumptions;

1. Since most of the experiments on He atom has been performed at high incident electron energy and low momentum transfer, we use plane wave for the incident and scattered electrons.
2. The initial state of the target has been described by various correlated and uncorrelated wave functions.
3. The final state wave function of correlated pair of the ejected electrons in the continuum state is described by approximate Brauner, Briggs and Klar (BBK) type wave function.
4. The second Born calculation is performed using closure approximation. As there are infinite numbers of intermediate states probable in the calculation of second Born approximation so it is not possible to incorporate all the intermediate states before the double ionization to take place. One way to ease the problem is to set the target energy difference $(\omega_n - \omega_{1s})$ for any intermediate state by an average value of excitation energy I_n and by using the closure property of the wave function ψ_n ($\sum_n |\psi_n\rangle\langle\psi_n| = 1$) (see, [17-18]).

Let us return to the first Born matrix element (16);

$$T_{fi}^{B1} = -\langle \phi_f | V | \phi_i \rangle, \tag{22}$$

where ϕ_f and ϕ_i are the final and initial state wave functions and V is the interaction between the incident electron and the target. Now, substituting the expression of V to equation (22), we get the form of T_{fi}^{B1} in the following form;

$$T_{fi}^{B1} = -\langle \phi_f | -\frac{Z}{r_i} + \frac{1}{r_{i1}} + \frac{1}{r_{i2}} | \phi_i \rangle. \tag{23}$$

The initial state, which contains the incident electron and two bound electrons for He and He like ions (or K-shell electrons), can be described by a product of the plane wave for the incident electron and target wave function for the two bound electrons by the following form;

$$|\phi_i (r_i, r_1, r_2)\rangle = \left| \frac{e^{i k_i \cdot r_i}}{(2\pi)^{3/2}} \psi_i (r_1, r_2) \right\rangle, \tag{24}$$

where, k_i and r_i are the momentum and position vector of the incident electron respectively. The position vectors of the two bound electrons are described by

r_1 *and* r_2 for the electrons 1 and 2 respectively with respect to the nucleus of the target which is assumed to be located at the origin. In literature, various correlated and uncorrelated wave functions are available for He atom as well as for He like ions. Since the (e, 3e) processes are very useful for the study of electron-electron correlation, we exploit the fact by taking various wave functions for the targets. We use following wave functions in our calculation;

(i) Slater type wave function:

$$\psi_i(r_1, r_2) = \frac{Z'^3}{\pi} \exp(-Z'(r_1 + r_2)), \ Z' = Z - \frac{5}{16} \tag{25}$$

(ii) Byron and Joachain wave function: This wave function was derived by Byron and Joachain (1966) [19]. The form of the wave function is separable and can be written as;

$$\psi_i(r_1, r_2) = \phi_1(r_1).\phi_1(r_2), \tag{26}$$

where, $\phi_i(r_i) = (A e^{-\alpha r_i} + B e^{-\beta r_i})$

and A=2.60505, B=2.08144, $\alpha = 1.41$ *and* $\beta = 2.61$.

(iii) Le Sech wave function: This wave function is highly correlated which includes radial as well as angular correlation and has been derived by Le Sech (1997) [20] and can be used for He like ions. The form of the Le Sech wave function is given as;

$$\psi_i(\mathbf{r_1}, \mathbf{r_2}) = N e^{-Z r_1} e^{-Z r_2} [\cosh(\lambda r_1) + \cosh(\lambda r_2)][1 + 0.5 \mathbf{r_{12}} e^{-a r_{12}}] \tag{27}$$

We have dropped the angular correlation factor $\mathbf{r_{12}}$ in computing the matrix element, needed for evaluation of FDCS by using analytically evaluated radial integrals. Therefore, the approximate form of the Le Sech wave functions as;

$$\psi_i(r_1, r_2) = N e^{-Z r_1} e^{-Z r_2} [\cosh(\lambda r_1) + \cosh(\lambda r_2)], \tag{28}$$

(iv) Hyllerass type wave functions: We use the Hylleraas type wave function [21] for He atom which is modified by Bonham and Kohl (1966) [22] with the following form;

$$\psi_i(r_1, r_2) = N[F(a,b)\{1 + A r_{12}^{n_1} e^{-\lambda r_{12}}\} + F(c,d)\{B + C r_{12}^{n_2} e^{-\mu r_{12}}\} + D F(e,f) \tag{29}$$

with $F(a,b) = [e^{-a r_1} e^{-b r_2} + e^{-a r_2} e^{-b r_1}]$,

Here, we use following correlated wave functions;

$$\psi_i(r_1, r_2) = N[F(a,b) + B F(c,d)],$$

where , $a = 1.399, b = 2.097, c = 1.63, d = 1.63$ *and* $N = \sqrt{1.442}$

Though we have chosen four different initial state wave functions for K-shell bound electrons we are describing theoretical formalism for Le Sech wave function only.   The formalism using other wave functions can be done in a similar way.

Final state wave function

The final state has four particles (i.e., three outgoing electrons in the field of the residual ion) in the continuum state interacting with each other by infinite long range Coulomb potential.   These various interactions make the exact treatment extremely difficult and some approximations have to be used.   One way to ease the problem is to consider the (e, 3e) process in the dipole regime (low momentum transfer condition) wherein the incident and scattered electrons are highly energetic as compared to the ejected electrons.   This allows us to describe the scattered electron like to the incident electron by plane waves, i.e. it doesn't include the Coulomb distortions on it due to other particles in the continuum states.   Thus we can reduce a four-body problem to a three-body problem.   The three-body problem has been extensively investigated in the literature for the study of the (e, 2e) and $(\gamma, 2e)$ processes.

We describe the final state by plane wave for the scattered electron and the 3 Coulomb body BBK type wave function which is orthogonalized with respect to the initial state wave function $\psi_i(r_1, r_2)$ of the following forms;

$$\left\langle \phi_f^- \right| = \left\langle \frac{e^{ik_i \cdot r_i}}{(2\pi)^3} \psi_f^\perp (r_1, r_2) \right| \tag{32}$$

$$= \left\langle \frac{e^{ik_i \cdot r_i}}{(2\pi)^3} \psi_f^- (r_1, r_2) \right| - \left\langle \frac{e^{ik_i \cdot r_i}}{(2\pi)^3} \right| \psi_f^- (r_1, r_2) \left| \frac{e^{ik_i \cdot r_i}}{(2\pi)^3} \phi_i (r_1, r_2) \right\rangle \left\langle \frac{e^{ik_i \cdot r_i}}{(2\pi)^3} \phi_i (r_1, r_2) \right|$$

The final state wave function $\psi_f^-(r_1, r_2)$ is the solution of the Schrodinger equation for the 3C-body system consisting of two ejected electrons and residual ion Z and can be written as;

$$\left[-\frac{\nabla_1^2}{2} - \frac{\nabla_2^2}{2} - \frac{Z_1}{r_1} - \frac{Z_2}{r_2} + \frac{1}{r_{12}} - E \right] \psi_f^- (r_1, r_2) = 0, \tag{33}$$

The BBK wave function, which is solution of above equation, treats all three particles in the continuum state on an equal footing and also satisfies the three-body Coulomb boundary condition.   The form of the BBK wave function [23] is as follows;

$$\psi_f^- (r_1, r_2) = \frac{1}{\sqrt{2}} \left[\psi_f (r_1, r_2) + \psi_f (r_2, r_1) \right], \tag{34}$$

with,
$$\psi_f (r_1, r_2) = M \, e^{ik_1 \cdot r_1} e^{ik_2 \cdot r_2} \, \chi(r_1, r_2), \tag{35}$$

and,
$$\chi(r_1, r_2) = \prod_{j=1}^{2} {}_1F_1 (i\alpha_j, 1; -i(k_j r_j + \boldsymbol{k}_j \cdot \boldsymbol{r}_j)) \tag{36}$$

$$\times {}_1F_1 (i\alpha_{12}, 1; -i(k_{12} r_{12} + \boldsymbol{k_{12}} \cdot \boldsymbol{r_{12}}))$$

The constant M is given by;

$$M = \prod_{i=1}^{3} \frac{\left\{ \exp\left(-\frac{\pi \alpha_i}{2}\right) \Gamma(1 - i\alpha_i) \right\}}{(2\pi)^3} \qquad (37)$$

where, α_1, α_2 and α_{12} are Somerfield parameters and given as;

$$\alpha_1 = -\frac{Z_1}{k_1}; \ \alpha_2 = -\frac{Z_2}{k_2} \ and \ \alpha_{12} = -\frac{Z_{12}}{|k_{12}|}, \qquad (38)$$

with, $$k_{12} = \frac{(k_1 - k_2)}{2}, \qquad (39)$$

The effective charges Z_1, Z_2 and Z_{12} describe the interaction among all the outgoing particles in the continuum state and its analytical form is given by [24]. With the exact BBK wave function choice in the final state, a huge amount of computational time is required due to the presence of six dimensional radial integrals with an inter-electronic correlation term r_{12} in the Coulomb wave function describing the interaction between the ejected electrons. This makes the evaluation of the integrals extremely difficult. However, simplification can be achieved if we choose an approximate BBK type wave function in which the inter-electronic Coulomb wave is replaced by Gamow factor which includes the correlation factor k_{12} between the ejected electrons. Grin et al (2000) [25] have suggested that the use of Gamow factor is sufficient in predicting the experimental results and can explore the possible mechanism. The approximate BBK type wave function is;

$$\psi_f^-(r_1, r_2) = \frac{C}{\sqrt{2}} \left[\phi_{k_1}(z_1, r_1) \phi_{k_2}(z_2, r_2) + \phi_{k_2}(z_2, r_1) \phi_{k_1}(z_1, r_2) \right], \qquad (40)$$

where, $\phi_{k_j}(z_j, r_j)$ is the Coulomb wave function of the ejected electron j whose momentum is k_j. The repulsive Gamow factor C is given by;

$$C = \exp(-\pi / k_{12}) \Gamma\left(1 - \frac{1}{k_{12}}\right), \qquad (41)$$

On substituting equations (24) and (32) in equation (23), we obtain;

$$T_{fi}^{B1} = -\left\langle \frac{e^{ik_i \cdot r_i}}{(2\pi)^3} \psi_f^\perp(r_1, r_2) \left| \frac{1}{r_{i1}} + \frac{1}{r_{i2}} \right| \frac{e^{ik_s \cdot r_i}}{(2\pi)^3} \psi_i(r_1, r_2) \right\rangle, \qquad (42)$$

In equation (42), due to orthogonalization of final state wave function with the initial state wave function, the term $\left\langle \psi_f^\perp \left| -Z \right| \psi_i \right\rangle$ vanishes. Now, the integration over the incident electron coordinate r_i can be easily performed using Bethe's result;

$$\int \frac{e^{ik.r}}{|r - r_a|} d^3r = \frac{4\pi}{k^2} e^{ik.r_a} \, . \tag{43}$$

We can write;

$$T_{fi}^{B1} = \frac{1}{2\pi^2 k^2} [\langle \psi_f^{\perp}(r_1, r_2) | (\exp(iK.r_1) + \exp(iK.r_2) | \psi_i(r_1, r_2) \rangle$$

$$- \langle \psi_f(r_1, r_2) \| \psi_i(r_1, r_2) \rangle \langle \psi_i(r_1, r_2) | (\exp(iK.r_1) + \exp(iK.r_2) | \psi_i(r_1, r_2) \rangle] \tag{44}$$

We can also express the form of the matrix element (42) as;

$$T_{fi}^{B1} = -\frac{1}{2\pi^2 K^2} \left[T_{e1} + T_{e2} - \frac{T_c}{2}.T_o \right] . \tag{45}$$

Hence, the matrix element T_{fi}^{B1} has the following four contributive terms;

$$T_{e1} = \langle \psi_f^-(r_1, r_2) | \exp(iK.r_1) | \psi_i(r_1, r_2) \rangle , \tag{46}$$

$$T_{e2} = \langle \psi_f^-(r_1, r_2) | \exp(iK.r_2) | \psi_i(r_1, r_2) \rangle . \tag{47}$$

$$T_c = \langle \psi_f^-(r_1, r_2) | \psi_i(r_1, r_2) \rangle , \tag{48}$$

and $T_o = \langle \psi_i(r_1, r_2) | \exp(iK.r_1) + \exp(iK.r_2) | \psi_i(r_1, r_2) \rangle .$ (49)

On substituting final state BBK type wave function and initial state wave function (here **Le Sech wave function**) we get following term

$$T_{fi}^{B1} = -\frac{C^*}{2\pi^2 K^2 \sqrt{2}} \frac{N}{2} \begin{bmatrix} \{f(k_1, K, r_1, \alpha) + f(k_1, K, r_1, \beta)\} \{f(k_1, 0, r_2, z)\} \\ + \{f(k_2, 0, r_2, \alpha) + f(k_2, 0, r_2, \beta)\} \{f(k_1, K, r_1, z)\} \end{bmatrix}$$

$$- \frac{C^*}{2\pi^2 K^2 \sqrt{2}} \frac{N}{2} \begin{bmatrix} \{f(k_2, K, r_1, \alpha) + f(k_2, K, r_1, \beta)\} \{f(k_1, 0, r_2, z)\} \\ + \{f(k_1, 0, r_2, \alpha) + f(k_1, 0, r_2, \beta)\} \{f(k_2, K, r_1, z)\} \end{bmatrix}$$

$$- \frac{C^*}{2\pi^2 K^2 \sqrt{2}} \frac{N}{2} \begin{bmatrix} \{f(k_1, K, r_1, \alpha) + f(k_1, K, r_1, \beta)\} \{f(k_2, 0, r_2, z)\} \\ + \{f(k_2, 0, r_2, \alpha) + f(k_2, 0, r_2, \beta)\} \{f(k_1, K, r_1, z)\} \end{bmatrix}$$

$$- \frac{C^*}{2\pi^2 K^2 \sqrt{2}} \frac{N}{2} \begin{bmatrix} \{f(k_2, K, r_1, \alpha) + f(k_2, K, r_1, \beta)\} \{f(k_1, 0, r_2, z)\} \\ + \{f(k_1, 0, r_2, \alpha) + f(k_1, 0, r_2, \beta)\} \{f(k_2, K, r_1, z)\} \end{bmatrix}$$

$$+ T_0 \frac{Z C^*}{2\pi^2 K^2 \sqrt{2}} \frac{N}{2} \begin{bmatrix} \{f(k_1, 0, r_1, \alpha) + f(k_1, 0, r_1, \beta)\} \{f(k_2, 0, r_2, z)\} \\ + \{f(k_2, 0, r_2, \alpha) + f(k_2, 0, r_2, \beta)\} \{f(k_1, 0, r_1, z)\} \end{bmatrix}$$

$$+ T_0 \frac{Z C^*}{2\pi^2 K^2 \sqrt{2}} \frac{N}{2} \begin{bmatrix} \{f(k_2, 0, r_1, \alpha) + f(k_2, 0, r_1, \beta)\} \{f(k_1, 0, r_2, z)\} \\ + \{f(k_1, 0, r_2, \alpha) + f(k_1, 0, r_2, \beta)\} \{f(k_2, 0, r_1, z)\} \end{bmatrix}$$

$$\tag{50}$$

Where,

$$f(k_i, K, r_i, \lambda) = M_1 \int e^{-i(K + k_i).r_i - \lambda r_i} \, {}_1F_1(-i\alpha_i, 1, i(k_i r + k_i.r_i) d^3r_i$$

$$\text{and } M_1 = \left\{ \exp\left(-\frac{\pi\alpha_1}{2} \right) \Gamma(1 + i\alpha_1) \right\} \left\{ \exp\left(-\frac{\pi\alpha_2}{2} \right) \Gamma(1 + i\alpha_2) \right\}$$

2.1 Second Born matrix element

The derivation and computation of second Born matrix element for TS2 mechanism is more involved problem due to its complicated form as compared to the first Born matrix element T_{fi}^{B1}. The second Born matrix element can be written in the following form (See, [8]);

$$T_{fi}^{B2} = -(8\pi^4)^{-1} \lim_{\eta \to 0^+} \sum_n \int \frac{d^3 k_b}{(k_n^2 - k_b^2 + i\eta)}$$

$$\times \left\langle e^{i k_s \cdot r_i} \psi_f \left| -\frac{Z}{r_i} + \frac{1}{r_{i1}} + \frac{1}{r_{i2}} \right| \psi_n e^{i k_b \cdot r_i} \right\rangle \times \left\langle e^{i k_b \cdot r_i} \psi_n \left| -\frac{Z}{r_i} + \frac{1}{r_{i1}} + \frac{1}{r_{i2}} \right| \psi_i e^{i k_i \cdot r_i} \right\rangle$$

$$(51)$$

where, $\mathbf{k_b}$ is the intermediate momentum of the scattered electron and $k_n^2 = k_i^2 - 2(\omega_n - \omega_{1s})$, ω_n being the eigenvalue of the He atom Hamiltonian corresponding to the eigenvalue ψ_n. The summation on n runs over both the discrete and continuum states (see, [26-27])). In the evaluation of the matrix element (51), we have to calculate the radial integrals over the coordinate space spanned by r_i, r_1 *and* r_2 and then we integrate the resulting integrals over the intermediate momentum $\mathbf{k_b}$ in the momentum space.

Now carrying out the Bethe's integration over the coordinate r_i, one can write;

$$T_{fi}^{B2} = -\frac{2}{\pi^2} \lim_{\eta \to 0^+} \sum_n \int \frac{d^3 k_b}{K_1^2 K_2^2 (k_n^2 - k_b^2 + i\eta)}$$

$$\times \left\langle \psi_f \left| -2 + e^{i \mathbf{K_1} \cdot \mathbf{r_1}} + e^{i \mathbf{K_1} \cdot \mathbf{r_2}} \right| \psi_n \right\rangle \times \left\langle \psi_n \left| -2 + e^{i \mathbf{K_2} \cdot \mathbf{r_1}} + e^{i \mathbf{K_2} \cdot \mathbf{r_2}} \right| \psi_i \right\rangle$$

$$(52)$$

where, $\mathbf{K_1} = \mathbf{k_i} - \mathbf{k_b}$ and $\mathbf{K_2} = \mathbf{k_b} - \mathbf{k_s}$ are the successive momentum transfers from the incident electron to the target due to two step mechanism between the incident electron and target electrons. As there are infinite numbers of intermediate states probable for ψ_n so it is not possible to incorporate all the intermediate states before the double ionization to take place. One way to ease the problem is to set the target energy difference $(\omega_n - \omega_{1s})$ by an average value of excitation energy I_n and by using the closure property of the wave function $\psi_n (\sum_n |\psi_n\rangle\langle\psi_n| = 1)$ in the equation (52), we get;

$$\overline{T}_{fi}^{B2} = -\lim_{\eta \to 0^+} \frac{2}{\pi^2} \int \frac{d^3 k_b}{K_1^2 K_2^2 (k_i^2 - k_b^2 - 2I_n + i\eta)} \tag{53}$$

$$\times \langle \psi_f | (-2 + e^{iK_1 \cdot r_1} + e^{iK_1 \cdot r_2})(-2 + e^{iK_2 \cdot r_1} + e^{iK_2 \cdot r_2}) | \psi_i \rangle,$$

It may be mentioned here that (as in the present kinematics the incident electron energy is large) the excess ejected electrons energy and the momentum transfer to the target are low so use of closure approximation is valid in the present kinematics (see Byron *et al* (1982) [26]). This approximation, denoted by a bar over T_{fi}^{B2}, is known as closure approximation.

The matrix element $\overline{T}_{fi}^{B2}$ is singular at $k_b = \overline{k}_n$ (where, $k_b = \sqrt{k_i^2 - 2I_n}$), so we consider the contribution from the pole at $k_b = \overline{k}_n$ of the integrand of the integral (53). The integral (53) after analytic evaluation over k_b, results in the following form;

$$\overline{T}_{fi}^{B2} = -\lim_{\eta \to 0^+} \frac{2i}{\pi} \int_0^{2\pi} \int_0^{\pi} \frac{\overline{k}_n \sin \theta_b \, d\theta_b \, d\phi_b}{K_1^2 K_2^2}$$

$$\langle \psi_f | (-2 + e^{iK_1 \cdot r_1} + e^{iK_1 \cdot r_2})(-2 + e^{iK_2 \cdot r_1} + e^{iK_2 \cdot r_2}) | \psi_i \rangle, \tag{54}$$

Now, the integration will be performed on the polar and azimuthual angles (θ_b, ϕ_b) of the intermediate scattered electron momentum $|k_b|$. Now, we can write the matrix element $\overline{T}_{fi}^{B2}$ as;

$$\overline{T}_{fi}^{B2} = -\lim_{\eta \to 0^+} \frac{2i}{\pi} \int_0^{2\pi} \int_0^{\pi} \frac{\overline{k}_n \sin \theta_b \, d\theta_b \, d\phi_b}{K_1^2 K_2^2}, \tag{55}$$

$$\{ f_0 + f_1 + f_2 + f_3 + f_4 + f_5 + f_6 + f_7 + f_8 \}$$

where;

$$f_o = -4 \langle \psi_f^- (r_1, r_2) | \psi_i (r_1, r_2) \rangle = -4T_o ,$$

$$f_1 = \langle \psi_f^- (r_1, r_2) | \exp(iK.r_1) | \psi_i (r_1, r_2) \rangle = T_{e1} ,$$

$$f_2 = \langle \psi_f^- (r_1, r_2) | \exp(iK.r_2) | \psi_i (r_1, r_2) \rangle = T_{e2} ,$$

$$f_3 = \langle \psi_f^- (r_1, r_2) | \exp(iK_1.r_1) \exp(iK_2.r_2) | \psi_i (r_1, r_2) \rangle ,$$

$$f_4 = \langle \psi_f^- (r_1, r_2) | \exp(iK_1.r_2) \exp(iK_2.r_1) | \psi_i (r_1, r_2) \rangle ,$$

$$f_5 = -2 \langle \psi_f^- (r_1, r_2) | \exp(iK_1.r_1) | \psi_i (r_1, r_2) \rangle ,$$

$$f_6 = -2 \langle \psi_f^- (r_1, r_2) | \exp(iK_1.r_2) | \psi_i (r_1, r_2) \rangle ,$$

$$f_7 = -2 \langle \psi_f^- (r_1, r_2) | \exp(iK_2.r_1) | \psi_i (r_1, r_2) \rangle ,$$

82   Atomic and Molecular Physics: *Introduction to Advanced Topics*

and, $f_8 = -2\langle \psi_f^-(r_1,r_2) | \exp(iK_2.r_2) | \psi_i(r_1,r_2) \rangle$.

After substituting the derived expressions form $f_0, f_1, f_2, f_3, f_4, f_5, f_6, f_7$ *and* f_8 from their respective equations in equation (55), we can write the computational form of the second order matrix element $\overline{T}_{fi}^{B2}$ in the following form;

$$
\begin{aligned}
\overline{T}_{fi}^{B2} = {}& -4\int_\Omega F \; \frac{C^*}{\sqrt{2}}\frac{N}{2}\left[\begin{array}{l}\{f(k_1,0,r_1,\alpha)+f(k_1,0,r_1,\beta)\}\{f(k_2,0,r_2,z)\}\\+\{f(k_2,0,r_2,\alpha)+f(k_2,0,r_2,\beta)\}\{f(k_1,0,r_1,z)\}\end{array}\right]\\[4pt]
& -4\int_\Omega F.\frac{C^*}{\sqrt{2}}\frac{N}{2}\left[\begin{array}{l}\{f(k_2,0,r_1,\alpha)+f(k_2,0,r_1,\beta)\}\{f(k_1,0,r_2,z)\}\\+\{f(k_1,0,r_2,\alpha)+f(k_1,0,r_2,\beta)\}\{f(k_2,0,r_1,z)\}\end{array}\right]\\[4pt]
& +\int_\Omega F.\frac{C^*}{\sqrt{2}}\frac{N}{2}\left[\begin{array}{l}\{f(k_1,K,r_1,\alpha)+f(k_1,K,r_1,\beta)\}\{f(k_1,0,r_2,z)\}\\+\{f(k_2,0,r_2,\alpha)+f(k_2,0,r_2,\beta)\}\{f(k_1,K,r_1,z)\}\end{array}\right]\\[4pt]
& +\int_\Omega F.\frac{C^*}{\sqrt{2}}\frac{N}{2}\left[\begin{array}{l}\{f(k_2,K,r_1,\alpha)+f(k_2,K,r_1,\beta)\}\{f(k_1,0,r_2,z)\}\\+\{f(k_1,0,r_2,\alpha)+f(k_1,0,r_2,\beta)\}\{f(k_2,K,r_1,z)\}\end{array}\right]\\[4pt]
& +\int_\Omega F.\frac{C^*}{\sqrt{2}}\frac{N}{2}\left[\begin{array}{l}\{f(k_1,K,r_1,\alpha)+f(k_1,K,r_1,\beta)\}\{f(k_2,0,r_2,z)\}\\+\{f(k_2,0,r_2,\alpha)+f(k_2,0,r_2,\beta)\}\{f(k_1,K,r_1,z)\}\end{array}\right]\\[4pt]
& +\int_\Omega F.\frac{C^*}{\sqrt{2}}\frac{N}{2}\left[\begin{array}{l}\{f(k_2,K,r_1,\alpha)+f(k_2,K,r_1,\beta)\}\{f(k_1,0,r_2,z)\}\\+\{f(k_1,0,r_2,\alpha)+f(k_1,0,r_2,\beta)\}\{f(k_2,K,r_1,z)\}\end{array}\right]\\[4pt]
& +\int_\Omega F.\frac{C^*}{\sqrt{2}}\frac{N}{2}\left[\begin{array}{l}\{f(k_1,K_1,r_1,\alpha)+f(k_1,K_1,r_1,\beta)\}\{f(k_2,K_2,r_2,z)\}\\\{f(k_1,K_1,r_1,z)\}+\{f(k_2,K_2,r_2,\alpha)+f(k_2,K_2,r_2,\beta)\}\end{array}\right]\\[4pt]
& +\int_\Omega F.\frac{C^*}{\sqrt{2}}\frac{N}{2}\left[\begin{array}{l}\{f(k_2,K_1,r_1,\alpha)+f(k_2,K_1,r_1,\beta)\}\{f(k_1,K_2,r_2,z)\}\\\{f(k_2,K_1,r_1,z)\}+\{f(k_1,K_2,r_2,\alpha)+f(k_1,K_2,r_2,\beta)\}\end{array}\right]\\[4pt]
& +\int_\Omega F.\frac{C^*}{\sqrt{2}}\frac{N}{2}\left[\begin{array}{l}\{f(k_1,K_2,r_1,\alpha)+f(k_1,K_2,r_1,\beta)\}\{f(k_2,K_1,r_2,z)\}\\\{f(k_1,K_2,r_1,z)\}+\{f(k_2,K_1,r_2,\alpha)+f(k_2,K_1,r_2,\beta)\}\end{array}\right]\\[4pt]
& +\int_\Omega F.\frac{C^*}{\sqrt{2}}\frac{N}{2}\left[\begin{array}{l}\{f(k_2,K_2,r_1,\alpha)+f(k_2,K_2,r_1,\beta)\}\{f(k_1,K_1,r_2,z)\}\\\{f(k_2,K_2,r_1,z)\}+\{f(k_1,K_1,r_2,\alpha)+f(k_1,K_1,r_2,\beta)\}\end{array}\right]\\[4pt]
& -2\int_\Omega F.\frac{C^*}{\sqrt{2}}\frac{N}{2}\left[\begin{array}{l}\{f(k_1,K_1,r_1,\alpha)+f(k_1,K_1,r_1,\beta)\}\{(k_2,0,r_2,z)\}\\+\{f(k_2,0,r_2,\alpha)+f(k_2,0,r_2,\beta)\}\{f(k_1,K_1,r_1,z)\}\end{array}\right]\\[4pt]
& -2\int_\Omega F.\frac{C^*}{\sqrt{2}}\frac{N}{2}\left[\begin{array}{l}\{f(k_2,K_1,r_1,\alpha)+f(k_2,K_1,r_1,\beta)\}\{(k_1,0,r_2,z)\}\\+\{f(k_1,0,r_2,\alpha)+f(k_1,0,r_2,\beta)\}\{f(k_2,K_1,r_1,z)\}\end{array}\right]\\[4pt]
& -2\int_\Omega F.\frac{C^*}{\sqrt{2}}\frac{N}{2}\left[\begin{array}{l}\{f(k_1,K_2,r_1,\alpha)+f(k_1,K_2,r_1,\beta)\}\{(k_2,0,r_2,z)\}\\+\{f(k_2,0,r_2,\alpha)+f(k_2,0,r_2,\beta)\}\{f(k_1,K_2,r_1,z)\}\end{array}\right]\\[4pt]
& -2\int_\Omega F.\frac{C^*}{\sqrt{2}}\frac{N}{2}\left[\begin{array}{l}\{f(k_2,K_2,r_1,\alpha)+f(k_2,K_2,r_1,\beta)\}\{(k_1,0,r_2,z)\}\\+\{f(k_1,0,r_2,\alpha)+f(k_1,0,r_2,\beta)\}\{f(k_2,K_2,r_1,z)\}\end{array}\right]
\end{aligned}
$$

$$-2\int_{\Omega}F\cdot\frac{C^*}{\sqrt{2}}\frac{N}{2}\begin{bmatrix}\{f(k_1,K_2,r_1,\alpha)+f(k_1,K_2,r_1,\beta)\}\{(k_2,0,r_2,z)\}\\+\{f(k_2,0,r_2,\alpha)+f(k_2,0,r_2,\beta)\}\{f(k_1,K_2,r_1,z)\}\end{bmatrix}$$

$$-2\int_{\Omega}F\cdot\frac{C^*}{\sqrt{2}}\frac{N}{2}\begin{bmatrix}\{f(k_2,K_2,r_1,\alpha)+f(k_2,K_2,r_1,\beta)\}\{(k_1,0,r_2,z)\}\\+\{f(k_1,0,r_2,\alpha)+f(k_1,0,r_2,\beta)\}\{f(k_2,K_2,r_1,z)\}\end{bmatrix}$$

$$-2\int_{\Omega}F\cdot\frac{C^*}{\sqrt{2}}\frac{N}{2}\begin{bmatrix}\{f(k_1,K_1,r_1,\alpha)+f(k_1,K_1,r_1,\beta)\}\{(k_2,0,r_2,z)\}\\+\{f(k_2,0,r_2,\alpha)+f(k_2,0,r_2,\beta)\}\{f(k_1,K_1,r_1,z)\}\end{bmatrix}$$

$$-2\int_{\Omega}F\cdot\frac{C^*}{\sqrt{2}}\frac{N}{2}\begin{bmatrix}\{f(k_2,K_1,r_1,\alpha)+f(k_2,K_1,r_1,\beta)\}\{(k_1,0,r_2,z)\}\\+\{f(k_1,0,r_2,\alpha)+f(k_1,0,r_2,\beta)\}\{f(k_2,K_1,r_1,z)\}\end{bmatrix}. \tag{56}$$

Where, $F=-\dfrac{2i}{\pi}\dfrac{\overline{k}_n Sin(\theta_b)d\theta_b\,d\phi_b}{K_1^2\,K_2^2}$ \hfill (57)

Once the radial integrals $f_0 - f_8$ have been calculated by the above-mentioned derived expressions, the resultant integrand terms will be further integrated on the polar angle θ_b for the range $[0,\pi]$ and on the azimuthal angle ϕ_b for the range $[0,2\pi]$ of the intermediate momentum $\mathbf{k_b}$ numerically using Gauss quadrature technique;

$$\int f(x)\,dx=\sum_i W(x_i)\,f(x_i), \tag{58}$$

where, $W(x_i)$ and $f(x_i)$ are respectively the weight points and functions.

Once the first order matrix element $\overline{T}_{fi}^{B1}$ and second order matrix element $\overline{T}_{fi}^{B2}$ are calculated using the computational forms as given in equations (50) and (56) respectively, we calculate FDCS in the second Born approximation using the relation;

$$\frac{d^5\sigma(E_1,E_2,\Omega_s,\Omega_1,\Omega_2)}{dE_1\,dE_2\,d\Omega_s\,d\Omega_1\,d\Omega_2}=(2\pi)^4\cdot\frac{k_s\,k_1\,k_2}{k_i}\cdot\left|T_{fi}^{B1}+T_{fi}^{B2}\right|\cdot\left|T_{fi}^{B1}+T_{fi}^{B2}\right|^*$$

3. RESULTS AND DISCUSSION

This is to be mentioned here that we have calculated FDCS in first and second Born approximation using all above mentioned target wave function for He and He like ions for various geometrical arrangements. For brevity, in this Chapter we are not presenting these results in details due to our more emphasis on theoretical formalism. The results of our calculation and others calculation as well as experimental results can be found in literature. The experimental data for (e, 3e) processes on Ar, Ne, Kr and He atoms have been reported in [12, 28-30], [31], [32] and [24, 33-36] respectively by Orsay group at France led by Prof. A. Lahmam-Bennani. The German group led by Dr. A. Dorn have also reported (e, 3e) measurements on He at impulsive regime in [37-39]. Prof. Coplan and his co-workers from Maryland (USA) have conducted few (e, 3e) measurements to investigate the direct information on the correlated initial state wave function [40-41]. Some simplified experiments in the (e, 3-1e) on Ar, He and hydrogen molecule have been performed in [12, 42-46]. In such type of experiments one detects either pair of (e_s, e_1) (e_s, e_2) (e_1, e_2) of the three out going electrons. The theoretical results using any of the following theoretical formalism; 3 Coulomb (3C), 6C, 2CG and Convergent Close Coupling (CCC), have been reported in [17, 24, 35, 45, 47-59]. It is worth mentioning that the above-mentioned list is not complete as we have cited only few of the (e, 3e) results reported till now.

Here we present the results of our calculation of FDCS in θ-variable mode in coplanar geometry wherein momenta of all outgoing electrons lie in the scattering plane defined by the incident and scattered electrons. In θ-variable mode, one of the ejected electrons and the scattered electron are detected in a fixed direction and FDCS is investigated as a function of the angle of the other ejected electron. Further, the experimental data of Orsay group has been measured in θ-variable mode in two ways; in θ_1 as well as in θ_2 modes.

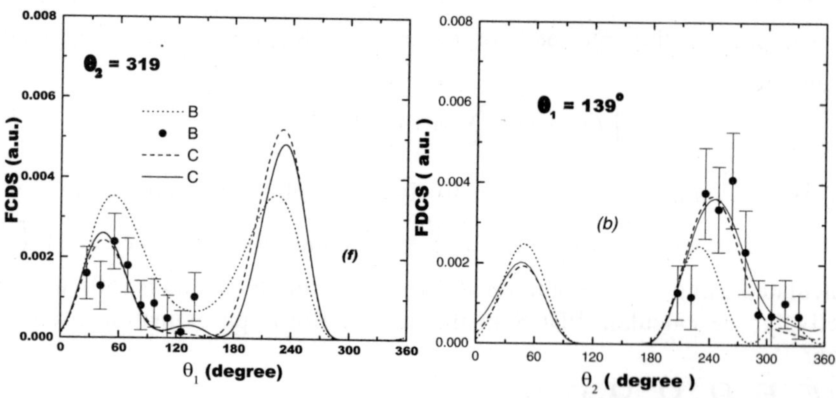

Figure 2: FDCS is plotted as a function of θ_2 in $\theta_1 - mode$ for He atom in coplanar geometry. The kinematics used in the depicted calculation is: $E_i = 5599\,eV$, $E_s = 5500\,eV$, $E_1 = E_2 = 10\,eV$, $\theta_s = 0.45°$ and θ_1 is given in each frame. We depict the results of FDCS in first Born approximation, second Born approximation with I_n = 0.0 a.u. and 0.9 a.u. by the dotted, dashed and solid curves respectively (Choubisa et al (2008) [52] and experimental data [24] by the symbol (•).

In the θ_1-mode, one of the two ejected electrons is kept fixed in the upper two quadrants ($27° \leq \theta_1 \leq 139°$) and the measured angular profile of FDCS is by varying θ_2 in lower two quadrants ($200° \leq \theta_2 \leq 360°$). We present in this subsection, the results of our calculation as well as the experimental data in Fig. 2.

Figures 2(a) and 2(b) represent an interesting situation in which one of the ejected electrons is ejected opposite to and along the direction of momentum transfer respectively and FDCS is investigated as a function of the angle of the other ejected electron angle. In both cases, the calculated FDCS in first Born approximation exhibits symmetry about the axis of momentum transfer (see dotted curve of Fig. 2). This reveals that symmetrical profile of FDCS about the axis of momentum transfer is a natural outcome of the first order theory and any deviation from the symmetry is an indication of second-order projectile-target interaction (i.e., TS2 mechanism). Unfortunately, due to lack of experimental data in full range, it is not possible here to see whether the symmetry is broken or not. It is very useful to investigate the contribution from second-order effect on the symmetry breaking of FDCS. Clearly, the second Born calculations of FDCS (solid and dashed curves) demonstrate a broken symmetry of the FDCS which exists for the first Born calculation (see, dotted curve) inherently. Unfortunately, we cannot confirm whether the symmetry of FDCS is broken or not in the present kinematics as the available experimental data is not in the full range (see, experimental data in the Figure 2). We suggest for more experimentation in the full range to confirm this effect.

Conclusion

In conclusion, we have discussed (e, 3e) processes on atoms (K-shell for many-electrons system) with special emphasis on theoretical formalism for the calculation of FDCS in first and second Born approximation. The mathematical expressions for first and second Born matrix elements have been written in explicit form. We have also compared our results with the available experimental data. We have discussed the importance of second order projectile-target interaction on the symmetry breaking on FDCS about the momentum transfer axis.

References

[1] T. G. Gooding and H. G. Pugh Nucl. Phys. **18** 46 (1960).
[2] U. Amaldi et al., Phys. Rev. Lett. **13** 341 (1964).
[3] U. Amaldi, et al., Phys. Lett. **22** 593 (1966).
[4] U. Amaldi et al., Phys. Lett. B **25** 24 (1967).
[5] H. Ehrhardt et al., Phys. Rev. Lett. **22**, 89 (1969)
[6] U. Amaladi et al., Rev. Sci. Instr. **40** 1001 (1969).
[7] I. E. McCarthy and E. Weigold *Phys. Rep.* C **27** 275 (1976).
[8] F. W. Jr. Byron and C. J. Joachain *Phys. Rep.* **179** 211 (1989).
[9] A. Lahmam-Bennani *J. Phys. B: At. Mol. Opt. Phys.* **24**, 2401 (1991).
[10] W. Nakel and C. T. Whelan Phys. Rep. **315**, 409 (1999).
[11] J. Briggs and V. Schmidt *J. Phys. B: At. Mol. Opt. Phys.* **33**, R1 (2000).
[12] A. Lahmam-Bennani et al., *Phys. Rev. Lett.* **63**, 1582 (1989).
[13] B. H. Bransden and C. J. Joachain, *Physics of Atoms and Molecules* (Pearson Education) (2003).
[14] J. Berakdar et al., *Phys. Rep.* **374,** 91 (2003).

[15] T. A. Carlson and M. O. Krause *Phys. Rev.* **140**, 1057 (1965)
[16] R. Choubisa, Ph.D. thesis, M L S University, Udaipur (2004)
[17] R. Choubisa et al., *J. Phys. B* **36,** 1731 (2003).
[18] H. R. J. Walters, J. Phys. B **13**, L749 (1980)
[19] F. W. Byron and C. J. Joachain *Phys. Rev.* **146,** 1-8 (1966).
[20] C. Le Sech *J. Phys. B* **30,** L47 (1997).
[21] A. Hylleraas Z Phys. **54,** 347 (1929).
[22] R. A. Bonham and D. A. Kohl *Chem. Phys.* **45,** 2471 (1966).
[23] M. Brauner et al., *J. Phys. B: At. Mol. Opt. Phys* **22,** 2265 (1989).
[24] A. Lahmam-Bennani et al., *Phys. Rev. A* **59**, 3548 (1999).
[25] M. Grin et al., *J. Phys. B* **33,** 131 (2000).
[26] F. W. Jr. Byron et al., *J. Phys. B* **15,** L293 (1982).
[27] F. W. Jr. Byron et al., *J. Phys. B* **18,** 3203 (1985).
[28] B. El. Marji et al., *J. Phys. B* **30,** 3677 (1997).
[29] C. C. Jia et al., *J. Phys. B* **35,** 1103 (2002).
[30] C. C. Jia et al., *J. Phys. B* **36,** L17 (2003).
[31] A. Lahmam-Bennani et al., *J. Phys. B* **25,** 2873 (1992).
[32] C. D. Schröter et al., *J. Phys. B* **31**, 131 (1998).
[33] I. Taouil et al., *Phys. Rev. Lett.* **81**, 4600 (1998).
[34] A. Lahmam-Bennani et al., *J. Phys. B* **34**, 3073 (2001).
[35] A. Lahmam-Bennani et al., *J. Phys. B* **35,** L215 (2002).
[36] A. Lahmam-Bennani et al., *Phys. Rev. A* **67,** 010701 (2003).
[37] A. Dorn et al., *Phys. Rev. Lett.* **82,** 2496 (1999).
[38] A. Dorn et al., *Phys. Rev. Lett.* **86,** 3755 (2001).
[39] A. Dorn et al., *Phys. Rev. A* **68,** 012715 (2002).
[40] M. J. Ford et al., *Phys. Rev. Lett.* **77,** 2650 (1996).
[41] B. El. Marji et al., *Phys. Rev. Lett.* **83,** 1574 (1999).
[42] A. Lahmam-Bennani et al., *J. Phys. B* **24**, 3645 (1991).
[43] A. Duguet et al., *J. Phys. B* **24** 675 (1991).
[44] B. El. Marji et al., J. Phys. B **29** L157 (1996).
[45] A. Lahmam-Bennani et al., *J. Phys.* **B35**, L59 (2002).
[46] P. Bolognesi et al., *Phys. Rev. A* **67,** 034701 (2003).
[47] C. Dal Cappello and H. Le Rouzo *Phys. Rev. A* **43,** 1395 (1991).
[48] Yu. V. Popov et al., J. Phys. B **27,** 1599 (1994).
[49] P. Defrance et al., *J. Phys. B* **33,** 4323 (2000).
[50] A. Kheifets et al., *J. Phys. B* **35,** L15 (2002)
[51] B. Nath and C. Sinha *J. Phys. B* **33,** 5525 (2002).
[52] R. Choubisa et al., *Pramana-J. Phys.* **60,** 1187 (2003).
[53] S. Jones and D. H. Madison *Phys. Rev. Lett.* **91,** 073201 (2003).
[54] S. Keller *J. Phys. B: At. Mol. Opt. Phys.* **36,** 755 (2003).
[55] A. S. Kheifets *Phys. Rev A* **69**, 032712 (2004).
[56] M. K. Srivastava *Phys. Rev. A* **70,** 062702 (2004).
[57] S. Elazzouzi et al., *J. Phys. B* **39,** 496 (2006).
[58] R. Choubisa and K K Sud *J. Phys. B* **41** 035202 (2008).
[59] R. Choubisa and K K Sud *J. Phys. B* **41** 208002 (2008).

Fast Ion-Atom Collision Processes: Some Recent Investigations

Lokesh C. Tribedi

*Tata Institute of Fundamental Research, Homi Bhabha Road, Colaba,
Mumbai-400005, India
e-mail: lokesh@tifr.res.in*

1 INTRODUCTION

The study of ion-atom collision physics involves various inelastic processes, such as, ionization, excitation and electron capture. The capture process can be radiative and non-radiative or Coulomb capture. Atomic collision processes find a lot of application in other fields, such as, astrophysical and laboratory plasma physics, atmospheric physics, biophysics, accelerator technology and laser-physics. In case of high energy collisions the ionization is one of the most important inelastic process. Electrons or photons have been the conventional probes in most of the collision experiments. However, with the advent of the high energy accelerators the fast heavy ions have been an useful tool for the atomic collision studies, which not only plays a complementary role regarding the collision mechanisms, but also provides many new insights into the collision dynamics. The availability of ion beams with varieties of velocities and also of different charge states allows one to study the collision mechanisms under the different perturbation strengths. With the wide variation of beam energies, by using different kind of accelerators, one can explore the collision physics from ionization-dominated regime (high energy) to capture-dominated regime (low energy). There have been a tremendous progress in the field of theoretical modelling and calculations for various types of atomic collision processes. While the collisions with fast electrons and photons provide a basic platform to test the theoretical mechanisms, the fast ions with predefined structures and charge-states are considered as to provide even more stringent test to the sophisticated models. Based on the ion-beam energy and charge states one can probe the strongly bound inner-shell processes besides the conventional outer shell processes. The physics of ion-atom collisions are readily extended to collisions involving the small molecules, large-biomolecules, clusters and solids. Besides the single particle excitations one can deal with several mechanisms influencing the collision processes, such as, e^- - e^- correlation, collective excitations, electron interference and two or many electron processes. The most commonly used (a few) basic theoretical techniques have been introduced for elastic scattering and electron capture in terms of the basic approximations and its range of validity etc. In many cases, instead of elaborate discussions, only relevant references are added. In addition, some of the recent studies on electron emission in fast ion-atom ionization, the interference effect, binary-encounter and soft collision processes, electron capture, scaling rule, Ist order theories on electron transfer and comparison with experimental data, recent experimental studies on 1s - 1s electron transfer have been outlined which are taken mostly from our recent research work.

2   SOME BASIC CONCEPTS

The basic aspects and definitions regarding the elastic, inelastic collisions, the threshold and channels can be found in comprehensive reviews and texts (see Ref. [1] for details regarding the following discussions). In general, one can find excellent reviews and texts on the theoretical and experimental techniques, for a detail understanding of atomic collision phenomenon [1, 2, 3, 4, 5, 6]. In a typical collision experiment a well collimated beam of particles with given energy is made to collide with a target atom in a single collision. If the incident and the target particles are denoted by A and B, respectively, then the several possibilities are prescribed as: A and B are scattered in which there is no change in their internal energies (or structures) i.e. A+B $\rightarrow$ A+B. This is known as elastic scattering. In elastic scattering the total kinetic energy remains the same before and after the interaction. The collision may also introduce a change in the internal energy or internal quantum state of A or B i.e. A+B $\rightarrow$ A+B*. The superscript '*' represents the excited state and such collision process is denoted as inelastic collision. In third case, the composite system A+B may split into two or more particles C and D i.e. A+B $\rightarrow$ C+D, or, more particles. This type of collision process is known as reaction. In a general reaction in which A+B composite system is decomposed into several (n) particles, one can write $K_f = K_i + Q$, where, K_f and K_i are final and initial kinetic energies (i.e. C.M. frame) and Q is the change of internal energies $(I_A + I_B)$-$(I_1+I_2+.....+I_n)$ before and after the collision. The reaction takes place if the condition: $K_f \geq 0$ is satisfied. If $Q \geq 0$ i.e. in exothermic reaction it is always allowed, at least energetically, to have the reaction. But in case $Q < 0$ the following condition has to be satisfied in order to make the (endothermic) reaction possible: $K \geq |Q|$. The quantity '$-Q = K_{th}$' is termed as threshold of the reaction. However, for practical application one has to note that in an experiment one mostly deal with the target B at rest initially, in the laboratory frame. Therefore at the laboratory frame the corresponding threshold energy can be given by: $K_{th}(Lab)$=$K_{th}(M_A + M_B)/M_B$, with M_A and M_B being the mass of A and B, respectively. One can see that in case of electron-atom collisions M_A=m_e << M_B and hence the laboratory and CM frame coincide, giving $K_{th}(Lab) = K_{th}$. As an example, if one considers the electron impact ionization of any atom (say 1s shell of He with I_p=24.6 eV) one has to simply compare the beam energy and the ionization potential (I_p) of 1s shell to decide the threshold energy of the beam required for such ionization, which, in this case (He) would be $K_{th}(Lab)$ 24.6 eV. As mentioned , in an elastic scattering the total K.E. remains the same before and after the collision and there is no change in the internal energy of the total system. In an inelastic collision the total internal energy changes. The excitation, deexcitation, ionization, electron transfer are some of the examples. In order to keep the total energy conserved the total K.E. of the system may increase or decrease. One of the motivation of studying the elastic scattering of heavy ions on atoms/molecules is the fact that it helps to probe the the interaction potential [2, 3, 7, 8] apart from its various other applications in plasma physics, atmospheric physics, solar astrophysics, propagation of electromagnetic radiation in atmosphere [9, 10]. One can show [2] that the fractional average energy-loss by projectiles (allowing for collisions at all possible angles) in elastic scattering (crude elastic sphere scattering) with a target atom at rest is given by : $\Delta E = 2M_p M_t/(M_p + M_t)^2$ (M_p and M_t being the masses of projectile and the target, respectively). For electron projectile this reduces to $\Delta E \sim 2m_e/M_t \leq 10^{-3}$. This imply that the energy experienced by an elastically scattered electron is negligibly small. However, one can see from the same equation that for heavy ion projectile the fractional energy loss is very large. For example, if $M_p \sim M_t$ then $\Delta E = 1/2$.

The time independent quantum mechanical theory of elastic scattering is well formulated and can be found in several references. It is not the aim of this article to describe it in detail. Only some results will be indicated without any proof. We assume that an infinitely wide beam of structureless projectiles approaching along the -Z axis and is scattered from a structureless center of force which is spherically symmetric. To obtain the amplitude of scattering (at angle θ), denoted as, $f(\theta)$ and hence the scattering cross section one needs to find a solution of the wavefunction which will have the asymptotic form:

$$\psi = \psi_{inc} + \psi_{scatt} \simeq e^{ikz} + \frac{e^{ikr}}{r}f(\Theta), \tag{1}$$

since the outgoing wave must have the form of of an outgoing spherical wave whose amplitude decreases as $1/r$ at large distance. This will also make sure that the radial current density may fall of as $1/r^2$ and the the number of scattered particles is conserved. The time independent wave equation then can be written as:

$$\Delta^2\psi + [\kappa^2 - U(r)]\psi = 0, \tag{2}$$

where $\kappa = \sqrt{(2\mu E)}$, E=total energy (> 0), and $U(r)$ is the reduced potential, given by $U(r) = 2\mu V(r)/\hbar^2$. The solution to this can be obtained in different ways. One of the common ways is to use the partial wave method. The soultion for the scattered wavefunction, differential scattering cross section and total cross section can then be given by,

$$\psi(r, \Theta) = \Sigma_0^\infty (2l + 1)(e^{2i\eta_l} - 1)P_l(\cos\Theta) \tag{3}$$

$$\frac{d\sigma}{d\Omega}(\Theta) = |f(\Theta)|^2 = \frac{1}{\kappa^2}|\Sigma_0^\infty(2l+1)e^{i\eta_l}\sin\eta_l P_l(\cos\Theta)|^2 \tag{4}$$

$$\sigma = \frac{4\pi}{\kappa^2}\Sigma_0^\infty(2l+1)\sin^2\eta_l. \tag{5}$$

The η_l is the phase shift of l-th partial wave. The limits of the sum in the above two equations are $l = 0$ to ∞. Therefore one has to calculate the associated phse-shift in order to obtain the integral scattering cross section. As a simple example of getting phase shift one typically uses a square well potential and then we may calculate the phase shift for hard sphere potential (i.e. $U(r) = \infty$ for $r < a$, and $U(r) = 0$ for $r > a$, where a is the radius of the sphere. It turns out that for low energy (i.e. for $l \sim 0$) $\sigma \sim 4\pi a^2$ which is 4-times the classical value. For higher energy collisions a larger no. of partial waves may contribute and one can show that $\sigma \sim 2\pi a^2$ which is 2-times the classical value. One can make a general comment that the partial wave method is most useful when a small no of partial wave contributes, a situation which generally occurs at low energies. There exists a large amount of data on the elastic scattering of the heavy particles in the low energy collisions of the type neutral-neutral or ion-neutral. The cross sections show structure as a function of beam energy or scattering angle. The related effects, such as, the Rainbow-scattering and Ramsauer-Townsend effect, orbiting resonances and symmetry oscillations are greatly discussed in literature which can be found in [3]. In case of resonances the phase shift varies rapidly in a certain energy region which causes a dramatic change in the corresponding partial cross section.

Though various approximations are used in such collisions, the mostly used one is the first-Born approximation which uses the first term in the Born series. The Born series for scattering amplitude physically represents multiple scattering series in which the projectile interacts repeatedly with the potential V and propagates freely between the two such interactions [1]. The Born series thus converges faster if the scattering potential is weak and the particle is fast enough so that it does not interact too many times with the potential. Therefore one can then use only the first Born approximation (FBA) in case of sufficiently high velocity. The first Born approximation for the scattered amplitude can be given by:

$$f(\Theta) = -\frac{2\mu}{\hbar^2}\int_0^\infty \frac{\sin kr}{kr}V(r)r^2 dr \tag{6}$$

where $k = 2\kappa\sin(\Theta/2)$ and represents the magnitude of momentum transfer. Now, if one calculates the Coulomb scattering cross section in the FBA the relevant potential $V(r) = Z_p Z_T e^2/r$ gives a mathematically meaningless integral to calculate $f(\Theta)$. However, to avoid this situation one then uses $V(r) = (Z_p Z_T e^2/r)e^{-\gamma r}$ (with $\gamma > 0$). Then after integrating in the RHS of above equation (7) one then uses $\gamma \to 0$ to get the final expression for the scattering cross section:

$$\frac{d\sigma}{d\Omega_{CM}} = \frac{Z_p^2 Z_T^2 e^4}{4\mu^2 v_0^4 \sin^4(\Theta/2)}d\Omega_{CM} \tag{7}$$

One can see that this equation although derived from an approximation method and using the approximate potential (with $\gamma \to 0$) one still obtains the exact classical result (Rutherford scattering). Also one may note that the term $e^{-\gamma r}$ helps to obtain the screened Coulomb potential. The similar potential functions are used in case of plasma with $1/\gamma$ being the Debye length. The screening of a nucleus by atomic electrons can also be represented by such potential. The Born approximation is, in general, can be applied for high energy collisions.

Bethe-Born approximation: Bethe proposed a modification to the FBA. This assumes an additional condition kr is small $(<< 1)$ where k is the momentum transfer and r is the range of the potential. This allows one to expand the e^{kr} term in the integrand giving the dipole, quadrupole contributions in the atomic transitions. In Coulomb-Born approximation the plane wavefunction is replaced by a Coulomb wavefunction which of course depends on the nuclear charge. It is generally used for the collisions involving electrons/ions as projectiles and target ions.

3   POTENTIAL FUNCTIONS

The interaction potential plays the most central role in determining the scattering mechanism. There are various kinds of functions used to represent realistically the interaction. At the same time, the mathematical forms are, however, chosen in order to satisfy the mathematically solvable. here we give some examples of the potential functions which depend on the separation (r) only. This discussion is only in brief and the details of it can be found in Ref.[2] (Section 1-7, 1-8). The hard sphere potential ($V(r) = \infty$ for $r < D$ where, D is the sum of the radii of the smooth elastic spheres and 0 for $r > D$) is one of the common potential which are applied in many simple cases. Repulsive or attractive potentials are generally represented as $\pm 1/r^n$. For Coulomb interaction $n = 1$. In particular, one may refer to the so called "Sutherland potential" which behaves like elastic sphere for $r < D$, and like attractive $-1/r^n$ type for $r > D$. Another type of function, known as Lennard-Jones potential has the form $(r^{-12} - r^{-6})$. One may note that inverse-sixth-power attraction represents an induced-dipole induced-dipole interaction. The attractive potential function $-1/r^4$ represents the potential of a (point charge) induced-dipole and represents the interaction of an charged particle and neutral molecule. Another potential which includes higher multipolar interactions, such as, induced-dipole induced-dipole ($-1/r^6$) and induced-dipole induced-quadrupole ($-1/r^8$) interactions, is the so called Buckingham potential. This is defined as $V(r) = \alpha e^{-\beta r} - \gamma/r^6 - \delta/r^8$. One can find more detailed discussions on this as well as the angle dependent potential functions in Ref. [11, 12, 13].

4   TWO CENTER ELECTRON EMISSION IN ION-ATOM ION-IZATION

4.1   Binary Encounter

Ionization is one of the most important reactions in high energy ion-atom collisions. The distinct features characterizing the ionization process are the soft collision, the electron capture in continuum (ECC) cusp and the binary encounter peak. The low energy electrons produced in large impact parameter collisions (soft collision) dominate the electron energy spectrum. Binary encounter (BE) electrons are the target electrons ionized through direct hard collisions with energetic projectiles and show up as a broad peak in the electron double differential distribution. Although the cross sections for BE process are very small compared to the soft collision electrons, the studies of the angular distributions of the BE electrons can enrich our understanding of the low impact parameter collision physics. A detailed understanding of the BE process is necessary to elucidate the ion-atom ionization mechanism.

Binary encounter process has been extensively studied recently both experimentally and theoretically [14, 15, 16, 17, 18, 19, 20, 21, 22, 23, 24, 25, 26, 27, 28, 29, 30]. In case of the light ion projectiles the most of the measurements are carried out for zero degree laboratory emission angle. A comprehensive understanding of the binary encounter process requires the study of the angular distribution of the electrons emitted. Only a few measurements on the angular distributions have been reported [14, 20, 23, 31] for light ions. However, there have been several measurements on the angular distribution studies using much heavier ions [19, 21, 22] and a double peak structure was observed in the BE electron spectrum.

Lee et al. [16] have shown that in case of $0°$ emission the BEE DDCS follows Z_p^2 dependence (where Z_p is the projectile atomic number) as expected on the basis of first order perturbation theory. But for dressed ions, Richard et al. [17] have shown that the DDCS of BE process *increases* with *decreasing* projectile charge state. However, in case of non-zero degree emission angles, the DDCS was found to decrease with decreasing charge state for θ larger than a critical angle θ_0 where θ_0 depends on the collision system. For example, it was found that $\theta_0 = 25°$ for 2.5 MeV/u $O^{q+}+O_2$ [14]; $35°$ 1 MeV/u $C^{q+}+H_2$ [23] and approximately $25°$ for 1 MeV/u $F^{q+}+H_2$ [20].

Recent theoretical and experimental studies are also been aimed to the peak position of the BE distribution. Using elastic two-body collision dynamics for heavy-ion impact on a *free* electron, the the peak energy of the BE electron can be given by:

$$E_{BE} = 4t\cos^2\theta_L \qquad (8)$$

where the cusp electron energy is given by,

$$t = 4t\cos^2\theta_L \qquad (9)$$

$t = (m_e/M_p)E_p$, where m_e and M_p are the electron and projectile mass, respectively. Since the electrons have binding energy in the initial state (ϵ_i), the BE peak is not centered exactly at the position corresponding to the prediction of the free electron scattering model. It has been shown [31, 32] that the position of the BE peak is shifted to lower electron energies by an amount which is proportional to $|\epsilon_i|$ ($\approx 2|\epsilon_i|$). For $Z_P > Z_T$ the polarization of the initial state by the projectile must be considered. According to the tunnelling model [32], the ionization and charge transfer (for $Z_P > Z_T$) may proceed through the potential barrier of finite width, formed by the projectile electric field superimposed on the target atomic potential.

4.2   Soft-Collision Electron Emission

In case of heavy-ion impact ionization the ionized electron is influenced by the two moving sources of Coulomb potentials, namely, the receding highly charged heavy projectile ion and the residual recoil ion. This is not so much important in case of low charged projectiles, such as, e^-, H^+ etc. Although, the projectile velocity considered here is much larger than the velocity of the electron in the atom, $(v_p/v_e \sim 10$ a.u.) it is seen that the first Born calculation (B1) fails to explain the energy and angular distributions of the ionized electrons. A series of experimental data have revealed that the forward emission is enhanced compared to the prediction of B1. Accordingly for the backward angles the double-differential cross sections are reduced. This forward focussing is more dominant for high-atomic number ions. This is a consequence of the fact that, B1 accounts only for the target centered effects and doesn't consider the effect of the receding projectile after the electron has been ionized. In order to account for the projectile center effects, a theoretical model based on the continuum distorted wave eikonal initial state (CDW-EIS) approximation has been developed [33]. This method is a first order in the distorted wave series and is shown to be adequate to describe the dynamics of the ionized electron in the combined Coulomb fields of the projectile and the target. The model was extended and fine-tuned by many workers over the period of time for multi electronic targets [34, 35]. Fainstein *et.al.*, have also extended it to molecular targets [36, 37] which has recently been applied to explain the interference effect (see below) in heavy ion induced ionization of H_2. However, there are many experimental investigations on the heavy ion induced continuum electron emissions from a few research-groups [31, 38, 39, 40, 41, 42, 69, 70, 45, 46, 47, ?, 50, 51, 52, 53, 54, 55, 56, 74]. It has been demonstrated that the measurements of the electron DDCS in case of He-target provides important information regarding the TCE while that molecular hydrogen provide crucial information regarding the interference effect, in addition. The CDW-EIS calculations provide excellent agreement with the electron DDCS measured in GeV energy collisions, as revealed from many of the GANIL-based works. However, in MeV energy (20-100 MeV) collisions for which the perturbation strength is relatively higher, the overall agreement with the CDW-EIS model is again good except for extreme backward angles where the theory underestimates the experimental results. Also the deviations from the data at the extreme forward angles are also noted. It is well known that the B1 calculations underestimate the cross sections at forward angles whereas it overestimates the cross sections at backward angles. Such a difference is quoted as a typical example of the effect of two-center mechanism. However, the B1 calculation provides a good agreement with measured data for angles close to 90°, which is dominated by binary encounter process, which is mainly a two-body process and single center model works farely well. Therefore,the experimental evidences are there to suggest the substantial deviation of the CDW-EIS model from the observations even at high velocity ($v_p \sim 10$ a.u.) collisions. The CDW-EIS is again a first order in the distorted wave series and there is further scope to improve it in order to provide a better wgareemnet with the series of experimental data available.

Such two-center effect is less relevant when electrons are used as projectiles, since they are singly charged particles [58]. The electron angular distributions are found to be more symmetric (giving almost dame cross section for the extreme forward and backward angles) compared to that for heavy-ion collisions However, the soft collision and the B.E. are dominant features besides the elastic peak in the ejected electron spectrum of atoms and molecules in collision with fast electron ($\sim keV$ beam. There are several studies available on BE process for heavy ion collision, as indicated above. Unlike heavy ions, for electron impact elaborate measurement on BE process is scarce. Only recently we have investigated the effect of e-e binary encounter in electron impact ionization of low Z (atomic number) targets, such as, He and H_2 [59].

5   YOUNG TYPE ELECTRON INTERFERENCE

5.1   Evidence of Oscillations in DDCS Spectrum

The low energy electron double differential spectrum from H_2,is very rich since it provides the evidence of interference effect. Although this effect has been known for many years to exist in case of photo-ionization [60, 61, 62] and electron capture [64, 63], it is interesting to note that the influence of interference on the ion impact ionization of molecule was not known until recently. It is only very recently that, the interference effect in the low energy electron spectra emitted from H_2 has been demonstrated, in collisions with high energy (60 MeV/u) Kr^{34+} ion impact [65, 66] and relatively low energy collisions [67, 68, 69, 70, 71, 72]. Since then much effort has been put both on theoretical as well as experimental fronts by many authors. The angular dependence of the interference pattern was subsequently understood both experimentally [66, 73, 74] and theoretically [75, 36]. Further, measurements were also carried out to see the effect of a second order interference effect in the ionization spectrum and was explained by invoking a double scattering mechanism [76, 77]. The role of interference effect on the double ionization of D_2 was also studied very recently by Landers *et al.,* [78] using a cold target recoil ion momentum spectroscopy technique. The influence of the target Compton profile on the interference oscillation and the associated energy-deoendent complications were also discussed [79, 68]to derive the oscillation. The influence of the interference in photoionization of molecules, such as, H_2, N_2 and H_2^+ targets

[80, 81, 82, 83, 84, 85, 86, 87] has been investigated very recently in detail.

Since the two H-atoms in a homonuclear-diatomic molecule H_2, are indistinguishable, their contributions to the ionization probability add coherently and an interference effect might be expected in the single ionization of H_2. Such electron emission from H_2 may be closely related to the well known Young's two-slit experiment which played a most crucial role in the the development of the quantum mechanics. Since the DDCS varies over several orders of magnitude over an energy range of 300 eV, it becomes very difficult to notice small variation of the order of 50% due to interference in the DDCS spectrum. Therefore, in order to amplify the visibility of the structure, the experimental DDCS for H_2 is divided by the DDCS for atomic H which was either calculated or measured *in situ* [67, 40]. The expression for DDCS ratio (R) can be written as:

$$\frac{d^2\sigma_{H_2}}{d\Omega_e d\epsilon_e} \Big/ \frac{d^2\sigma_{2H}}{d\Omega_e d\epsilon_e} = \left(1 + \frac{\sin(kcd)}{kcd}\right). \tag{10}$$

where k is related to the electron energy $\varepsilon = \frac{\hbar^2 k^2}{2m}$ and c is known as frequency parameter which is emission angle (θ) dependent. According to some authors $c = \cos\theta$. However, it was later later found that for forward angles c can be represented as $\cos\theta$. This was obtained by averaging over all possible molecular orientations[60, 65]. However, several authors [65, 66]have shown that, in order to derive the interference oscillations, one can also use the different values for Z_{eff}, such as 1.05, derived from the binding energy or 1.19, originated from the variational parameter treatment of the wave function for H, to compensate the effect of Compton profile mismatch. However, the oscillation itself was observed in the DDCS ratio (i.e. the ratios of DDCS for H_2 to that for 2H) even earlier [40].

The interference mechanism in electron elastic scattering from molecular H_2 was introduced as early as 1977 both theoretically and experimentally [88, 89]. In electron impact ionization studies, the presence of oscillations due to coherent emission was first predicted [90] and then experimentally observed in DDCSs [91] and, later on, in the angular distribution of electron emission [92, 93, 94] for H_2 in coplanar kinematics by investigating the relative change of binary and recoil peak intensities caused by partial constructive or destructive interferences. In triple differential cross section (TDCS) studies, the single-center He target was also used as a reference to compare the fluctuation of intensity due to interference effects in the corresponding H_2 spectrum [92, 93]. Recently, Young type interference patterns were reported in the double electron capture process from molecular hydrogen [95].

5.2   Second Order Interference

Following the initial investigations by Stolterfoht *et al.* [65] and Misra *et al.* [67] many aspects of the interference phenomenon have been studied [76, 96, 73, 69, 78, 99, 100] in the charged particle impact ionization of H_2. Besides the experimental observations there have been a substantial progress in the theoretical models [36, 37, 97, 98]. However, if higher order contributions (e.g. second order amplitudes) are considered, one expects to have a higher frequency component to be present in the oscillation, besides the Ist order oscillation. Such a formalism was originally introduced by Messiah [103] in the context of electron scattering from a molecule. Although there is no analogue of higher order interference term in the Young's interference using photons, the double scattering mechanisms are known to give rise many interesting features in atomic collision process, such as Thomas mechanism of double-scattering mechanism in electron capture. A close inspection of the DDCS ratios [96], shows an indication of a high frequency component riding over the first order oscillatory structure and the periodic deviation from the Cohen-Fano type [60] fit function was attributed, by Stolterfoht *et al.*, to the second order process. The second order component arises due to the interference between the ionized electron from one H-center and the re-scattered electron from the second H-center. However, such interpretation of doubling of frequency in terms of a second scattering model, although present in the model calculation by Nagy et al. [101], does not get a support from the very recent theoretical work by Sisourat et al. [102]. In order to resolve this and to have a further understanding of these secondary structures, recently, Misra *et al.* [77] carried out experiments with improved statistics. They also derived the necessary theoretical formalism which includes the contribution from both the processes (i.e., first and second order effects). The general interaction matrix for a particle, interacting with two scattering centers having interaction potentials V_1 and V_2 can be written as an expansion in the Born series as [103]:

$$T = (T_1 + T_2) + (T_2 G_0 T_1 + T_1 G_0 T_2) + T(3) \qquad (11)$$

where, T_1, T_2 and T(3) are defined as:

$$T_1 = V_1 + V_1 G_0 V_1 + V_1 G_0 V_1 G_0 V_1 + ..., \qquad (12)$$

$$T_2 = V_2 + V_2 G_0 V_2 + V_2 G_0 V_2 G_0 V_2 + ... \qquad (13)$$

$$T(3) = (T_1 G_0 T_2 G_0 T_1 + T_2 G_0 T_1 G_0 T_2) + ... \qquad (14)$$

The first two terms in Eq. 11 correspond to the single scattering process. The next two terms correspond to a double scattering process, i.e., an incoming particle first gets scattered off one of the centers (T_1) and propagates (G_0) towards the other center with before it gets scattered off the second center (T_2) and emerges with the final momentum k. The last two terms (i.e. T(3)) correspond to a triple scattering, i.e., the particle gets scattered twice in between the two centers before it gets out with the final momentum k. Following Messiah's treatment of multiple scattering from two scattering centers [103], the expansion in Eq. 11 relates the transition amplitudes of the two scatterer problem to the amplitude of the single scatterer problem.

The total DDCS ratio is shown to be (see [77] for details):

$$R = D + F \left(1 + \frac{\sin(kcd)}{kcd} \right) + S \left(1 + \cos(2kd) \frac{\sin(kcd)}{kcd} \right), \qquad (15)$$

where D, F and S are free parameters, representing the direct, first and second order contributions, respectively. i) In the first order term the frequency parameter c has been used [76]). The second order contribution has an oscillation frequency that is twice as that of the first order contribution. Therefore, the double scattering mechanism essentially introduces additional path length which is responsible for the doubling of the oscillation frequency.

Finally they fit the experimentally observed oscillatory structure to the analytical formula derived. A double scattering component in the oscillation was clearly visible for the collision system with bare 4 MeV/u F-ions colliding with H_2. This also revealed that the amplitude of the second-order effect was as high as about 10% of that for the Ist order. This is larger contribution compared to that observed for projectile-ions with high atomic number(such as Kr) at GeV energies [96]. It is commented that it would be interesting to study this effect even for lower-atomic number projectiles. However, this experimental results with bare F-ions as projectile and analysis strongly supports the initial interpretation of Stolterfoht *et al.* [96] in terms of a second order scattering mechanism.

5.3   Asymmetry Parameter and Interference Effect

The long range Coulomb interactions between the electron-target and electron-projectile in the final state influence the evolution of electron wavefunction. The signature of the two-center effect in ion-atom collision induced ionization is the large forward-backward asymmetry in electron emission spectrum even for atomic H target [38, 31, 39, 40]. Apart form the TCE, for multi-electron target-atoms the non-Coulomb potential may cause some asymmetry. Therefore, if the TCE is not considered, then for H-atom target (i.e. with only Coulomb potential) one does not expect any angular asymmetry. However one observes a large asymmetry in collisions involving atomic H [40]. The asymmetry is also electron-energy dependent which also indicates its dependence on the impact parameter. Therefore the angular distribution is also influenced which can be seen in the large difference in the DDCS-values (σ) for small forward and large backward angles. Let's he quantity $\alpha(k)$ as,

$$\alpha(k, \theta) = \frac{\sigma(k, \theta) - \sigma(k, \pi - \theta)}{\sigma(k, \theta) + \sigma(k, \pi - \theta)}, \tag{16}$$

where, electron energy $\varepsilon_k = k^2/2$ (in a.u.) and θ is chosen to be low forward angle, 30°. However, by expanding the $\sigma(k, \theta)$ in terms of the Legendre's polynomials, it was shown by Fainstein *et. al.*, [104] that, the $\alpha(k)$ would represent the angular asymmetry parameter if $\theta = 0$. Since angular distributions vary slowly near 0 and π [40] the measured $\alpha(k, 30°)$ approximately represents the angular asymmetry parameter.

The experimentally derived values of $\alpha(k)$, (i.e. $\alpha(k, 30°)$) show a smooth monotonically increasing trend as a function of electron velocity, for an atomic target such as He [see Ref. [69, 70]. This behavior is expected based on the TCE-consideration which is qualitatively well represented by the CDW-EIS model [35]. On the contrary, for fast (80 MeV) C^{6+}-ions colliding with a molecular target, H_2, the asymmetry parameter shows an oscillatory structure superimposed on a smoothly varying function. To understand this effect we use the molecular CDW-EIS calculation [36, 37]. The main feature of this model is to represent the initial bound state of the active electron by a two-center molecular wavefunction. Within the impact parameter approximation, the transition amplitude reduces to a coherent sum of atomic transition amplitudes corresponding to individual molecular centers. This model, however, automatically reproduces the Young type interference effect in the electron emission from H_2 and its dependence on the emission angle such that the frequency of oscillation is higher for backward angles compared to the complementary forward ones. This difference in the frequency, for forward and backward angles, causes the oscillatory structure in the $\alpha(k)$.This again implies that the interference process built-in molecular CDW-EIS model using molecular wave function gives rise to the oscillations in the asymmetry parameter for H_2. Recently the studies on the oscillations in the asymmetry parameter was also observed in 95 MeV $F^{9+}+H_2$ collisions as well as in 10 keV electron impact ionization of H_2 [74, 58]. A comparative study showed that the angular asymmetry parameter in case of electron collision shows more prominent feature of oscillation than that for heavy-ion impact. This is due to the fact that the two-center effect does not play a major role in electron emission in case of fact electron-impact. Importantly in the technique one needs only the relative DDCS values for H_2 target for the forward and backward angles. This implies that it does not need any DDCS data for atomic H ionization which is not easy, rather challenging task experimentally. In particular this technique can be applied to other multielectronic molecules, such as, O_2 and N_2 to get information on such interference since it is very difficult to get the atomic cross section bothe experimentally and also theoretically.

6   ELECTRON TRANSFER

The ionization, electron capture and excitation are among the most important inelastic processes in ion-atom collisions, especially in the intermediate energy region. When a highly charge ion collides with a target atom, among other processes, the ion can capture one or more bound electrons into any unoccupied bound-state. This results in a ground state or excited state of the projectile ion. The capture can be non-radiative which is known as Coulomb capture or mechanical capture since the energy momentum conservation can take place through the change of projectile/recoil ion energy and momentum. In an radiative electron capture, on the other hand, a photon is emitted in order to have energy conservation. We only discuss the Coulomb capture here. At intermediate energy, the projectile energy (v_p) is approximately equal to the orbital velocity (v_e) of the active electron and the strengths of these processes are of same orders of magnitude and a coupling among these different channels become also important. The ion beams from medium or high energy tandem accelerator are used for fast collisions for which $v_p/v_e \gg 1$, for loosely bound outer-shell electrons. However, such fast ions of energy tens of MeV or more can be used for the study of strongly bound inner-shell electrons for some selected target-systems. Ionization and electron transfer involving deeply bound inner shells play a major role in producing vacancies in these shells in heavy ion-atom collision. In some cases, depending on the symmetry parameter (Z_p/Z_t) of collision systems the transfer channel could be much larger than the direct Coulomb ionization. There are numerous studies on the total electron capture for initially loosely bound electrons and several empirical scaling laws [105, 106] have been proposed to predict the capture cross sections which fall rapidly with the projectile velocity. On the other hand, for the projectiles with energies $\sim$ hundreds of MeVs the cross sections for deeply bound electron transfer (such as, σ_{K-K}) process has a maximum since the projectile velocity (v_p) is approximately same as the orbital velocity (v_e) of the active electron, as in the present studies.

7   Scaling Law

Many times the experimental data can not be exactly reproduced by *ab initio* theoretical models due to the inherent difficulties regarding the approximations used for the interaction potential as well as the wavefunctions. In such cases, in order to get an approximate value of the electron transfer cross sections, especially, for capture from the loosely bound outer shells of an atom to a dressed projectile-ion of charge state q, can be obtained by empirical scaling formula derived by different workers. One such scaling formula was predicted by Schlachter [105] for electron capture by a high energy projectile with energy E (in keV/u) and is given by,

$$\tilde{\sigma} = \frac{1.1 \times 10^{-8}}{\tilde{E}^{4.8}}[1 - exp(-0.037\tilde{E}^{2.2})][1 - exp(-2.44 \times 10^{-5}\tilde{E}^{2.6})] \qquad (17)$$

where $\tilde{\sigma}$ and $\tilde{E}$ are defined as:

$$\tilde{\sigma} = \frac{\sigma Z_T^{1.8}}{q^{0.5}}, \qquad (18)$$

$$\tilde{E} = \frac{E}{Z_T^{1.25} q^{0.7}}. \qquad (19)$$

It was observed that this scaling law fits data well for varieties of collision systems provided, $E \geq 10$ and $q \geq 3$.

7.1   First Order Models on Electron Capture

The state selective electron transfer cross sections involving deeply bound initial and final states can not be described by such empirical laws and mechanism of such transfer process in strongly perturbative collision is not yet understood completely, although such elementary collision processes are being studied for many decades. Reference [4] serves as an excellent text (see chapter 8). The additional complication arises due to the fact that one has to deal with a three-body problem involved in such process, such as, p+H$\rightarrow$H+p. The orthogonality of the wave functions in the initial and final states are not automatically guaranteed. The initial and final state binding energies of the transferred electron, the symmetry parameter $S_z = Z_P/Z_T$ and the reduced velocity $v_r = v_p/v_e$ of the collision system are the relevant parameters which are generally used to describe the transfer process. The binding energy matching between initial and final states provide a favourable condition for electron transfer as predicted by the first order calculations. This was first worked out by Oppenheimer-Brinkman-Kramer in 1928 (see [4]) and further generalized by Nikolaev (OBKN) (1967) [107]. They derived a simple analytical expression for electron transfer from n_i(target) to n-th state of projectile by using "almost" orthogonal wavefunctions and by neglecting the internuclear potential.

$$\sigma_{OBK}(n_i - n) = \frac{2^1 8v^8 (Z_T Z_P)^5}{5n^3 n_i^5 [v^4 + 2v^2(Z_T^2/n_i^2 + Z_P^2/n^2) + (Z_T^2/n_i^2 - Z_P^2/n^2)^2]^5}(\pi a_0^2)$$

(20)

In case of capture from 1s of target ($n_i = 1$ to n-th state of projectile and if $n \gg 1$ then one can get:

$$\sigma_{OBK}(1s - n \gg 1) \simeq \frac{2^1 8v^8 (Z_T Z_P)^5}{5n^3 (v^2 + Z_T^2)^{10}}(\pi a_0^2) \tag{21}$$

In the high velocity limit this expression reduces to:

$$\sigma_{OBK}(1s - n \gg 1, v \rightarrow \infty) \simeq \frac{2^1 8v^8 (Z_T Z_P)^5}{5n^3 v^{12}}(\pi a_0^2) \tag{22}$$

One may note that the capture cross section peaks if $Z_T^2/n_i^2 = Z_P^2/n^2$ i.e. if the initial and final binding energies are equal. This is known as the famous binding energy matching condition. Especially at low energy this condition becomes most important to provide resonance transfer. In the high velocity limit the $\sigma_{OBK} \sim v^{-12}$ i.e the cross sections drop drastically with velocity. In addition the cross sections to capture a K-shell electron to higher shell (n) falls with n as $1/n^3$ i.e. capture to n=2 state is less than that for 1s-capture by a factor of eight. In particular for 1s-1s transfer,

$$\sigma_{OBK}(1s - 1s, v \rightarrow \infty) \simeq \frac{2^1 8v^8 (Z_T Z_P)^5}{5v^{12}}(\pi a_0^2) \tag{23}$$

The OBK formalism is known to overestimate the experimental data, by a factor of 3-4, for electron transfer even for the simplest collision system: proton on H_2. Bates and Dalgarno (1952) [108], and, Jackson and Schiff [109] introduced the inter-nuclear potential in order to improve the agreement between the experiment and the calculations. They argued in favour of retaining this term in the Hamiltonian to calculate the capture cross section in the Ist-order theory since it will partially compensate the non-orthogonality of the wave-functions of the initial and final states. This mechanism provides capture cross section which is commonly known as the first-Born cross section: $\sigma_B \simeq 0.661\ \sigma_{OBK}$, at high velocity. In addition the second-order Born calculation was carried out by Drisko (1955) [4, 110]which provided the following relation at high velocity which is quite interesting:

$$\sigma_{B2} \simeq (0.3 + 5\pi v/2^{12})\sigma_{OBK} \tag{24}$$

Since σ_{OBK} has a v^{-12} dependence, the σ_{B2} varies as v^{-11} in contrast to that for σ_{OBK} which predicts the v^{-12} dependence. Note that the Schlachter's scaling formula uses a $v^{-9.6}$ dependence which could explain many of the experimental data but not based on any *ab initio* quantum mechanical models (such as, B1 or OBKN or B2).

7.2   1s-1s Transfer and 1s-Ionization

We give only a few example of 1s-1s electron transfer which is commonly known as K-K transfer. The single K-K electron transfer cross sections have been measured in a few cases in the past and mostly using solid targets [111, 112, 113, 114, 115, 116, 117, 118] in which the evolution of vacancy configurations due to multiple collisions inside target complicate the data analysis. A three component model is used (see references in [119]) to extract the cross sections at the limit of "zero"-thickness i.e. for single collision condition. Therefore it is desirable to have measurements on these processes in which the single collision condition is satisfied i.e. using low pressure gas target. However, such experimental data are also available for a limited number of collision systems in which the single K-K electron transfer cross sections are reported for the intermediate velocity range $(0.1 \le v_r \le 1.2)$ where these cross sections are expected to be near the maximum.

In case of ionization of strongly bound K-shell electrons by heavy projectiles the Ist Born calculations are known to be unsuccessful to predict total cross sections. In order to improve the situation, in one approach, Brandt and Lapicki and co-workers [120, 121] had developed a model known as ECPSSR based on perturbed stationary state approximation. In fact in case of K- and L-subshell ionization it has become conventional to use ECPSSR model which is a Ist order Born calculation and modified to include the corrections due to enhanced binding energy, Coulomb (C) deflection, energy loss (E) and any relativistic (R) effects [120]. We will compare ECPSSR calculations with our experimental data obtained with different symmetry parameters.

The electrons emitted in heavy-ion induced ionization are subject to long range Coulomb interactions with the recoil-ions and projectiles. Theoretical models based on continuum distorted wave (CDW) have been developed [122, 123] in order to explain such two center effect on ionization. In the CDW the initial and final unperturbed target wave functions are distorted by a projectile continuum factor. In one of its simplified version known as CDW-EIS originally developed by Crothers and McCann [33] the final state is chosen as in the CDW but the initial distorted state is represented as a bound state multiplied by a projectile eikonal phase (eikonal initial state). As mentioned above that the CDW-EIS has been quite successful in explaining the angular distributions of the electron DDCS in fast ion-atom ionization [31, 38, 53, 40]. However, in collision systems for which the electrons are much more strongly bound ($v_r \leq 1$), it is not clear whether the CDW-EIS calculations can explain the ionization data for such highly non-perturbative collision system. Such examples are found in our recent works [124, 125, 126] for gaseous (Ar, Kr) as well as fullerene (C_{60}) targets.

It is well known that the Ist order calculations based on the OBKN approximation overestimates the cross sections by a large factor. In perturbed stationary state approach Lapicki and McDaniel (1980) [127] have included the second Born term and corrections due to the enhanced binding energy and Coulomb deflection in the OBKN formalism in the same way as was done in ECPSSR formalism for ionization. Although this formalism is not an *ab initio* one but the simplicity of using analytical expression in this method and its ability to predict the cross sections for asymmetric collisions is worth mentioning. This semi-empirical technique has been quite successful in predicting the inner-shell transfer cross sections.

The two-center semiclassical close-coupling method [128, 129], based on atomic orbital expansion [130], is found to be quite successful in explaining the state selective electron transfer cross sections at least for loosely bound outer shell electrons. In this model the motion of the projectile is approximated by a classical trajectory and the target electrons are treated quantum mechanically. For treating electron capture from the inner shells, an independent electron model is used and the active electron is described by a model potential fitted so that the binding energies of the inner-shell electrons are reproduced. Although the possible role of outer shell electrons, (the so called Pauli exchange effect), is not included explicitly in theory it may be partially accounted for in using model potential. In the close-coupling calculation all the atomic states up to n=2 on both centers have been included.

A detail measurements on the K-ionization, K-K and (and partly K-L) electron transfer processes are reported by Dhal et al. [124, 125] for varieties of collision systems, such as, bare and H-like C, O, F, Si, S and Cl colliding with Ar and Kr targets. The measurements are pursued for different values of symmetry parameters between 0.25 and 0.5. A large enhancement in the double K-K transfer channel has been observed recently [124] in case of nearly symmetric collision (S_z=0.78) system of Si projectile on Ar target. The K-K electron transfer and the K-ionization cross sections derived for bare and H-like C, O, F, Si ions on Ar, F, S, Cl ions on Kr are used to provide stringent test for the Ist order perturbative, continuum distorted wave and close coupling calculations. It was found that the CDW-EIS model fails to reproduce the K-ionization cross sections at intermediate or lower energies and largely underestimates the data for relatively symmetric collisions. The deviation seems to be much larger for the heavier target atoms like Kr as compared to Ar. The perturbed stationary state calculations, ECPSSR, of Brandt and Lapicki overestimate the data except for most asymmetric collisions for which a good agreement was found. In the case of the K-K electron transfer process the close coupling calculations are found to deviate for the asymmetric collisions and give a very good agreement for nearly symmetric collisions. The perturbed stationary state calculations of Lapicki and McDaniel, on the other hand, explain the K-K electron transfer data for asymmetric systems with lighter target and deviates for near symmetric collisions.

Summary

In summary, we have discussed some basic definitions and formalisms for elastic scattering process in ion-atom collisions. It has been indicated that by using a screened Coulomb potential one can get the exact classical Rutherford formula for elastic scattering. The concept of threshold has been indicated. The most commonly used potential functions are introduced. The techniques of partial wave expansion and Born approximations are outlined in the context of elastic scattering. In addition a detail account on the electron double differential cross section measurements has been provided. In this context the binary encounter mechanism and continuum electron emission in soft collisions have been discussed. The influence of the two center mechanism on ionization has been detailed giving references from recent experimental results. The theoretical techniques based on the continuum distorted wave approximation which accounts for the two Coulomb center effects in the initial and final states has been outlined. In particular, the recent rapid progress in the investigation of the electron interference in electron emission from a molecular double slit i.e. H_2 has been introduced. The influence of such atomic scale interference on the forward-backward asymmetry parameter in heavy-ion collisions is discussed. It has been shown how a double scattering mechanism gives rise a double frequency component of the interference oscillation. The electron capture in fast ion-atom collision has been discussed including a scaling formula derived at least for the outer-shell electrons and first order quantum mechanical theories. In particular, the OBKN method, the first Born and the second Born techniques are introduced, in brief, to explain the electron transfer involving inner-shells. The basic features in terms of the dependence of the transfer process on projectile velocity, the binding energies of the involved electron-states and the initial and final principle quantum numbers of the states involved are indicated. It is mentioned how the perturbed stationary state approach terms (including several corrective terms) and close coupling methods may give increasingly good agreement with the experimental data on 1s-1s transfer. In most of the cases the references are made to the existing experimental data which provides suitable tests to the theoretical methodology.

References

[1] B.H. Bransden and C.J.Joachain, *Physics of Atoms and Molecules*, Pearson Education, 2nd Ed. 2001

[2] Earl W. McDaniel, *Atomic Collisions-Electron and Photon Projectiles*, Wiley, New York 1989.

[3] Earl W. McDaniel, *Atomic Collisions-Heavy particle Projectiles*, Wiley, New York.

[4] M.R.C McDowell and J.P.Coleman, *Introduction to the theory of ion-atom collisions*, M.R.C. Mcdowell and J.P.Coleman, Noth-Holland Publishing Co., Amsterdam-London, 1970.

[5] A. Messiah, *Quantum Mechanics*, North-Holland, Amsterdam, 1970, Vol. II

[6] Patric Richard, Editor, *Methods of experimental physics*, Academic, New York **17** 193 (1980).

[7] R.D. Levine and R.B. Bersstein, Molecular Reaction Dynamics and Chemical Reactivity, Oxford University Press, New York, 1987.

[8] Scoles G Ed. Atomic and Molecular beam methods, Vol-1, Oxford University Press, New York, 1988.

[9] B.R. Junker in Applied Atomic Collision Physics, H.S.W. Massey, E.W. MCDaniel and Bederson Series Eds. 5 vols, Academic Press, New York, 1982, p379.

[10] E.W. McDaniel and L.A. Viehland Phys. Rep. 110, 333 (1984).

[11] J.O. Hirschfelder, C.F. Curtiss and R.B. Bird, *Molecular Theory of Gases and Liquids*, Wiley, New York (1964), pp. 31-35.

[12] R.B. Bernstein (Ed.) *Atom-Molecule Collision Theory*, Plenum, New York (1979).

[13] T. Kihara, *Intermolecular Forces* Wiley, New York (1978).

[14] N. Stolterfoht, D. Schneider, R. Burch, H. Wieman, and J. S. Risley, Phys. Rev. Lett. **59**, 59 (1974).

[15] M. E. Rudd and J. H. Macek, Case Stud. At. Phys. **3**, 47 (1972).

[16] D.H. Lee, P. Richard, T.J.M. Zouros, J.M. Sanders, J.L. Shinpaugh, and H. Hidmi, Phys. Rev. A**41**, 4816 (1990).

[17] P. Richard, D.H. Lee, T.J.M. Zouros, J.M. Sanders, J.L. Shinpaugh, and H. Hidmi, J. Phys B **23**, L637 (1990).

[18] T. B. Quinteros *et al*, J. Phys B **24**, 1377 (1990).

[19] C. Kelbch, R. E. Olson, S. Schmidt, H. Schmidt-Bocking, and S. Hagmann, J. Phys B **22**, 2171 (1989).

[20] C. Liao, P. Richard, S.R. Grabbe, C.P. Bhalla, T.J.M. Zouros and S. Hagmann, Phys. Rev. A**50**, 1328 (1994).

[21] S. Hagmann *et al.*, J. Phys B **25**, L827 (1992).

[22] W. Wolff, J. L. Shinpaugh, H. E. Wolf, R. E. Olson, U. Bechthold, and H. Schmidt-Bocking, J. Phys B **26**, L65 (1993).

[23] A. D. González, P. Dahl, P. Hvelpund, and K. Taulberg, J. Phys B **25**, L573 (1992).

[24] A. D. González, P. Dahl, P. Hvelpund, and P. D. Fainstein J. Phys B **26**, L135 (1993).

[25] H. I. Hidmi, P. Richard, J.M. Sanders, H. Schoene, J. P. Giese, D.H. Lee, T.J.M. Zouros, and S. L. Varghese, Phys. Rev. A **48** 4421 (1993).

[26] C. P. Bhalla and R. Singhal, J. Phys B **24**, 3187 (1990).

[27] R. E. Olson, C. O. Reinhold, and D. R. Schultz, J. Phys B **23**, L455 (1990).

[28] D. R. Schultz and R. E. Olson J. Phys B **24**, 3409 (1991).

[29] C. O. Reinhold, D. R. Schultz, and R. E. Olson, J. Phys B **23**, L591 (1990).

[30] T.J.M. Zouros, K. L. Wong, S. Grabbe, H. I. Hidmi, P. Richard, E. C. Montenegro, J.M. Sanders, C. Liao, S. Hagmann, and C. P. Bhalla, Phys. Rev. A **53** 2272 (1996).

[31] J.O.P. Pedersen, P. Hvelplund, A. Petersen and P. Fainstein, J. Phys. B**24**, 4001 (1991).

[32] Pablo D. Fainstein, Victor H. Ponce, and Roberto D. Rivarola Phys. Rev. A 45, 6417 (1992)

[33] D. S. F. Crothers and J. F. McCann, J. Phys. B **16**, 3229 (1983).

[34] P D Fainstein, V H Ponce and R D Rivarola, J. Phys. B: At. Mol. Opt. Phys. 24 3091 (1991).

[35] L. Gulyás, P. D. Fainstein and A. Salin, J. Phys. B **28**, 245 (1995).

[36] M. E. Gallasi, R.D. Rivarola, P.D. Fainstein and N. Stolterfoht, Phys. Rev A **66**, 052705 (2002).

[37] M. E. Galassi, R. D. Rivarola and P. D. Fainstein, Phys. Rev. A **70**, 032721 (2004).

[38] N. Stolterfoht, et al. Phys. Rev. A **52**, 3796 (1995).

[39] S. Suárez, C. Garibotti, W. Meckbach, and G. Bernardi, Phys. Rev. Lett. **70**, 418 (1993).

[40] Lokesh C. Tribedi, P. Richard, W. DeHaven, L. Gulyás, M. E. Rudd, J. Phys. B **31**, L369 (1998).

[41] Lokesh C. Tribedi, P. Richard, L. Gulyás, M. E. Rudd, and R. Moshammer, Phys. Rev A, **63** 062723 (2001).

[42] Lokesh C. Tribedi, P. Richard, L. Gulyás and M.E. Rudd, Phys. Rev A, 63 062724 (2001).

[43] D. Misra et al. Phys. Rev. A. **74**, 060701(R) (2006).

[44] D. Misra, A. Kelkar, U. Kadhane, Ajay Kumar, Y. P. Singh, PD Fainstein and Lokesh C. Tribedi, Phys. Rev. A **75**, 052712 (2007).

[45] D. Misra, K.V. Thulasiram, W. Fernandes, A.H. Kelkar, U. Kadhane, A. Kumar, Y. Singh, L. Gulyas and L.C. Tribedi, Nucl. Instrum. Methods Phys. Res. B **267**, 157 (2009).

[46] T. W. Shyn, Phys. Rev. A **45**, 2951 (1992).

[47] M. W. Gealy, G.W. Kerby III, Y.-Y. Hsu, and M.E. Rudd, Phys. Rev. A **51**, 2247 (1995).

[48] G. W. Kerby III, M.W. Gealy, Y.-Y. Hsu, and M.E. Rudd, Phys. Rev. A **51**, 2256 (1995).

[49] Y. -Y. Hsu, M.W. Gealy, G.W. Kerby III, and M. E. Rudd, Phys. Rev. A **53**, 297 (1996);

[50] Y.-Y. Hsu, M.W. Gealy, G.W. Kerby III, M. E. Rudd, D. R. Schultz and C. O. Reinhold, Phys. Rev. A **53**, 303 (1996).

[51] Lokesh C. Tribedi, P. Richard, Y. D. Wang, C. D. Lin and R. E. Olson, Phys. Rev. Letters 77, 3767 (1996).

[52] Lokesh C. Tribedi, P. Richard, D. Ling, Y. D. Wang, C. D. Lin, R. Moshammer, G. W. Kerby III, M. W. Gealy and M. E. Rudd, Phys. Rev. A 54, 2154 (1996).

[53] Lokesh C. Tribedi, P. Richard, Y. D. Wang, C. D. Lin, L. Guly'as and M.E. Rudd, Phys. ReV A 58, 3619 (1998).

[54] D Misra, A H Kelkar, U Kadhane, A Kumar, P D Fainstein and L C Tribedi, b(2007).

[55] Deepankar Misra, A.H. Kelkar, P.D. Fainstein and Lokesh C. Tribedi, Journal of Physics (IoP, UK): Conf. Ser. 80 012013 (2007).

[56] L. C. Tribedi and D. Misra, in *Current Topics in Atomic, Molecular and Optical Physics, Invited Lectures of TC2005*, pp-209-228, World Scientific, Singapore: Ed. by Chandana Sinha and Shibshankar Bhattacharyya (2005)

[57] S. Chatterjee, D. Misra, A. H. Kelkar, P.D. Fainstein, and L. C. Tribedi, J. Phys. B: At. Mol. Opt. Phys. 43 (2010) 125201.

[58] S. Chatterjee, D. Misra, A. H. Kelkar, L. C. Tribedi, C. R. Stia, O. A. Fojón, and R. D. Rivarola, Phys. Rev. A **78**, 052701 (2008).

[59] S. Chatterjee, A. Agnihotri, C. R. Stia, O. A. Fojn, R.D. Rivarola and Lokesh C Tribedi, Phys. Rev. A **82**, 052709 (2010).

[60] H. D. Cohen and U. Fano, Phys. Rev **150**, 30 (1966).

[61] M. Walter and J. S. Briggs, J. Phys. B **32**, 2487 (1999).

[62] R. Dorner et al., Phys. Rev. Lett., **81** 5776 (1998).

[63] S. E. Corchs, R.D. Rivarola, J. H. Mcguire, and Y.D. Wang, Phys. Scr. **50**, 469 (1994).

[64] T. F. Tuan and E. Gerjuoy, Phys. Rev **117**, 756 (1960).

[65] N. Stolterfoht et al., Phys. Rev. Lett., **87** 023201 (2001).

[66] N. Stolterfoht et al., Phys. Rev. A., **67** 030702(R) (2002).

[67] Deepankar Misra, Umesh Kadhane, Yeshpal Singh, P. D. Fainstein, P. Richard and Lokesh C. Tribedi. Phys. Rev. Lett. **92** 153201 (2004).

[68] Deepankar Misra, U. Kadhane, Y. P. Singh, L. C. Tribedi, P. D. Fainstein and P. Richard, Phys. Rev. Lett., **95** , 079302 (2005).

[69] Deepankar Misra, A. Kelkar, U. Kadhane, Ajay Kumar, Lokesh C. Tribedi and P. D. Fainstein, Phys. Rev. A. **74**, 060701(R) (2006).

[70] D. Misra, A. Kelkar, U. Kadhane, Ajay Kumar, Y. P. Singh, PD Fainstein and Lokesh C. Tribedi, Phys. Rev. A **75**, 052712 (2007).

[71] S. Hossain, A. S. Alnaser, A. L. Landers, D. J. Pole, H. Knutson, A. Robison, B. Stamper, N. Stolterfoht, and J. A. Tanis, Nucl. Instr. Meth. Phys. Res. B **205**, 484 (2003).

[72] S. Hossain, A. Landers, N. Stolterfoht and J. A. Tanis, Phys. Rev. A **72**, 010701(R) (2005).

[73] J. A. Tanis, J. -Y. Chesnel, B. Sulik, B. Skogvall, P. Sobocinski, A. Cassimi, J. -P. Grandin, L. Adoui, D. Hennecart, and N. Stolterfoht, Phys. Rev. A **74**, 022707 (2006).

[74] S. Chatterjee, D Misra, A H Kelkar, P D Fainstein and L C Tribedi, J. Phys. B: At. Mol. Opt. Phys. **42**, 125201 (2010).

[75] L. Nagy, L. Kocbach, K. Pora and J. P. Hansen, J. Phys. B **35**, L453 (2002).

[76] N. Stolterfoht et al., Phys. Rev. A. **67**, 030702(R) (2003).

[77] Deepankar Misra, Aditya H. Kelkar, Shyamal Chatterjee, and Lokesh C. Tribedi, Phys. Rev. A **80**, 062701 (2009).

[78] A. L. Landers et al., Phys. Rev. A., **70** 042702 (2004).

[79] J. A. Tanis, S. Hossain, B. Sulik, and N. Stolterfoht, Phys. Rev. Lett., **95** , 079301 (2005).

[80] Daniel Rolles, Markus Braune, Slobodan Cvejanovi, Oliver Gessner, Rainer Hentges, Sanja Korica, Burkhard Langer, Toralf Lischke, Georg Prumper, Axel Reinköster, Jens Viefhaus, Björn Zimmermann, Vincent McKoy, and Uwe Becker, Nature **437**, 711 (2005).

[81] X-J Liu, N. A. Cherepkov, S. K. Semenov, V. Kimberg, F. Gelmukhanov, G. Prümper, T. Lischke, T. Tanaka, M. Hoshino, H. Tanaka, and K. Ueda, J. Phys. B: At. Mol. Opt. Phys. **39** 4801 (2006).

[82] K. Kreidi, D. Akoury, T. Jahnke, Th. Weber, A. Staudte, M. Schöffler, N. Neumann, J. Titze, L. Ph. H. Schmidt, A. Czasch, O. Jagutzki, R. A. Costa Fraga, R. E. Grisenti, M. Smolarski, P. Ranitovic, C. L. Cocke, T. Osipov, H. Adaniya, J. C. Thompson, M. H. Prior, A. Belkacem, A. L. Landers, H. Schmidt-Böcking, and R. Dörner, Phys. Rev. Lett. **100**, 133005 (2008).

[83] D. Akoury, K. Kreidi, T. Jahnke, Th. Weber, A. Staudte, M. Schöffler, N. Neumann, J. Titze, L. Ph. H. Schmidt, A. Czasch, O. Jagutzki, R. A. Costa Fraga, R. E. Grisenti, R. Diez Muino, N. A. Cherepkov, S. K. Semenov, P. Ranitovic, C. L. Cocke, T. Osipov, H. Adaniya, J. C. Thompson, M. H. Prior, A. Belkacem, A. L. Landers, H. Schmidt-Böcking, R. Dörner, Science **318**, 949 (2007).

[84] R. Della Picca, P. D. Fainstein, M. L. Martiarena, N. Sisourat and A. Dubois, Phys. Rev. A **79**, 032702 (2009).

[85] O. A. Fojón, J. Fernández, A. Palacios, R. D. Rivarola and F. Martín, J. Phys. B **37**, 3035 (2004).

[86] J. Fernández, O. Fojón, A. Palacios and F. Martín, Phys. Rev. Lett. **98**, 043005 (2007).

[87] J. Fernández, O. Fojón and F. Martín, Phys. Rev. A **79**, 023420 (2009).

[88] D.K. Jain and S.P. Khare Phys. Lett. **A63**, 237 (1977).

[89] A. Jain, A.N. Tripathi and M.K. Srivastava, Phys. Rev. A **20**, 2352 (1979).

[90] C. R. Stia, O. A. Fojón, Ph. Weck, J. Hanssen and R. D. Rivarola, J. Phys. B: At. Mol. Opt. Phys. **36** L257 (2003).

[91] O. Kamalou, J.-Y. Chesnel, D. Martina, J. Hanssen, C. R. Stia, O. A. Fojón, R. D. Rivarola, and F. Frémont, Phys. Rev. A **71**, 010702(R) (2005).

[92] D. S. Milne-Brownlie, M. Foster, Junfang Gao, B. Lohmann, and D. H. Madison, Phys. Rev. Lett. **96**, 233201 (2006).

[93] E. M. Staicu Casagrande, A Naja , F Mezdari , A Lahmam-Bennani , P Bolognesi , B Joulakian , O Chuluunbaatar , O Al-Hagan , D H Madison , D V Fursa , I Bray, J. Phys. B: At. Mol. Opt. Phys. **41** 025204 (2008).

[94] O. A. Fojón, C. R. Stia, and R. D. Rivarola, AIP Conference Proceedings **811**, 42 (2006).

[95] Deepankar Misra, H. T. Schmidt, M. Gudmundsson, D. Fischer, N. Haag, H. A. B. Johansson, A. Kallberg, B. Najjari, P. Reinhed, R. Schuch, M. Schoeffler, A. Simonsson, A. B. Voitkiv, and H. Cederquist, Phys. Rev. Lett. **102**, 153201 (2009).

[96] N. Stolterfoht et al., Phys. Rev. A **69**, 012701 (2004).

[97] L. Sarkadi, J. Phys B **36**, 2153 (2003).

[98] C. R. Stia, O. A. Fojón, P. F. Weck, J. Hanssen and R. D. Rivarola, J. Phys. B **36**, L264 (2002).

[99] K. Støchkel *et al.*, Phys. Rev. A **72**, 050703(R) (2005).

[100] D. S. Milne-Brownlie *et al.* Phys. Rev. Lett. **96**, 233201 (2006).

[101] K. Póra and L. Nagy, Nucl. Instr. and Meth. B. **233**, 293 (2005).

[102] Nicolas Sisourat, Jrmie Caillat, Alain Dubois, and Pablo D. Fainstein Phys. Rev. A **76**, 012718 (2007).

[103] A. Messiah, *Quantum Mechanics*, North-Holland, Amsterdam, Vol. II , pp. 848-852 (1970).

[104] P. D. Fainstein, L. Gulyás, F. Martin, and A. Salin, Phys. Rev. A**53**, 3243 (1996).

[105] A.S. Schlachter, J.W. Stearns, W.G. Graham, K.H. Berkner, R.V. Pyle and J.A. Tanis, Phys Rev A 27, 3372 (1983).

[106] H. Knudsen, H.K. Haugen and P.Hvelplund, Phys Rev A23, 597 (1981)

[107] V.S. Nikolaev, Zh. E ksp. Teor. Fiz. **51**, 1263, 1966, Sov. Phys. JETP **24**, 847 (1967).

[108] D.R. Bates and A. Dalgarno, Proc. Phys. Soc A **65**, 919 (1952).

[109] J.D. Jackson and H. Schiff, Phys. Rev.**89**, 359 (1953).

[110] R.M. Drisko, Thesis, Carnegie Institute of technology (1955) [unpublished].

[111] J. Hall et al. Phys Rev A **33**, 914 (1986); and *ibid* **28**, 99 (1983).

[112] K. Wohrer, Chetioui, J.P. Rozet, A. Jolly and C. Stephan, J.Phys. B**17**, 1575 (1984)

[113] L.C. Tribedi, K.G. Prasad, P.N. Tandon, Z. Chen and C.D. Lin, Phys. Rev A **49**, 1015 (1994).

[114] J.A. Tanis, S.M. Shafroth, J.E. Willis, Morwat, Phys. Rev. Lett.**45**, 1547 (1980).

[115] L. C. Tribedi, K. G. Prasad and P. N. Tandon Z. Phys.D. **27**, 143 (1993).

[116] L.C. Tribedi, K.G. Prasad, P.N. Tandon Phys. Rev A **47**, 3739 (1993).

[117] L. C. Tribedi, K. G. Prasad and P. N. Tandon, Z.Phys. D **24**, 215 (1992)

[118] Lokesh C. Tribedi, K. G. Prasad and P. N. Tandon, Phys. Rev. A 51, 3783 (1995)

[119] T.J. Gray in Methods of experimental physics, Ed. P Richard (Academic, New York) **17** 193 (1980).

[120] W. Brandt and G. Lapicki, Phys. Rev. A 23, 171 (1981) and refs. therein.

[121] G. Basbas, W. Brandt and R. Laubert, Phys. Rev. A **17** 1655 (1978).

[122] I.M. Chesire, Proc. Phys. Soc. **84**, 89 (1964).

[123] D.ź Belkic, J. Phys. B **11**, 3529 (1978).

[124] B.B. Dhal, L.C. Tribedi, U. Tiwari, P.N. Tandon, Teck Lee, C.D. Lin, et al., J.Phys B. **33** 1069, (2000).

[125] B.B. Dhal, L. C. Tribedi, U. Tiwari, K.V. Thulasiram, P.N. Tandon,T. G. Lee, C.D. Lin, L. Gulyás, Phys. Rev. A, **62** 022714 (2000).

[126] U. Kadhane, D. Misra, Y.P. Singh, L.C. Tribedi, Phys. Rev. Letters., **90**, 093401 (2003).

[127] G. Lapicki and F.D. McDaniel, Phys Rev A22, 1896 (1980).

[128] Jiyun Kuang and C.D. Lin, J. Phys. B29, 1207 (1996).

[129] W. Fritsch and C.D. Lin Phys. Rep. **202**,1 (1996)

[130] D.R. Bates and McCarroll, Proc. Roy. Soc. London A245, 175 (1958).

Study of Collisions Involving Exotic Particles

Chandana Sinha

Theoretical Physics Department,
Indian association For the Cultivation Of Science, Kolkata- 700032, India
email: chand_sin@hotmail.com, *tpcs@iacs.res.in*

1. INTRODUCTION

In this article some important aspects of the exotic atoms are considered. Positronium (Ps) and antihydrogen made up of particles and antiparticles are relevant to the Atomic Collision Physics. Certain collision processes involving matter – antimatter are briefly discussed. The following section (2) deals with the Ps atom and its formation and sections (3) and (4) describe the formation of antihydrogen in the absence and in the presence of an external laser field.

2. POSITRON AND POSITRONIUM PHYSICS

One of the most important discoveries in science is the existence of positron (e^+), the antiparticle of electron by Anderson in 1932 [1]. The theoretical prediction [2] of the existence of the exotic atom Positronium (Ps), the bound state of two leptons, a positron and an electron led to the experimental discovery of Ps in 1953 by Deutsch [3]. The interesting property of this atom is that it is an isotope of hydrogen, whose charge and mass centers coincide. It is itself its antiparticle. Depending on the relative spin orientation of its constituents (e & e^+), Ps may be formed in the ground state as an Ortho positronium ($^3S_1$, triplet state, parallel spin) or Para positronium ($^1S_0$, singlet state, antiparallel spin). Ps is an instable atom, its two constituents annihilate each other to produce gamma ray photons after a certain mean life time again depending on the relative spin of the constituents. Ortho Ps has a mean life time against annihilation in vacuum of ~142 nanosecond (10^{-9} s) and the leading mode of decay is three photon (always odd number of photons) while the para Ps has a mean life time ~125 picoseconds (10^{-12} s) and the leading mode is two photon (always even number of photons).

The odd and even number of photons is governed by the charge conjugation and parity (CP) symmetry as follows.

CP is the product of two symmetries: C for charge conjugation which transforms a particle into its antiparticle, and P for parity, which creates the mirror image of a physical system. The strong interaction and electromagnetic interaction seem to be invariant under the combined CP transformation operation, but this symmetry is slightly violated during certain types of weak decay. The charge conjugation parity (CP) is given by $(-1)^{l+s}$, l and s being the orbital and spin angular momentum quantum number. For ground state, Ortho Ps has l=0 and S=1 and has therefore odd CP while for para Ps with S=0 and l=0 has even CP. The interchange of positive and negative charge reverses the sign or polarization of the associated electric field. Thus the CP of photon is $(-1)^n$, n being the no of photons. Since CP is conserved in electromagnetic interactions according to the Standard Model of Particle Physics, Ortho Ps must decay into an odd no of photons while the Para Ps must decay into an even no. of photons.

The energy levels of the Ps atom are similar to that of the H atom but only scaled by the reduced mass factor. The reduced mass of the Ps is

$$\mu = \frac{m_e m_{e^+}}{m_e + m_{e^+}} = \frac{m_e}{2}$$

The Bohr radius a_0 $(Ps) = 2a_0(H)$: $E_I(Ps) \approx 0.5 * E_I(H)$, E_I being the ionization energy of the atom (6.8 eV for Ps). Thus for a given state of Ps, the average $e - e^+$ distance is twice the $e - p^+$ distance for the corresponding H atom. However a precise calculation of the Ps energy levels uses the Bethe Salpeter equation, the similarity between the Ps and H allows for a rough estimate. Ps atom is highly polarizable, the dipole polarizability being 36 a_0^3, eight times that of hydrogen. The large magnetic moment of the positron, 657 times that of the proton makes the magnitudes of the spin–orbit and spin–spin interactions comparable and as such the fine-structure of Ps is very different from that of H atom. Measurements of the $2^3S_1 \rightarrow 2^3P_2$ fine structure transition were first made by Mills *et al* [4].

Free annihilation

A free e–e^+ pair cannot decay into a single photon, because energy and momentum cannot be conserved simultaneously in such a reaction. However, occurrence of single photon annihilation is possible when the pair is in the field of a nucleus as the latter can absorb the excess momentum. Annihilation of the pair into two photons is allowed kinematically. Since the annihilation rate is inversely proportional to the relative velocity of the electron and the positron, the annihilation mostly occurs when the particles have a small kinetic energy in the center of mass (CM) system so that the total energy available to the photons is approximately equal to the rest energy of the system, $2mc^2$, m being the mass of the e/e^+. The two photons are emitted in opposite directions each of them having the same energy, mc^2. If $\psi(\vec{r})$ is the e-e$^+$ wave function in the CM system, normalized to a plane wave of unit amplitude at large distance of separation $(\vec{r} \to \infty)$, the probability of finding the electron and the positron at the same position which is the measure of annihilation probability, is given by $|\psi(0)^2|$. For annihilation of a free pair $\psi(\vec{r})$ must be taken to be the solution of the Schrodinger equation that describes the scattering by the Coulomb interaction [5] between the e and e$^+$, i.e.,

$$\left(\frac{\hbar^2}{2m} \nabla_r^2 + \frac{e}{r} + E \right) \psi(\vec{r}) = 0$$

The solution of this equation is the hyper-geometric function [5] and

$$|\psi(0)|^2 = \frac{2\pi\alpha c / v}{1 - \exp(-2\pi\alpha c / v)}$$

The Coulomb attraction between the e and e$^+$ increases the annihilation rate at low velocities. At large velocities, when $2\pi\alpha c / v < 1$, the Coulomb attraction can be ignored.

The density $|\psi_{nl}(0)|^2$ at r = 0, that occurs in the expression for the annihilation rate for the Ps atom vanishes except for states with l=0 and for these states

$$|\psi_{nl}(0)|^2 = \frac{1}{(2a_0)^3 \pi n^3}$$

The characteristic collision time for Ps atom collision is extremely short compared to the lifetime against annihilation and this implies that when calculating the cross section for a Ps atom reaction, the possibility of annihilation can be ignored and the collision for stable particles can be employed.

Formation of Ps

The Ps atom is generally formed when an incident positron captures an electron from a target atom or ion. The threshold energy for Ps formation is the difference between the first ionization energy of the target atom 6.8 eV e.g., for positron collisions with H and He atoms the threshold energies are 6.8 and 17.8 eV respectively.

Theory

Positronium (Ps) Formation in Positron – Helium Atom System

$$e^+ \left(\vec{r}_1\right) \; + \; He(\vec{r}_2, \vec{r}_3) \;\longrightarrow\; \left(e^+ e\right) \; + \; He^+ \left(\vec{r}_3\right)$$

The incident positron and the two atomic electrons are denoted as particles 1, 2 and 3 respectively, and assume that the helium nucleus is infinitely heavy (as compared to the incident positrons) and is chosen to be the centre of the coordinates. The Hamiltonian for the positron-helium system, in atomic units, is [6]:

$$H \;=\; -\tfrac{1}{2}(\nabla_1^2 + \nabla_2^2 + \nabla_3^2) + \frac{2}{r_1} - \frac{2}{r_2} - \frac{2}{r_3} + \frac{1}{r_{23}} - \frac{1}{r_{13}} - \frac{1}{r_{12}}.$$

We now approximate the wave function of the system as [6]

$$\Psi(r_1, r_2, r_3) \;=\; \psi_0(r_2, r_3) F_0(r_1) \chi_1(1, 23)$$

$$+\, \phi_0(r_3)\omega(r_{12})[G^P(s_{12})\chi_1(3, 12) + G^O(s_{12})\chi_2(3, 12)]$$

$$+\, \phi_0(r_2)\omega(r_{13})[G^P(s_{13})\chi_1(2, 31) - G^O(s_{13})\chi_2(2, 31)],$$

where χ_1 and χ_2 are the appropriate spin functions, ψ_0, ϕ_0 are the wave functions of the helium atom and ionized helium in the ground state respectively and ω that of the positronium, with

$$s_{12} \;=\; \tfrac{1}{2}(r_1 + r_2), \qquad r_{12} \;=\; r_1 - r_2.$$

F_0 describes the motion of the positron while G^P and G^O represent he motion of the Ps in the para and ortho states respectively relative to the helium nucleus. We now substitute eqn.(2) in the time independent Schrodinger equation

$$(H - E)\Psi = E\Psi$$

and assuming ψ_0 to be exact, we obtain the following set of integro-differential equations following the Hartree-Fock variational principle: We first substitute eqn. (1) in the Schrodinger equation and then we multiply both sides of the equation (from the left) by the complex conjugate of the functions ψ_0, $\Phi_0(\vec{r}_3)$ $\omega(r_{12})$ and then integrating over $\vec{r}_2$ and $\vec{r}_3$ we obtain respectively the equation for $F_0(\vec{r}_1)$ and $G^{P,O}(s_{12})$.

$$\tfrac{1}{2}(\nabla_1^2 + k_1^2) F_0(r_1)$$

$$= \iint \psi_0^*(r_2, r_3)\left(\frac{2}{r_1} - \frac{1}{r_{12}} - \frac{1}{r_{13}}\right)\psi_0(r_2, r_3)F_0(r_1)\, dr_2\, dr_3$$

$$-\tfrac{1}{2}\iint \psi_0^*(r_2, r_3)(H - E)\{\phi_0(r_3)\omega(r_{12})[G^P(s_{12}) + \sqrt{3}G^O(s_{12})]$$

$$+\, \phi_0(r_2)\omega(r_{13})[G^P(s_{13}) + \sqrt{3}G^O(s_{13})]\}\, dr_2\, dr_3 \tag{3}$$

$$\tfrac{1}{4}(\nabla^2_{s_{12}} + k^2_2)G^P(s_{12})$$

$$= -\tfrac{1}{2}\iint \phi^*_0(r_3)\omega^*(r_{12})(H-E)\{\psi_0(r_2,r_3)F_0(r_1)$$

$$-\phi_0(r_2)\omega(r_{13})[-G^P(s_{13}) + \sqrt{3}G^O(s_{13})]\}\,dr_3\,dr_{12} \qquad (4)$$

$$\tfrac{1}{4}(\nabla^2_{s_{12}} + k^2_2)G^O(s_{12})$$

$$= -\tfrac{1}{2}\iint \phi^*_0(r_3)\omega^*(r_{12})(H-E)\{\sqrt{3}\psi_0(r_2,r_3)F_0(r_1)$$

$$-\phi_0(r_2)\omega(r_{13})[\sqrt{3}G^P(s_{13}) + G^O(s_{13})]\}\,dr_3\,dr_{12} \qquad (5)$$

where the wavenumbers k_1 and k_2 are given by

$$\tfrac{1}{2}k^2_1 = E - E_0, \qquad \tfrac{1}{4}k^2_2 = E + \tfrac{1}{4} + 2,$$

E_0 being the ground-state energy of helium. In the last two equations the static interaction terms vanish because of the coincidence of the centre of charge and centre of mass of Ps. On subtracting equation (4) multiplied by a factor of $\sqrt{3}$ from equation (5) one obtains an equation for $G^O - \sqrt{3}G^P$ which is coupled with itself only, there being no coupling with F_0, whence one may conclude that $G^O - \sqrt{3}G^P = 0$, *i. e.* $G^O = \sqrt{3}G^P$. In other words one can say that the capture probability of a positron by an atomic electron in the triplet state (ortho-Ps) is three times that in the singlet state (para –Ps). Thus, equations (3) and (4) can be recast as

$$\tfrac{1}{2}(\nabla^2_1 + k^2_1)F_0(r_1)$$

$$= \iint \psi^*_0(r_2,r_3)\left(\frac{2}{r_1} - \frac{1}{r_{12}} - \frac{1}{r_{13}}\right)\psi_0(r_2,r_3)F_0(r_1)\,dr_2\,dr_3$$

$$-4\iint \psi^*_0(r_2,r_3)(H-E)\omega(r_{12})\phi_0(r_3)G^P(s_{12})\,dr_2\,dr_3$$

$$\tfrac{1}{4}(\nabla^2_{s_{12}} + k^2_2)G^P(s_{12})$$

$$= -\tfrac{1}{2}\iint \phi^*_0(r_3)\omega^*(r_{12})(H-E)[\psi_0(r_2,r_3)F_0(r_1)$$

$$-2\phi_0(r_2)\omega(r_{13})G^P(s_{13})]\,dr_3\,dr_{12}.$$

Using the Greens function technique, the First Born (FBA) matrix element for para PS formation is obtained as

$$f^{B}_{21}(\mathbf{k}', \mathbf{k}) = -\frac{1}{2}\left(\frac{\mu_2}{2\pi}\right) \iiint \phi^{*}_0(r_3)\omega^{*}(r_{12}) \exp(-i\mathbf{k}' \cdot \mathbf{s}_{12})\left(\frac{2}{r_1} - \frac{2}{r_2} + \frac{1}{r_{23}} - \frac{1}{r_{13}}\right)$$

$$\times \psi_0(r_2, r_3) \exp(i\mathbf{k} \cdot \mathbf{r}_1) \, d\mathbf{r}_{12} \, d\mathbf{s}_{12} \, d\mathbf{r}_3$$

with $\mu_2 = 2.$ (6)

The para-Ps formation cross section σ^{para}_{Ps} is expressed by the following relation

$$\sigma^{para}_{Ps} = 2\left(\frac{v'}{v}\right) \int [f^{B}_{21}]^2 \, d\Omega = 2\pi \left(\frac{k'}{k}\right) \int_0^\pi \left|f^{B}_{21}\right|^2 \sin\theta \, d\theta \quad (7)$$

v and $\acute{v}$ being respectively the velocities in the incident and final channels. The total Ps formation cross section is given by

$$\sigma_{Ps} = 4\sigma^{para}_{Ps}.$$

Ps as projectile

Ps (e^+e) has the structure of a hydrogenic atom and is the lightest known atom, its mass being $\sim 10^{-3}$ that of any conventional atoms and due to its low mass, recoil effects of Ps may be as much as $e^{\pm}$ scattering. From the theoretical point of view, Ps – atom scattering poses to be a very difficult problem. Unlike electron and positron scattering, the projectile Ps, atom has internal degrees of freedom (apart from those of the target) which must be taken into account leading to a significant complication. Since the center of charge of the Ps coincides with its center of mass, the static potential for a Ps – target system vanishes and its neutrality leads to zero first order polarization. As a consequence, the electron exchange mechanism seems to be the main driving force at low incident energies, apart from the correction expected from polarization and Van der Waals force. However, the proper inclusion of the exchange effect is very difficult since it involves electron swapping between two different center, the Ps and the target.

With the advent of mono-energetic energy tunable positronium (Ps) beams a new area of atomic collision process has been opened up. Ps is special in that it is a light neutral projectile and therefore can interact with various forms of matter ranging from electron and protons to atoms, molecular, solids and plasmas and can provide important information about the target medium.

Scattering of the neutral ortho Ps atom (being relatively more stable than para Ps) with different gas (atomic/molecular) targets, having an internal charge and mass symmetry is one of the hot topics [7] nowadays both experimentally [8] and theoretically [9-19]. The recent development of energy controllable Ps beams may further progress [11] the detailed study of the mechanisms governing the processes of Ps – atom / molecule (A) collisions as listed below:

(i) $Ps + A \rightarrow Ps + A$ elastic scattering

(ii) $Ps + A \rightarrow PS^* + A$

(iii) $Ps + A \rightarrow Ps + A^*$ projectile/target excitation

(iv) $Ps + A \rightarrow Ps^* + A^*$

(iv) $Ps + A \rightarrow PSA^+ + e^-$

(v) $Ps + A \rightarrow e^- + e^+ + A$ projectile/target ionization

(vi) $Ps + A \rightarrow e^+ + A^-$

(vii) $Ps + A \rightarrow Ps^- + A^+$

The scattering of the neutral Ps differs from that of its constituents e^- and e^+ in several aspects. In the case of the charged particles at low energies, the static and the polarization effects are dominant factors in determining the size of the interaction with atoms, with the exchange / correlation effects between the incident particles and the atomic electron contributing accordingly [6]. While, for the Ps atom, the mean static interaction is zero and the influence of polarization from the induced dipole is also limited. In fact the description of the Ps – atom scattering is quite complex due to their own internal structures.The Ps – atom interaction for a rare gas target is known to be repulsive at short distances since the Pauli exclusion principle does not allow the electron of the incident Ps to occupy the lowest energy electron states. Further away from the atom, the Ps atom interaction should be attractive due to the van der Waals interactions that play a major role as it provides the leading order term in the effective range expansion for the $l \geq 2$ phase shifts close to threshold [19].

Now regarding the practical applications, the Ps atom is considered to be an ideal probe to solid surfaces for determining their structures mainly because it can only undergo elastic reflection from the outer surface layer of a solid. Because of the large break up probability of the Ps (above its binding energy 6.8 eV), the multiple scattering effect from inner layer atoms of the solid is expected to be negligible for the Ps (unlike the low energy electron and positron) and as such the low energy Ps collision should be confined to the outer most surface layers. However, neutral atoms and molecules like He, H_2 also interact mainly with the surface atoms, but the available low energy beams are not energetic enough to probe small scale surface structure. Thus the Ps provides a great deal of advantage over charged as well as neutral heavy projectiles as a probe to study the structure of atoms and molecules and the surface properties of solids and the knowledge of different scattering parameters for Ps atom, molecule or ion collisions could be highly useful for such studies. Further, by virtue of very light mass of Ps (3 orders of magnitude smaller than hydrogen), its interaction with various forms of matter, ranging from electrons, protons, alkali ions to atoms, molecules, solid surfaces and plasmas can provide important information about the target medium . Ps atom is also used as a probe to Tokomak fusion plasma in order to deposit the e^+ s in the hot plasma. The subsequent transport of e^+ s out of the plasma provides a mechanism for the anomalous transport of the particles and energy by scattering from the plasma turbulence. Important aspects in Ps - atom / molecule scattering is that the first Born amplitude for such process vanishes if the initial and the final Ps states have the same parity, regardless of how the initial and the final target states may differ. The first Born scattering amplitude (direct) factorizes into a product of Ps form factor and a target form factor.

3. ANTIHYDROGEN

Why Antiparticles and Antimatters are so important?

According to the standard model each of the fundamental particles, the quarks and leptons making up the material Universe has an equivalent antimatter partners. It is generally believed that at the time of the *Big Bang* antiparticles and particles were created in equal numbers. But the question is that why antimatter is then so rare today is one of the greatest challenges to the Scientists. Our Universe seems to be made up of normal matter only. Some underlying asymmetry (matter-antimatter) should be there that has led the Universe to prefer matter to the antimatter. This asymmetry, called the *baryon asymmetry*, could be attributed to violation of the *CP symmetry* relating matter and antimatter. The exact mechanism of this violation still remains a mystery. This has motivated Scientists all over the world to work on the properties of the antiparticles. Probably the recent activation of the *LHC at CERN* could throw some light in this direction. Anti-hydrogen, the antimatter partner of the Hydrogen atom is the simplest Anti atom made up of antiparticles only, as it is the bound state of a positron (antiparticle of electron) and an antiproton (antiparticle of proton).

The recent experiments [20-37] at CERN (in collaboration with different parts of the world) have been able to produce cold ($\sim$ meV) $\overline{H}$ using a nested Penning trap via the TBR process, which normally favors its formation in highly excited states and then through cascading they could finally reach to the ground state.

Antihydrogen formation in ground state, essential for the high precision spectroscopic studies (in order to test the CPT invariance) as well as for the antimatter gravitation studies, is now drawing increasing attention both from the Experimentalists and the Theorists. It is really a challenging task to the Scientists to find an efficient mechanism for a significant production of cold ground state Antihydrogen ($\overline{H}$) atoms mainly because, the violation of the CPT (Charge, Parity and Time reversal) would require a new physics beyond the Standard Model. As such, this area of research now a days belongs to one of the thrust areas of research. In fact cold $\overline{H}$ atom is an ideal system for studying the fundamental symmetries in physics. As has been proved experimentally, the *Three Body Recombination* (TBR) process takes the leading role in this direction. Further, it was predicted experimentally that the laser assisted TBR might further enhance the field free $\overline{H}$ yield. Moreover, since in designing most of the collisional experiments now a days, an external laser/ magnetic field is necessary for trapping, cooling, confinement or even for collimation, it is highly desirable and worthwhile to study theoretically the TBR process in presence of an external laser field.

Apart from these, the study of $\overline{H}$ formation finds important applications in propulsion [38] system, nuclear weapons and many other fields to be described below.

The fundamental questions in physics regarding antimatter- matter interactions can be studied from the properties and production of an anti-matter.

Motivation for antimatter study

Why Antihydrogen.

1st, antihydrogen, the simplest antimatter is an ideal system for such studies because of its stability than any other exotic atoms, such as positronium, muonium, protonium etc.

2nd, Production of *cold and trapped* antihydrogen atom & studies of its properties and interactions could provide a critical test of the fundamental symmetries in physics, e.g., the 2 postulates of Standard Model in Particle Physics

(1) The CPT invariance theorem

(2) The gravitational weak equivalence principle for anti matter. CPT invariance can be tested through the high precision spectroscopic measurement of the 1s to 2s two photon transition of anti H.

Important application in propulsion system

Once antimatter is produced & stored it can be used in propulsion [38] by releasing it into a chamber and allowing it to annihilate with normal matter which produces tremendous power in the form of energetic sub atomic particles.

Nuclear weapons

Anti-hydrogen triggered thermonuclear explosives are compact & have extremely reduced fallout & hence such devices enhance proliferation of nuclear weapons and further diffuse distinction between conventional explosives & low yield nuclear weapons.

Other practical applications under consideration

For igniting inertial confined fusion pellets [39]. Space based Power Generator, Cancer Therapy. Next Question is Why Cold & Trapped Anti Hydrogen? Trapped & cold anti hydrogen in ground state is essential in order to achieve the long term goal of performing high precision spectroscopy as well as the gravitational studies to test CPT. If the Anti-H atoms can be trapped in their ground state it should be possible to compare spectroscopy of Anti-H with ordinary H spectroscopy that has reached astronomical precision. This high precision comparison is supposed to be the most accurate tests of CPT symmetry. Any violation of CPT symmetry would require new physics beyond the Standard Model of particle physics.

Some important reactions leading to antihydrogen

(1) *3 Body Charge Transfer in an Ortho Ps + $\bar{p}$ system* [40-48].

$$\bar{p} + Ps\ (\ nlm\ )\ \rightarrow\ \overline{H}\ (\ n'l'm'\ )\ +\ e$$

(2) *Radiative Recombination* [49-51]

$$e^+ + \bar{p} \rightarrow \overline{H} + \gamma$$

e^+ is passing by $\bar{p}$, it can emit photon & radiatively recombine.

(3) *If e^+ density is high enough there can be a three body recombination (TBR)* [20-37, 52-55]

$$\bar{p} + e^+ + e^+ \rightarrow \overline{H}\ (\ n\ ,\ l\ ,\ m\ )\ +\ e^+$$

The extra energy in the above processes is carried off by the
(1) Photon in RR Process;
(2) Electron (e) in the Three Body Charge Transfer Reaction (CT)
(3) e^+ in the TBR process.

The cross section for recombination (TBR) is several orders of magnitude larger if an electron or positron carries off extra energy. Each of the processes have Certain Advantages & Disadvantages:

Advantages of the 3 body charge transfer (CT)

(i)   Large cross sections in the order of 10^{-15} cm^2.
(ii) Provides slow and confined H necessary for the ultimate goal of high precision spectroscopic studies of H.
(iii) Cross section can be further enhanced by order of magnitude if lasers are used to pump the Ps into highly excited states.

Disadvantages of the 3 body charge transfer (CT)

Ps is neutral and short lived so that it cannot be confined in the same volume as antiprotons.

Advantages of the RR process
(i) Recombination can be further stimulated by intense laser.

$$e^+ \;+\; \bar{p} \;+\gamma \rightarrow \overline{H} \;+\; \gamma^{\,'}$$

(ii) $\overline{H}$ is typically formed in deeply bound state, essential for high precision spectroscopic studies.

Disadvantages of the RR process

(i)   The cross sections rate is small.
(ii) The time required to radiate a photon is typically longer than the duration of collision between antiproton & positron.

TBR in the trapped plasma of anti proton and positrons poses to be the *most efficient* H production reaction at low and intermediate energies.

Advantages of the present TBR process

The TBR reaction (3) poses to be the most efficient by orders of magnitude as compared to other two processes [(1) and (2)] particularly low incident energies mainly because the spectator particle (i. e., the extra e +) efficiently carries off the excess energy (for conservation of energy) and momentum released in the recombination (unlike the other $\overline{H}$ production reactions). The Reactants are stable charged particles which can be held in a trap, first for cooling to meV and then subsequently for the recombination to occur. This reaction has dependence on both the density and the temperature of the positron plasma and hence cross sections can be controlled by varying these parameters.

Theory

We consider an ensemble of weakly correlated e^{+}'s and the e^+ plasma density is assumed to be low enough so that e^+ - e^+ interaction can be treated as a perturbation. The $\overline{P}$ is considered to be of infinite mass and is treated as stationary target (i.e. laboratory frame) which corresponds to the experimental situation of a cold and trapped $\overline{P}$.

The effect of exchange between the active and the spectators e^+ s which is supposed to be quite dominant at low incident energies,is incorporated in a proper way in the present prescription.

The three body recombination process for the formation of Anti-hydrogen atom:

$$\bar{p} + e^+ + e^+ \rightarrow \overline{H} + e^+ . \tag{1}$$

The prior form of the transition amplitude T_{if} for this process is given by

$$T_{if} = \left\langle \Psi_f^{-}(\vec{r}_1,\vec{r}_2) \, (1+\vec{P}) \,\middle|\, V_i \,\middle|\, \psi_i(\vec{r},\vec{r}_2) \right\rangle \tag{2}$$

where $\vec{P}$ denotes the exchange operator corresponding to the interchange of the positrons in the final channel. V_i in Eqn. (2) is the initial channel perturbation which is the part of the total interaction not diagonalised in the initial state and ψ_i is the corresponding asymptotic wave function . The final channel wave function Ψ_f^{-} satisfies the three body Schrodinger equation obeying the incoming wave boundary condition;

$$(H - E) \, \Psi_f^{-} = 0 \tag{3}$$

The total Hamiltonian (H) of the system can be written as

$$H = - \frac{1}{2}\nabla_1^{\,2} - \frac{1}{2}\nabla_2^{\,2} - \frac{1}{r_1} - \frac{1}{r_2} + \frac{1}{r_{12}} \tag{4}$$

Where $\vec{r}_1$ and $\vec{r}_2$ represent the position vectors of the active e^{+} (to be transferred) and the spectator e^{+}s respectively. The atomic unit (a.u.) is used throughout the work.

The initial channel asymptotic wave function ψ_i in Eqn. (2) satisfies the following Schrodinger equation :

$$\left(- \frac{1}{2}\nabla_1^{\,2} - \frac{1}{2}\nabla_2^{\,2} - \frac{1}{r_1} - \frac{1}{r_2} - E \right) \psi_i = 0 \tag{5}$$

and is given by

$$\psi_i = N_j \, e^{i\vec{k}_j \cdot \vec{r}_j} \,\, {}_1F_1[\, i\alpha_j, \, 1 , - i\,(k_j r_j - \vec{k}_j \cdot \vec{r}_j)] \tag{6}$$

with $N_j = \exp\left(\dfrac{-\pi\,\alpha_j}{2} \right) \Gamma\left(1 - i\alpha_j \right) ; j = 1, 2 \; ; \; \alpha_j = -\dfrac{1}{k_j} ;$

$\vec{k}_j$ denotes the incident momentum of the active or the spectator e^{+}s respectively. The approximated final state wave function Ψ_f^{-} is chosen in the framework of eikonal approximation as follow:

$$\Psi_f^{-} = \phi_f\,(r_1)\, e^{i\vec{k}_{2f} \cdot \vec{r}_2} \exp\left[i\eta_f \int_z^{\infty}\left(\frac{1}{r_{12}} - \frac{1}{r_2} \right) dz' \right] \tag{7}$$

with $\eta_f = \dfrac{1}{k_f}$, k_f being the final momentum of the spectator e^+; $\phi_f(r_1)$ represents the bound state wave function of the $\overline{H}$ atom .

$$\phi_{f1s}(\vec{r}_1) = N_{1s} \, \exp\,(-\lambda_H r_1) \qquad (8)$$

with $N_{1s} = \lambda_H^{3/2}/\sqrt{\pi}$ and $\lambda_H = 1$.

The 2s state wave function of the $\overline{H}$ atom is

$$\phi_{f2s}(\vec{r}_1) = N_{2s} \, \exp\,(-\lambda_H r_1)\,(1 - \lambda_H r_1) \qquad (9)$$

with $N_{2s} = \lambda_H^{3/2}/\sqrt{\pi}$ and $\lambda_H = 1/2$

The 2p state wave function of the $\overline{H}$ atom is

$$\phi_{f2p,x,y,z}(\vec{r}_1) = N_{2p} \, r_{1,x,y,z} \exp\,(-\lambda_H r_1) \qquad (10)$$

with $N_{2p} = \lambda_H^{5/2}/\sqrt{\pi}$ and $\lambda_H = 1/2$

where $r_{1,x,y,z}$ denotes the x, y or z component of the coordinate $\vec{r}_1$. However, it should be mentioned that according to our choice of the coordinate system $\phi_{2p,y} = 0$.

The integration variable z' in Eqn. (7) is the z component of the vector $\vec{r}_2$. The phase integral occurring in Eqn. (7) is given by

$$(r_2 + r_{2z})^{i\eta_f} (r_{12} + r_{12z})^{-i\eta_f}$$

where z_2 and z_{12} are the z components of the respective vectors $\vec{r}_2$ and $\vec{r}_{12}$. Thus in the frame work of Coulomb Modified Eikonal approximation (CMEA) the expression for T_{if}^{prior} in Eqn. (2) is given as:

$$T_{if}^{prior} = -\frac{\mu_f}{2\pi} \int \exp(i\,\vec{k}_1 \cdot \vec{r}_1) \,{}_1F_1[i\,\alpha_1\,,1,i\,(k_1 r_1 - \vec{k}_1 \cdot \vec{r}_1)]\,e^{i\vec{k}_{2i}\cdot\vec{r}_2}$$

$${}_1F_1[i\,\alpha_2,1\,,i(k_{2i}r_2 - \vec{k}_{2i}\cdot\vec{r}_2)]\,\left(\frac{r_2 + z_{2z}}{r_{12} + z_{12}}\right)^{-i\eta_f}\frac{1}{r_{12}}\,e^{-\lambda_H \bar{n}}\,e^{-i\vec{k}_{2f}\cdot\vec{r}_2}$$

$$\exp\!\left(-\frac{\pi\,\alpha_1}{2}\right)\Gamma\,(1 - i\,\alpha_1)\exp\!\left(-\frac{\pi\,\alpha_2}{2}\right)\,\Gamma(1 - i\,\alpha_2)d\vec{r}_1\,d\vec{r}_2 \qquad (11)$$

In order to evaluate Eqn. (11) , we start with the integral of the following form

$$I = -\frac{\mu_f}{2\pi} \int \int \int \exp(i\,\vec{k}_1.\vec{r}_1)\,_1F_1[i\alpha_1,1,i(k_1r_1 - \vec{k}_1.\vec{r}_1)]e^{i\vec{k}_{2i}.\vec{r}_2}$$

$$_1F_1[i\alpha_2,1,i(k_{2i}r_2 - \vec{k}_{2i}.\vec{r}_2)]\frac{1}{r_{12}}\left(\frac{r_2 + z_{2z}}{r_{12} + z_{12}}\right)^{-i\,\eta_f} e^{-\lambda_H\,\bar{n}}\,e^{i\,\vec{\varepsilon}.\vec{n}}\,e^{-i\vec{k}_{2f}.\vec{r}_2}$$

$$\exp(-\frac{\pi\alpha_1}{2})\Gamma(1 + i\alpha_1)\exp(-\frac{\pi\alpha_2}{2})\Gamma(1 + i\alpha_2)d\vec{r}_1 d\vec{r}_2 \tag{12}$$

Where $\vec{\varepsilon}$ is an arbitrary vector introduced in order to generate the x and z components of the amplitude for charge transfer to 2p state through the relation

$$I_{x,z} = -i \lim_{\varepsilon\,x,\,z\,\to\,0}\frac{\partial\,I}{\partial\varepsilon_{x,z}}$$

However, for charge transfer from 1s or 2s state we put $\vec{\varepsilon} \to 0$ in Eqn. (12). For evaluation of the integral in Eqn. (12) we use the following contour integral representation of the eikonal phase factors [56] as well as the coulomb functions [57] of the confluent hyper-geometric function. The eikonal phase factor is of the form,

$$y^{\pm(i\eta-n)} = \frac{(-1)^{n+1}}{2i\sin(\mp\pi\,i\eta)\Gamma(\mp i\eta \pm n)}\int_c (-\lambda)^{\mp i\eta\pm n-1}\exp(-\lambda y)dy \tag{13a}$$

where the contour c has a branch cut from 0 to ∞ [61]; the confluent hyper-geometric function:

$$_1F_1(i\alpha,1,z) = \frac{1}{2\pi\,i}\int_{\Gamma_1}^{(0^+,1^+)} p(\alpha,t)\exp(zt)dt \tag{13b}$$

With $p(\alpha,t) = t^{-1+i\alpha}(t-1)^{-i\alpha}$, Γ_1 is a closed contour encircling the two points 0 and 1in anti - clockwise [57]. At the point where the contour crosses the real axis to the right side of 1, arg (t) and arg (t-1) are both zero. Further, we also follow Fourier transform for the evaluation of equation (12);

$$\exp(-\lambda r) = \frac{\lambda}{\pi^2}\int\frac{\exp(\,i\vec{p}.\vec{r}\,)}{(p^2 + \lambda^2)^2}d\vec{p} \tag{14a}$$

$$\frac{1}{r} = \frac{1}{2\pi^2}\int\frac{\exp(\,i\vec{p}.\vec{r}\,)}{p^2}d\vec{p} \tag{14b}$$

$$\frac{\exp(-\lambda r)}{r} = \frac{1}{2\pi^2}\int\frac{\exp(\,i\vec{p}.\vec{r}\,)}{(p^2 + \lambda^2)}d\vec{p} \tag{14c}$$

Using the Equations 14(a) - (c) and then performing the space integration over $\vec{r}_1, \vec{r}_2$ in Eqn. (1) we obtain,

$$I_0 = 8 \int_{c_1} \int_{c_2} \int_{\Gamma_i} p(\alpha_1, t_1) dt_1 \, p(\alpha_2, t_2) dt_2 d\lambda_a d\lambda_b (-\lambda_a)^{-i\eta_f^{-1}} (-\lambda_b)^{i\eta_f^{-1}}$$

$$\int \frac{d\vec{s}}{(2\pi i)^2 (s^2 + \varepsilon_2'^2)(|\vec{s} + \vec{q}_1|^2 + \lambda_H'^2)^2 (|\vec{s} - \vec{q}_2|^2 + \varepsilon_1'^2)^2} \tag{15}$$

Where $\lambda_H' = \lambda_H - i k_1 t_1$, $\quad \varepsilon_1' = \lambda_a + \chi_1 - i \, k_{2i} \, t_2$, $\quad \varepsilon_2' = \chi_2 + \lambda_b$, $\vec{q}_1 = \vec{k}_1 - \vec{k}_1 t_1 + \vec{m}_2 + \vec{\varepsilon}$, and $\quad \vec{q}_2 = \vec{k}_{2i} - \vec{k}_{2i} t_2 - \vec{k}_{2f} + \vec{m}_1 - \vec{m}_2$ where $m_1 = i \lambda_a \hat{k}_{2f}$, $m_2 = i \lambda_b \hat{k}_{2f}$. The final results of the space integrations over $\vec{r}_1$ and $\vec{r}_2$ in Eqn. (15) can be generated by suitable parametric differentiation from the basic parent integral I_0 in Eqn. (15). The $\vec{s}$ integral occurring in Eqn. (15) can be performed analytically following Sinha and Sil [58] to obtain a type integral

$$I_0 = \frac{16\pi^2}{(2\pi i)^2} \int_{\Gamma_i} \int_0^\infty \int_{c_1} \int_{c_2} \frac{p_1(\alpha_1 t_1) p_1(\alpha_2 t_2)(-\lambda_a)^{-i\eta_i - 1}(-\lambda_b)^{i\eta_i - 1}}{(A + B\lambda_a + C\lambda_b + D\lambda_a\lambda_b)} dv d\lambda_a d\lambda_b dt_1 dt_2 \tag{16}$$

where the contour c_1 and c_2 refer to the same as that in eqn. (13). In Eqn.(16), A, B, C and D are functions of the $\vec{k}_1$, $\vec{k}_{2i}$, $\vec{k}_{2f}$, ε_1', ε_2', λ_H' and the integration variable v. Now the λ_a and λ_b integrations are performed analytically [59] to obtain

$$I_0 = -4\pi^2 \times 16\pi^2 \int_{\Gamma_i} p_1(\alpha_1, t_1) p_2(\alpha_2, t_2) \int_0^\infty \frac{1}{A} \left[\frac{B}{A}\right]^{i\eta} \left[\frac{C}{A}\right]^{-i\eta} {}_2F_1\left[-i\eta_1, i\eta_1, 1, z'\right] dv dt_1 dt_2 \tag{17}$$

where $z' = 1 - \dfrac{AD}{BC}$ and ${}_2F_1$ denotes the hyper-geometric function with argument z'. We are thus finally left with a three- dimensional integral over v, t_1 and t_2. The v integral has been evaluated numerically [60, 61] by using Gauss-Legendre quadrature method and the complex integration over t_1 and t_2 have been converted to a real one dimensional integral over the range 0 to 1 and evaluated numerically by using Gauss Lagurre quadrature method. The actual integrals occurring from the expression of I in Eqn. (12) are then derived from I_0 by parametric differentiations.

The differential cross sections for (1s – 1s), (1s – 2s) and (1s- 2p) capture processes are given as;

$$\frac{d\sigma_{1s}}{d\Omega} = \frac{k_f}{k_1 k_2}\left|A_{(1s-1s)}(\theta)\right|^2 \tag{18}$$

$$\frac{d\sigma_{2s}}{d\Omega} = \frac{k_f}{k_1 k_2}\left|A_{(1s-2s)}(\theta)\right|^2 \tag{19}$$

$$\frac{d\sigma_{2p}}{d\Omega} = \frac{k_f}{k_1 k_2}\left[\,\left|A_{(1s-2p),x}(\theta)\right|^2 + \left|A_{(1s-2p),z}(\theta)\right|^2\,\right] \tag{20}$$

where A's are the corresponding transition matrix elements. Finally the differential cross section for the process (1) is given by,

$$\frac{d\sigma}{d\omega} = \frac{k_f}{k_1 k_2}\left[\,\frac{1}{4}\,\left(\left|\,f+g\,\right|^2\right) + \frac{3}{4}\,\left(\left|\,f-g\,\right|^2\right)\,\right] \tag{21}$$

Where f and g correspond to the direct and the exchange amplitude respectively.

4. LASER ASSITED ANTI HYDROGEN FORMATION

Since nowadays in designing most of the collisional experiments, an external field is involved, it is therefore quite interesting and of worth to study the TBR process in the presence of an external laser field. Further, in the recent experiments producing $\overline{\mathrm{H}}$ an external magnetic field is involved. This has motivated us in treating the TBR process in the presence of external fields. To our knowledge, this is the first reported [62] theoretical work along this line.

In the present calculation a single particle collision is considered for the TBR reaction and as the density of the constituents particles are not incorporated. Moreover, in most of the theoretical TBR calculations, only the reaction rates are considered whereas the present study concentrates on the TBR cross sections only.

The laser polarization is chosen to be either parallel or perpendicular to the incident momentum.

Theory

The present study deals with the following laser assisted (LA) three body recombination process (TBR) in the ground and excited states:

$$\bar{p} + e^+ + e^+ + N\gamma(\omega,\,\vec{\varepsilon}) \rightarrow \overline{\mathrm{H}}(\overline{\mathrm{H}}^*) + e^+ \tag{22}$$

Where N stands for the multi photon absorption and emission and $\gamma(\omega,\,\vec{\varepsilon})$ denotes the laser photon with angular frequency ω and field strength

$\vec{\varepsilon}_0$, the laser being chosen as continuum wave (CW). The first incoming positron is the active one to form the anti-hydrogen atom in the final state while the second positron acts as a spectator that carries away the excess energy and momentum released in the recombination process.

The collision dynamics (TBR) is treated quantum mechanically in the frame work of Coulomb modified eikonal approximation (CMEA) while the laser field is treated classically in the dipole approximation using Coulomb gauge. The laser field dresses the two incoming positrons (active and the spectator) as well as the outgoing e^+ states non perturbatively by using eikonal modified Volkov wave functions [59,62] while the dressing of the final bound state of the $\overline{\text{H}}$ atom (ground or excited) is treated perturbatively by solving the time dependent Schrodinger equation. The laser field is chosen to be much less than the atomic units of field strength (5×10^{11} V/m) in order to avoid the effect of field ionization and the frequency of the laser field is kept much below the binding energy of $\overline{\text{H}}$ (soft photon limit). Both the differential and the total $\overline{\text{H}}$ formation cross sections are studied for single as well as multi photon exchange. The expression for the transition matrix element for the process (22) is the same as in Eqn.(2). The energy conservation relation for the TBR process is given by,

$$\frac{k_1^2}{2} + \frac{k_2^2}{2} \pm N\omega = \frac{k_f^2}{2} + \varepsilon_{\overline{H}} \tag{23}$$

where $\vec{k}_1$ and $\vec{k}_2$ are the incident momentum of the active and the spectator positrons (e^+s) respectively; $\vec{k}_f$ being the final momentum of the outgoing spectator e^+, $\varepsilon_{\overline{H}}$ is the binding energy of $\overline{\text{H}}$. '+N' refers to absorption and '$-N$' refers to the emission of photons. The total Hamiltonian (H) of the system may be written as,

$$\text{H} = -\frac{1}{2}(i\vec{\nabla}_1 + \vec{A})^2 - \frac{1}{2}(i\vec{\nabla}_2 + \vec{A})^2 - \frac{1}{r_1} - \frac{1}{r_2} + \frac{1}{r_{12}} \tag{24}$$

where $\vec{r}_1$ and $\vec{r}_2$ represent the position vectors of the active $e^+(\vec{r}_1)$ and the spectator $e^+(\vec{r}_2)$. The laser (CW) field is chosen to be a single mode linearly polarized, spatially homogeneous electric field represented by $\vec{\varepsilon}(t) = \vec{\varepsilon}_0 \sin(\omega t + \xi)$; ξ being the initial phase of the laser field; the corresponding vector potential in the Coulomb gauge being $\vec{A}(t) = \vec{A}_0 \cos \omega t$ with $\vec{A}_o = \vec{\varepsilon}_0 / \omega$; ξ is chosen to be zero in the present work.

The initial channel asymptotic wave function ψ_i in Eqn. (2) satisfies the following Schrödinger equation:

$$(-\frac{1}{2}(i\vec{\nabla}_1 + \vec{A})^2 - \frac{1}{2}(i\vec{\nabla}_2 + \vec{A})^2 - \frac{1}{r_1} - \frac{1}{r_2} - E)\psi_i = 0 \tag{25}$$

and is given by,

$$\psi_i = \chi_{k_1}\chi_{k_2} \tag{26}$$

where χ_{k_i} ; $i=1, 2$ refers to the Coulomb Volkov (CV) solution [59,62] and is given by,

$$\chi_{k_i} = N_i e^{i(\vec{k}_i.\vec{r}_i + \vec{k}_i.\vec{\alpha}_o \sin\omega t - E_{k_i}t)}\,_1F_1(i\eta_i,1,-i(k_ir_i - \vec{k}_i \cdot \vec{r}_i)) \tag{27}$$

with

$$N_i = \exp(-\pi\eta_i/2)\Gamma(1-i\eta_i) \; ; \; \eta_i = -\frac{1}{k_i} \; ; \vec{\alpha}_0 = \vec{\varepsilon}_0/\omega^2. \tag{28}$$

and $\vec{k}_i$ denotes the momentum of the active $(\vec{k}_1)$ or the spectator e^+ $(\vec{k}_2)$. The final channel wave function Ψ_f^- satisfies the three-body Schrödinger equation obeying the incoming wave boundary condition:

$$(H - E)\Psi_f^- = 0 \tag{29}$$

where the dressed final channel wave function Ψ_f^- is approximated in the framework of eikonal approximation as follows:

$$\Psi_f^- = \chi_{k_f}(\vec{r}_2,t)\phi_{\overline{H}}^d(\vec{r}_1,t) \tag{30}$$

where $\chi_{k_f}(\vec{r}_2,t)$ represents the eikonal modified Volkov state of the outgoing scattered positron and is given by

$$\chi_{k_f} = \exp i(\vec{k}_f.\vec{r}_2 + \vec{k}_f.\vec{\alpha}_0 \sin\omega t - E_{k_f}t) \times \exp[i\eta_f \int_z^\infty (\frac{1}{r_{12}} - \frac{1}{r_2})dz'] \tag{31}$$

with $\vec{\alpha}_0 = \vec{\varepsilon}_0/\omega^2$; $\eta_f = \dfrac{1}{|\vec{k}_f - \vec{A}(t)|}$

The dressed ground state wave function of the $\overline{H}$ in the final channel is constructed using the first order perturbation theory in the Coulomb gauge and is given as,

$$\phi_{\overline{H}}^{d}(\vec{r}_1,t) = \frac{1}{\sqrt{\pi}} e^{-iW_0^{\overline{H}}t} e^{-\lambda_f r_1} [1 + i\vec{A}(t).\vec{r}_1] \tag{32}$$

where $W_0^{\overline{H}}$ is the energy of the ground state $\overline{H}$.

In presence of the laser field, the angular momentum l is no more a good quantum number and as such the excited states (2s, 2p) lose their identity in presence of the field. In fact, since the dipole operator has a non-vanishing matrix element between the 2s and $2p_0$ states, these two states are mixed by the dipole perturbation and as such the dressed $\overline{H}$ formed in the excited state (2s or 2p) could be expressed as a superposition of the 2s and $2p_0$ state (linear combination) as follows:

$$\phi_{\overline{H}}^{d}(\vec{r}_1,t) = \frac{1}{\sqrt{2}}(\psi_1(r_1)e^{-i/\hbar(E_{n=2}-\Delta E)t} \pm \psi_2(r_1)e^{-i/\hbar(E_{n=2}+\Delta E)t}) \tag{33}$$

with

$$\psi_1 = \frac{1}{\sqrt{2}}(\psi_{200} + \psi_{210})$$

$$\psi_2 = \frac{1}{\sqrt{2}}(\psi_{200} - \psi_{210}) ; \tag{34}$$

$\Delta E = 3\varepsilon/Z$ denotes the Stark shift in a.u., Z being the charge of the target atom. In equation (34) ψ_1 represents the lower energy state while ψ_2 represents the upper energy state. We designate hence forth the lower state as 2s and the upper energy state as 2p. It may be mentioned here that since the laser field is always chosen along the direction of the polar axis, only m=0 state contributes to this laser assisted process since by the dipole selection rule Δm=0.

Finally, after performing the time integration [52, 59] in eqn. (2), the transition matrix element T_{if} reduces to,

$$T_{if} = -i\frac{1}{(2\pi)^{1/2}} \sum_N \delta(E_{k_f} - E_{k_1} - E_{k_2} + N\omega)J_N(\vec{K}.\vec{\alpha}_0)I \tag{35}$$

where J_N is the Bessel function of order N; $\vec{K} = \vec{k}_1 + \vec{k}_2 - \vec{k}_f$; the integral I being the space part of the transition matrix element. It should be noted that in performing the time integration analytically, we approximate (for the sake of simplicity) the quantity $\vec{A}(t)$ in the eikonal phase term in Eqn. (31) by its $t = 0$ value A_0. We have checked that the eikonal phase factor is not very sensitive with respect to the time dependence of the vector potential and as such this is not supposed to be a very crude approximation.

The total laser assisted cross section (TCS) for a given value of N (no of photon exchange) can be obtained by integrating the differential cross section (Eqn.(21)) over the solid angle.

$$\sigma_N = \int (\frac{d\sigma}{d\Omega})_N \, d\Omega \tag{36}$$

and the total multi photon cross section is given by

$$\sigma = \sum_N \sigma_N \tag{37}$$

For $\vec{\varepsilon}_0 \parallel \vec{k}_i$, both $\vec{\varepsilon}_0$ and $\vec{k}_i$ are along the polar axis Z, while for $\vec{\varepsilon}_0 \perp \vec{k}_i$, since $\vec{\varepsilon}_0$ is chosen to be the polar axis, $\vec{k}_i$ is $\perp^r$ to the Z axis. For the parallel geometry the differential cross section (DCS) is merely θ_f (scattering angle) dependent while for the perpendicular geometry the cross section depends both on θ_f and ϕ (azimuthal angle) of the outgoing spectator positron. The DCS results are, as a rule, presented in terms of the scattering angle (i.e. the angle between $\vec{k}_i$ and $\vec{k}_f$) and, conventionally, $\vec{k}_i$ or $\vec{q}$ ($\vec{k}_i - \vec{k}_f$) is chosen to be the polar axis. However, since the present theory is based on the choice that the polar axis is along the field direction $\vec{\varepsilon}_0$, the analysis requires a transformation of the coordinate system so that, in the final coordinate system, $\vec{k}_i$ becomes the z axis (new coordinate system).

The laser modification is found to be quite significant both qualitatively and quantitatively particularly for the ground state.

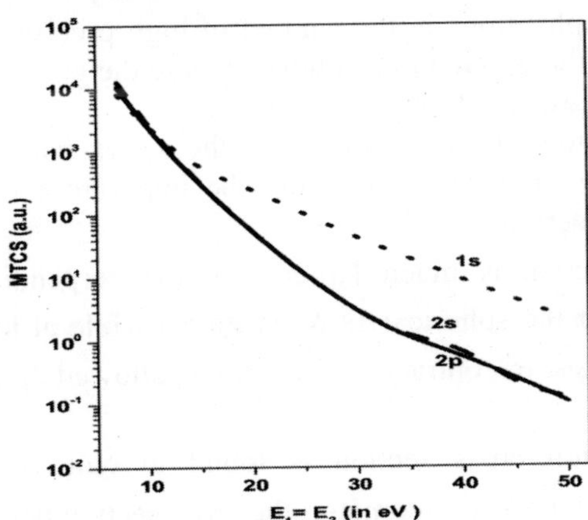

Fig.1: Partial laser assisted total cross sections (MTCS) (units of $\pi a_0^{\,2}$) for antihydrogen formation in 1s, 2s, 2p states for a wider range of incident energy (0 - 50 eV) where $E_1 = E_2$. Solid line for 2p state, dashed line for 2s state, dotted line for 1s state.

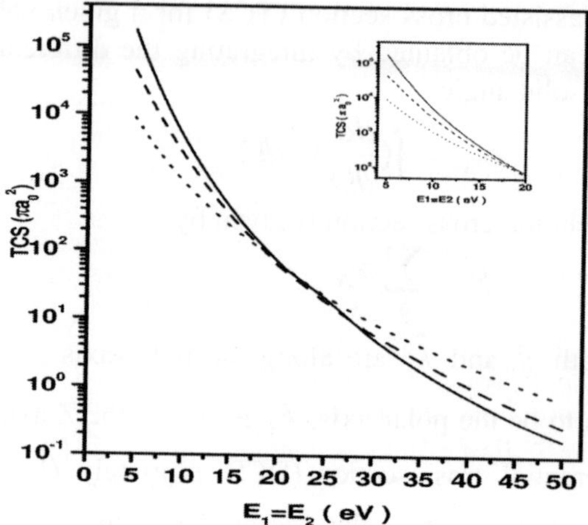

Fig.2: same as Fig.1 but for the field free case (FF)   Inset: same TCS but for smaller range of incident energy.

Conclusion

The salient features of the present study are as follows

In presence of the laser field, the ground state multi photon cross sections (MTCS) are enhanced by orders of magnitude with respect to the FF [55] cross sections while the excited states are mostly suppressed. Further, the laser assisted ground state MTCS falls off much more slowly particularly at lower energies than the excited states for unequal energy sharing $(E_1 > E_2)$. These findings might have some implications in the context of high precision spectroscopy (to test CPT) as well as the gravitational studies where the ground state is essential in significant quantities, at extreme low energies.

The single photon cross sections, on the other hand are significantly suppressed w.r.t. the FF results indicating the importance of the multiphoton effects in the TBR reaction.

The $\|'$ cross section is much larger than the perpendicular one at low incident energies (for the spherical 1s & 2s states) while at higher energies the $\perp'$ geometry overtakes. In contrary, for the dipole allowed 2p state, the situation is just the reverse.

The $\overline{\mathrm{H}}$ formation cross section is found to quite sensitive w.r.t. the azimuthal angle (ϕ) for the $\perp'$ geometry, the cross section being maximum for a nonzero ϕ value for a given kinematics.

For a more efficient production of $\overline{\mathrm{H}}$, the unequal $(E_1 > E_2)$ distribution of energy between the active and the spectator positrons could be suggested rather than the equal one $(E_1 = E_2)$ for both the FF and the LA cases.

Unlike the FF case, the LA ground state MTCS is highly favored throughout the energy range for a higher value of E_2.

Emission cross section (N<0) is always higher than the absorption cross section (N > 0) as is expected in an exothermic reaction.

The present model is not supposed to be suitable for the experimental extreme low energy (~ meV) regime and a more sophisticated theory is needed.

References

[1]   C D Anderson, *Science* **76** (1933) 238.

[2]   S. Mohorovic, *Astron. Nachr* **235** (1934) 94.

[3]   M Deutsch, *Phys. Rev,* **82** (1951) 455.

[4]   A.P.Mills., Jr.,S. Berko and K.F. Canter, *Phys. Rev Lett.*, **34** (1975) 1541.

[5]   N. F. Mott and H.S.W. Massey, "The Theory Of Atomic Collisions", 3rd Edn.,Oxford Press (1965).

[6]   A .S.Ghosh, N.C.Sil and P Mandal, *Phys. Rep.* **87** (1982).

[7]   N. Zafar, G Larichhia and M Charlton, *Hyperfine Interactions* **84** (1994) 355.

[8]   S Armitage *et al*, *Phys. Rev. Lett.,* **89** (2002) 173402.

[9]   H.S.W. Massey and C.B.O. Mohr, *Proc. Phys. Soc. London Sect. A* **67** (1954) 695.

[10]   L. Sarkadi, *Phys. Rev. A* **68** (2003) 032706.

[11]   C. Starett. M .T. McAlinden and H.R.J. Walters, *Phys. Rev. A* **72** (2005) 012508.

[12]   J. E. Blackwood, M. T. McAlinden and H.R.J.Walters, *Phys. Rev. A* **65** (2002) 032517.

[13]   P.K.Sinha, A.Basu and A.S.Ghosh, *J. Phys. B* **33** (2000; see also other references cited there in) 2579.

[14]   P. K. Biswas and S. K. Adhikari, *J. Phys. B* **33** (2000) 1575.

[15]   Hasi Ray, *J. Phys. B* **35** (2002; see other references cited there in) 3365.

[16]   S Roy, D Ghosh and C Sinha *J. Phys. B* **38** (2005) 2145.

[17]   S. Roy and C. Sinha, *Eur. Phys. J. D* **47** (2008) 327.

[18]   S Roy and C Sinha, *Phys. Rev. A* **80** (2009) 022713.

[19]   J Mitroy and M.W. J. Bromley, *Phys. Rev. A* **68** (2003) p 035201.

[20]   G. Gabrielse, S. L. Rolston, L. Haarsma and W. Kells, *Phys. Lett. A* **129** (1988) 38.

[21]   G. Gabrielse, *Advances in Atomic Molecular and Optical Physics* **50** (2005) 155.

[22]   M. H Holzscheiter and M. Charlton, *Rep. Prog. Phys.* **62** (1999) 1.

[23]   M.Amoretti *et al*, *Nature* **419** (2002) 456.

[24]   M. Amoretti *et al*, *Phys. Rev . Lett.* **91** (2003) 055001.

[25]   M. Amoretti *et al*, *Phys. Lett . B* **578** (2004) 23.

[26]   M. Amoretti *et al*, *Phys. Lett. B* **583** (2004) 59.

[27]   N. Madsen *et al*, *Phys. Rev . Lett.* **94** (2005) 033403.

[28]   M. Amoretti *et al*, *Phys. Rev. Lett.* **97** (2006) 213401.

[29]   G. Gabrielse *et al*, *Phys. Rev. Lett.* **89** (2002) 213401.

[30]   G. Gabrielse *et al*, *Phys. Rev. Lett.* **89** (2002) 233401.

[31]   G. Gabrielse *et al*, *Phys. Lett. B* **548** (2002) 140.

[32]   G. Gabrielse *et al, Phys. Rev. Lett.* **93** (2004) 073401.

[33]   G. Gabrielse *et al, Phys. Rev. Lett.* **98** (2007) 113002.

[34]   G. Gabrielse *et al, Phys. Lett. B* **507** (2001) 1.

[35]   A. Speck *et al, Phys. Lett. B* **597** (2004) 257.

[36]   T. Pohl, H. R. Sadeghpour and G. Gabrielse, *Phys. Rev. Lett.* **97** (2006) 143401.

[37]   G. Andresen *et al , Phys. Rev. Lett.* **98** (2007) 023402.

[38]   R. L. Forward, *Journal of the British Interplanetary Society* **35** (1982) 387.

[39]   A. Gsponer and J. P. Hurni, e-print *arXiv: physics/0507125v2.*

[40]   J. W. Humberston *et al, J. Phys. B* **20** (1987) L25.

[41]   J. W. Darewych, *J. Phys. B* **20** (1987) 5917.

[42]   M. Charlton, *Phys. Lett. A* **143** (1990) 143.

[43]   S. Tripathi, C.Sinha and N. C. Sil, *Phys. Rev. A* **42** (1990) 1785.

[44]   J. Mitroy and A. T. Stelbovics, *Phys. Rev. Lett.* **72** (1994) 3495.

[45]   J. Mitroy and G. Ryzhikh, *J. Phys. B* **30** (1997) L371.

[46]   S.. Tripathi, R. Biswas and C. Sinha, *Phys. Rev. A* **51** (1995) 3584.

[47]   D. B. Cassidy *et al, J. Phys. B* **32** (1999) 1923.

[48]   A. Chattopadhyay, C. Sinha, *Phys. Rev. A* **74** (2006) 022501.

[49]   S. M. Li, Z. J. Chen, Q.Q. Wang, Z.F. Zhou, *Eur. Phys. J. D.* **7** (1999) 39.

[50]   S. M. Li, Y. G. Miao, Z. F. Zhou, J. Chen, Y. Y. Liu, *Phys. Rev. A* **58** (1998) 2615.

[51]   S. Bivona, R. Burlon, G. Ferrante and C. Leone, *Optics Express* **14** (2006) 3715.

[52]   M. E. Glinsky and T. M. O'Neil, *Phys. Fluids B* **3** (1991) 1279.

[53]   F. Robicheaux, *J. Phys. B* **40** (2007) 271.

[54]   F. Robicheaux, *Phys. Rev. A* **70** (2004) 022510; ibid **73** (2006) 033401.

[55]   S. Roy, S. Ghosh Deb and C Sinha, *Phys. Rev. A* **78,** (2008) 022706.

[56]   I. S. Gradshteyn and I. M. Ryzhik; *Tables and Integrals* (Academic New York, 1980).

[57]   A Messiah; *Quantum Mechanics* (North- Holland, Amsterdam, 1966), **Vol. 1**, p. 481.

[58]   C. Sinha and N. C. Sil, *J. Phys. B* **11** (1978) L333.

[59]   A. Chattopadhyay and C. Sinha, *Phys.Rev A* **72** (2005) 053406.

[60]   B. Nath and C. Sinha, *J. Phys. B* **33** (2000) 5525.

[61]   D. Ghosh and C. Sinha, *Phys. Rev. A* **69** (2004) 052717.

[62]   S. Ghosh Deb and C. Sinha, *Euro. Phys. Lett.,* **88** (2009) 23001.

Photodissociation lifetime studies on gas-phase biomolecules

G. Aravind

*Department of Physics, Indian Institute of Technology - Madras,
Chennai 600 036, India
e-mail: aravind@physics.iitm.ac.in*

1   INTRODUCTION

Interaction of photoactive chromophores with photons is the key to understand
the dynamics of the protein that contains the chromophore within it. Transitions that occur in the chromophore upon photoabsorption is by itself difficult
to study with quantum mechanical calculations owing to their large size. In the
real biological environment the interactions between the chromophore and the
protein containing it such as the dipole interactions, hydrogen bonds, steric constraints could influence the photoabsorption of the chromophore. Calculations
that take into account such interactions and the protein structure becomes far
more formidable. However, studying the intrinsic photoabsorption properties of
the chromophore and then considering the environmental interactions as perturbations is an alternative. Gas-phase studies [1–8] on chromophores throw light
on their intrinsic properties and calls for novel experimental techniques such
as the electro-spray ion production and iontraps. The photoabsorption properties, photodetachment, and photodissociation lifetime studies on gas-phase
chromophores convey vital informations on the photoresponse of the biosystem
for eg.,understanding our colour vision, phototaxis in biosystems. Photodissociation lifetime studies are important to understand radiation damage, signal
transmission in biosystems and the internal states of the biochromophore. In
this article we shall discuss photodissociation lifetime studies on adenosine 5'
monophosphate (AMP) nucleotide to address radiation damage and the novel
experimental techniques that are employed in such gas-phase studies.

Before we dvelve into the experimental technique and the information derived in photodissociation lifetime studies, we shall breifly describe the AMP
molecule. The adenosine 5' monophosphate (AMP) nucleotide is found in RNA
and consists of a phosphate group, sugar ribose, and the nucleobase adenine.
It is an intermediate molecule formed during the process of energy production in the form of adenosine triphosphate (ATP) in our biological system. The
structure of AMP is critically determined by H bonds at the phosphate group
with the base as well as the Ribose [9, 10]. The cyclized form of AMP plays
an important role in intracellular signal transduction in many different organisms. The significance of the nucleotides in fundamental biological processes
motivates the study of its structural and functional properties. The structure
and dynamics of nucleotides have been studied via collision induced dissocia-

tion (CID) using techniques such as Fourier transform ion cyclotron resonance mass spectrometry [10]. Dissociation studies on deprotonated dinucleotides are known to provide sequence information in nucleotides [9]. Yang *et al.* [11] have studied electron-capture and electron-detachment dissociation of oligoribonucleotides. There have also been theoretical and experimental studies on the photophysics related to the π-π^* transition at the nucleobases in gas [12–16] and solution phase [17–22]. Guan et al. [23] reported the first observation of electron photodetachment from multiply charged nucleotides. Recently, Marcum et al. [24] studied UV photodissociation of deprotonated 2-deoxyriboadenosine-5-monophosphate in the gas phase and analysed the photofragments. Recent UV photodissociation lifetimes studies on the protonated and deprotonated AMP ions by Aravind *et al.* [27] would be the subject of this article.

Nucleobases resonantly absorb UV radiation due to the π-π^* transition at the aromatic ring. Resonant absorption of about 5 eV in these biomolecules triggers several reaction pathways and ultrafast dissociation occurring at a timescale faster than that for dissipation of excess energy to the environment would render these biomolecules vulnerable to UV radiation. Ultrafast internal conversion of the photon energy into vibrational excitation has been studied using femtosecond transient absorption spectroscopy [17,18]. Ionizing radiation could also damage these biomolecules with the generated energetic electrons and the effect of electron collisions has been studied previously at Electrostatic Ion Storage ring at Aarhus (ELISA) [28]. Photodestruction of DNA could cause hazards such as skin cancer and it is therefore important to understand the intrinsic photoresponse of nucleotides, the fragmentation pathways, and the associated lifetimes.

Photodissociation lifetime studies on the gas phase deprotonated and protonated AMP have been carried out in the millisecond regime at ELISA. Henceforth the deprotonated and protonated forms will be mentioned as anion and cation, respectively. The study revealed that a large fraction of the AMP anions as well as the cations photodissociated too fast (within few tens of microseconds) to be detected at the applied delayed micro channel plate detector (MCP) [29] and only a depletion of the stored ion beam was registered. The depletion was found to be largest for cations and was thought possibly to be due to non-ergodic decay pathways. Non-ergodic dissociation pathways in NC bond cleavage in electron capture dissociation experiments is known before. Non-ergodic fragmentation would imply damage of the nucleotide before it could quench the excitation energy in its environment. However, there could be prominent fast dissociation pathways in the sub-microsecond regime that could not be perceived using ELISA.

We now discuss an experiment to probe the submicrosecond regime photodissociation of AMP anions and cations using a linear time of flight spectrometer (TOF) [30,31]. More than one dissociation lifetimes for both the anion and cation in the submicrosecond regime is observed in this experiment. UV excitation of the nucleobase in the AMP anion was not observed to cause photodetachment, leaving dissociation as the dominant decay pathway since the flourescence quantum yield is known to be as low as 10^{-4} [20].

2   EXPERIMENTAL METHODS

As mentioned in the introduction, gas-phase experiments with biomolecules employ novel techniques such as the electrospray ionization and iontrap. Electrospray ionsource enable the production of deprotonated and protonated ions rather easily than the radical ions. The chromophore ions of interest are usually found in deprotonated or protonated form in the biosystems. Detailed description of this versatile source is in the literature [25] and there is still research on the precise mechanism behind the ion production in this source. The density of ions produced from this source is rather insufficient for photon-interaction experiments. Iontraps [26] are employed for this purpose to accumulate ions from the source and then transport them as a bunch at desired rate to the interaction region. Iontraps are also useful to cool the stored hot ions before transporting them to the interaction region. The experiment described in this article employs electrospray ion source, multipole ion trap and an elegant time-of-flight technique that could deduce submicrosecond regime dissociation lifetimes. The

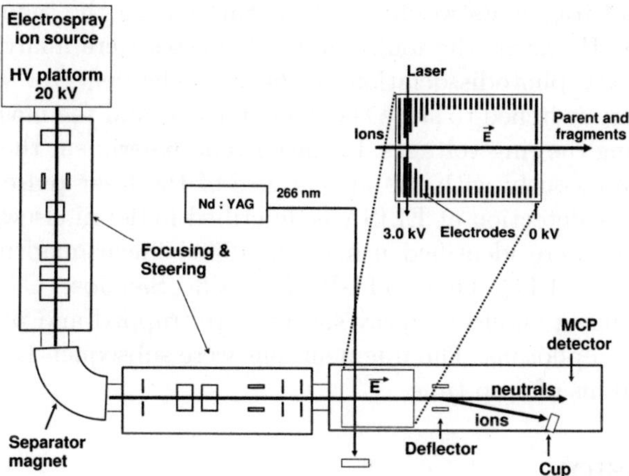

Figure 1: Schematic representation of the experimental set-up with the ion-beam path (without the 3^0 ion path deviation) and the laser set-up. The inset shows the spectrometer with a stack electrodes that maintain a uniform electric field. The electric field region is about 26 cm while the distance of the interaction spot to the MCP detector is 1.6 meters.

AMP anions and cations in the gas phase were produced by electrospraying AMP dissolved in ammoniated water/methanol and acetic acid/methanol, respectively. The ions were trapped using a Paul trap with helium as buffer gas. The trapped ions were extracted at 20 Hz repetition rate and accelerated to 20 keV. A magnet was employed to mass select the AMP ions which were then guided to the interaction region with a 3^0 deviation in the path for elimination of neutrals formed via background gas collisions. The ions are collided at right angles with UV photons produced from the fourth harmonic of a Nd:YAG laser flashing at half the ion bunch extraction rate to enable background subtraction.

Neutral fragments fly about 1.6 m before they hit a MCP detector. The ions are deflected away from the MCP using an electrostatic deflector and only the time of flight (TOF) of the neutral fragments created between the interaction point and the deflector is recorded.

A uniform electric field region of about 26 cm contains the interaction region. A stack of electrodes, as shown in the inset of Fig. 1 were employed for creating a homogenous electric field inside the spectrometer. The ions entering the spectrometer àre accelerated or decelerated depending on the charge and electric field. The interaction spot is about 5 cm downstream the entrance of the spectrometer and the neutral fragments formed at different times after the interaction have different velocities as the parent ion velocity varies across the spectrometer. The dissociation lifetime is hence tagged onto the TOF of the neutral fragments. All ions that fragment outside the spectrometer pile up to form a single TOF peak. For the AMP ions with 20 keV energy and about 3 kV spectrometer voltage (Vspec), dissociation times up to 2 microseconds are visible in this experiment.

As it can be seen, this setup cannot be used to mass analyse the fragments. Mass analysis of fragments would yield information on the sequential dissociation processes. However, the ionic fragment masses were analyzed at ELISA by performing UV photodissociation at 266 nm. The deflector voltages in the storage ring were switched to store the daughter ions and the mass analysis was done by scanning the ring voltages. In the current experiment the voltages were switched simultaneously with the appearance of the laser pulse. The method of daughter-mass detection at ELISA is described in detail elsewhere [32]. The fragment masses were identified using a modified linear-quadrupole-ion-trap-mass spectrometer (LTQ, Thermo Fisher Scientific, San Jose, CA) at Lyon [33]. Ions produced using an electrospray source were trapped and photoexcited by 5 shots of 266 nm photons. The fragment ions were subsequently mass analyzed after about 100 ms of trap time.

3   DISCUSSION

Figure 2 shows the TOF of neutral fragments from the UV photodissociation of AMP anions recorded with the spectrometer voltage (Vspec) off. The width of the TOF stems from the kinetic energy release, the finite interaction volume, and the spread in parent ion energy. The kinetic-energy release (KER) is estimated to be 40 meV. Figure 3 shows the TOF spectrum shifting to shorter times with Vspec raised to +3 kV, which is due to the fact that ions gain energy at the entrance of the spectrometer. The TOF peak seen at about 15.15 μs corresponds to photofragments formed outside the spectrometer. The fragments formed inside the spectrometer yield the TOF peak around 14.4 μs with a decaying tail towards the peak at 15.15 μs. The TOF section corresponding to fragmentation inside the spectrometer has a fast and a slower component. Photodetachment could deexcite the system by ejecting an energetic electron. The simulation with zero kinetic energy release in Fig. 3 shows the absense of a photodetachment pathway, indicating randomization of the absorbed energy without electron detachment. The data clearly thus indicates if there are slow and prompt processes upon photoexcitation.

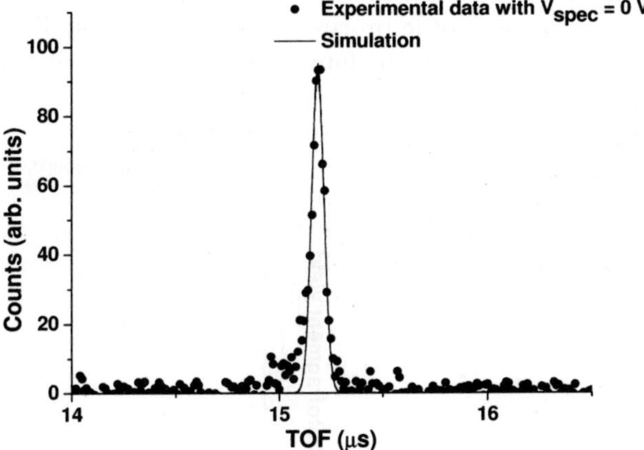

Figure 2: The TOF of the neutrals from the photodissociation of AMP anions with the Vspec switched off. The solid line is the monte carlo simulation for the time of flight (see text).

Fragmentations that occur inside the spectrometer was found to be a single-photon process as shown in Fig. 4.

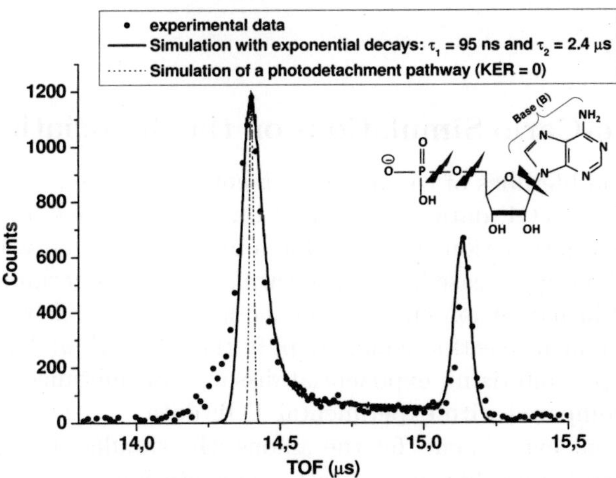

Figure 3: The TOF of the neutrals from the photodissociation of AMP anions with Vspec raised to 3 kV. The solid and dotted lines are the Monte Carlo simulations of the TOF for photodissociation and photodetachment pathways (KER = 0), respectively.

Figure 5 shows the TOF of photofragments from AMP cations measured with Vspec at +3 kV. Again, the peak with a tail corresponds to fragmentation inside the spectrometer. The spectrum shows a large fraction (94 %) of the cations fragmenting within the spectrometer correlating with the large ion depletion for cations measured at ELISA [29]. These fragments were also found to be associated with a single-photon process as shown in Fig. 4. The TOF (Fig. 5) apparently shows components in the hundreds of nanoseconds timescale.

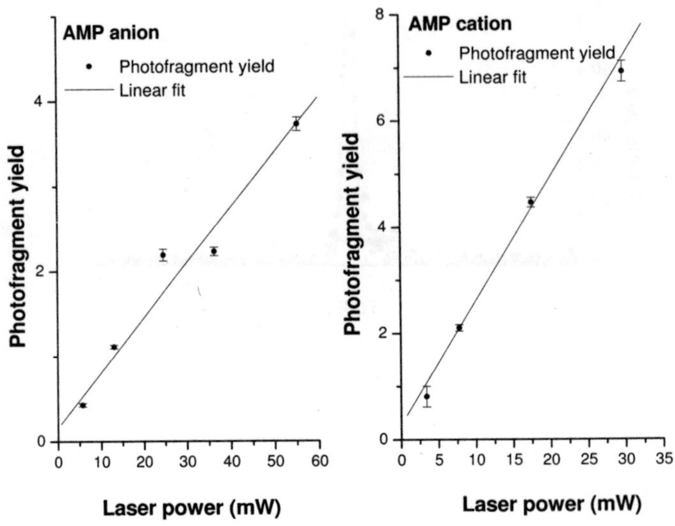

Figure 4: Laser-power dependence for the yield of fragments from AMP anion and cations that are produced within the spectrometer.

3.1   Monte-Carlo Simulations on the dissociation lifetimes

The dissociation lifetimes of the photofragmentation process could be deduced from the recorded TOF data using Monte Carlo simulations. In this we would generate random interaction events within a finite interaction volume considering the kinetic energy released, which in turn should be determined with Vspec off, the spread in parent ion energy, and the laser beam width. Assumption of an isotropic fragment ejection could simplify the simulation. The simulation is performed by paramterizing exponential dissociation lifetimes. The simulated TOF is then compared with experimental TOF data.

In the present experiment, for the anions, the simulated TOF yielded two exponential dissociation lifetimes: a fast fragmentation pathway with $\tau_1 = 95$ ns and a slower component with dissociation lifetime of $\tau_2 = 2.4$ μs as shown in Fig. 3. About 54 % of the ions fragmenting within the spectrometer have 95 ns lifetime.

The simulation for the cation (Fig. 5) shows about 74 % of ions, that fragment within the spectrometer, to have dissociation lifetime of about 640 ns and the remaining about 85 ns. Both components are too fast to be seen at the "delayed" MCP at ELISA. The large depletion of cations at ELISA previously lead to the conclusion that it could be a non-ergodic process [29]. As we can see above, the submicrosecond TOF studies reveal that the fast dissociation is indeed not prompt, but it has a lifetime much shorter than what can be mea-

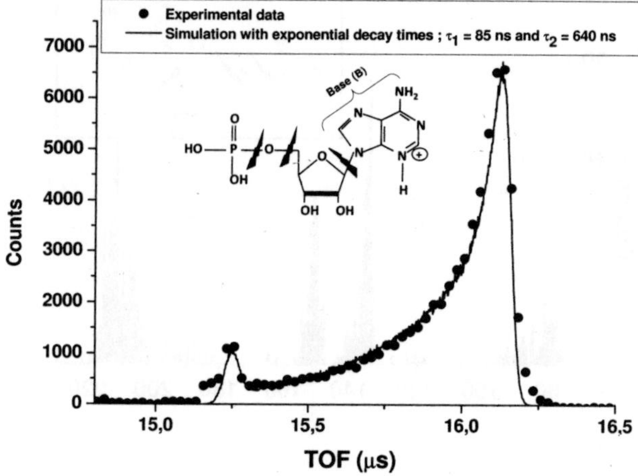

Figure 5: The TOF of the neutrals from the photodissociation of AMP cations with the Vspec raised to 3 kV. The solid line is the Monte Carlo simulation of the TOF for the photodissociation pathways.

sured in ELISA. Clearly the processes are ergodic and fast enough ($\tau_2 = 0.64$ μs) to go undetected at the delayed detector at ELISA [29].

3.2 Fragment masses of the AMP anion

As mentioned earlier, mass analysis of the fragments could be done at ELISA. The is done by switching the deflector voltages along with the laser trigger and only fragments produced faster than about 15 μs would be stored in the ring. The daughter ions are then collected after storing them for two revolutions. Fig. 6 shows the daughter mass spectra obtained at ELISA (top panel) along with the high resolution data (bottom panel) obtained using the mass spectrometer in P. Dugourd's laboratory at Lyon [33]. It is instructive to compare the present results on photofragments (see Fig. 6) and that of Marcum *et al.* [24]. The fragment spectra implies variety of fragmentation pathways. The two dissociation lifetimes that we deduced are the slow and fast dissociation lifetime regimes of these pathways.

UV photoabsorption at the nucleobase leads to π-π^* excitation which could undergo ultrafast conversion to the ground electronic state via a conical intersection. The excess energy would then randomize, eventually leading to various fragmentation pathways. A prompt dissociation channel from the excited electronic state is also possible but with minor contribution since we see no sign of it in the TOF. Prompt fragmentation pathways would be fatal as biomolecules would then fragment before the excess energy is exchanged with the surrounding environment.

Several mechanisms were proposed to explain collision induced dissociation of AMP anions [9, 34, 35]. The AMP anion, which is of biological relevance, is known to dissociate prominantly at the phosphate-ribose bond yielding

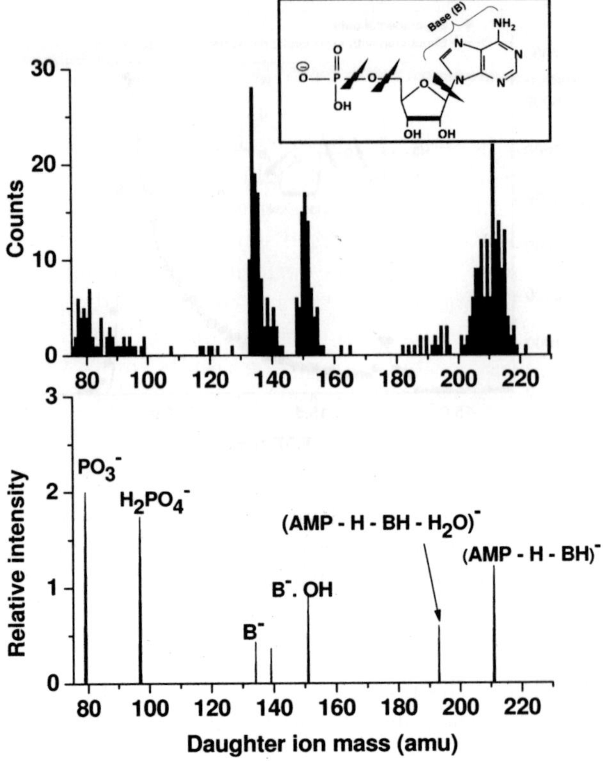

Figure 6: The daughter-ion mass spectrum obtained from UV photodissociation of the AMP anion at 266 nm performed at ELISA is shown in the top panel. The assignment of the ionic fragments is provided from a high resolution mass spectrometer [33] (bottom panel).

PO_3^- [9] in CID. The direct cleavage of the glycosidic C-N bond, following the photoexcitation of the base could be the fast (95 ns) dissociation component observed in the current experiment. The other fragments that are formed would require nuclear rearrangements after relaxation of the excited base and hence is likely to be the observed slower dissociation component (2.4 μs). Ho *et al.* [9] discuss an intermediate hydrogen-bond structure, formed after excitation, between the phosphate at the 3'-position and the base, with the negative charge on the base (B^-). This hydrogen-bonded structure could either undergo proton transfer to B^- expelling a neutral base (BH) or breakage of the hydrogen bond yielding a deprotonated base (B^-). B^- could be held with the remnant neutral to form an ion-dipole complex and the lifetime of dissociation to form B^- would depend on the shallow barrier holding the complex [9]. Cleavage of the strong hydrogen bond could be the rate determining step in this process. Such a process is also plausible through an intermediate with a H-bond between the base and the phospate in 5'-position.

Summary

Application of advanced techniques in atomic and molecular physics to biosystems were discussed by highlighting the significance of gas-phase studies on biosystems. UV Photodissociation lifetime studies on biologically relevant molecules namely the AMP was discussed in the light of radiation damage. Application of techniques such as Electrospray ion source, Iontrap and linear TOF for deducing sub-microsecond lifetimes were discussed. UV photodissociation of gas phase AMP cations and anions was performed at 266 nm. Dissociation lifetimes were observed for both ions in the sub-microsecond regime corresponding to ergodic dissociation. The photofragmentation was determined to be due to a single photon process for both ions. The results imply that the AMP ions can dissipate excess energy to the biological environment before they are damaged. The various dissociation lifetimes, observed over a wide timescale, imply complex dissociation mechanisms.

References

[1] D. P. Little, J. P. Speir, M. W. Senko, P. B. O'Connor and F. W. McLafferty *Anal. Chem.* (1994) **66**, 2809.

[2] R. A. Zubarev, N. A. Kruger, E. K. Fridriksson, M. A. Lewis, D. M. Horn, B. K. Carpenter, and F. W. McLafferty *J. Am. Chem. Soc.* (1999) **121**, 2857.

[3] S. B. Nielsen, A. Lapierre, J. U. Andersen, U. V. Pedersen, S. Tomita, and L. H. Andersen *Phys. Rev. Lett.* (2001) **87**, 228102-1.

[4] P. Hvelplund, B. Liu, S. B. Nielsen, S. Panja, J-C. Poully, K. Støchkel. *Int. J. Mass. Spectrom.* (2007) **263**, 66.

[5] L. Lammich, M. Å. Petersen, M. B. Nielsen, and L. H. Andersen *Biophys. J.*(2007) **92**, 201.

[6] L. H. Andersen, H. Bluhme, S. Boyé, T. J. D. Jørgensen, H. Krogh, I. B. Nielsen, S. B. Nielsen, and A. Svendsen *Phys. Chem. Chem. Phys.* (2004) **6**, 2617.

[7] L. H. Andersen, I. B. Nielsen, M. B. Kristensen, M. O. A. El Ghazaly, S. Haacke, M. B. Nielsen, and M. Å. Petersen. *J. Am. Chem. Soc.* (2005) **127**, 12347.

[8] J. A. Wyer, A. Ehlerding, H. Zettergren, M-B. S. Kirketerp, and S. B. Nielsen. *J. Phys. Chem. A* (2009) **113**, 9277.

[9] Y. Ho and P. Kebarle *Int. J. Mass Spectrom. Ion Processes* (1997) **165/166**, 433.

[10] M. T. Rodgers, S. Campbell, E. M. Marzluff, and J. L. Beauchamp *Int. J. Mass Spectrom. Ion Processes* (1995) **148**, 1.

[11] J. Yang and K. Håkansson *J. Am. Soc. Mass Spectrometry* (2006) **17**, 1369.

[12] B. B. Brady, L. A. Peteanu, and D. H. Levy. *Chem. Phys. Lett.* (1988) **147**, 538.

[13] N. J. Kim, G. Jeong, Y. S. Kim, J. Sung, S. K. Kim, and Y. D. Park *J. Chem. Phys.* (2000) **113**, 10051.

[14] A. L. Sobolewski and W. Domcke *Eur. Phys. J. D* (2002) **20**, 369.

[15] N. Ismail, L. Blancafort, M. Olivucci, B. Kohler, and M. A. Robb *J. Am. Chem. Soc.* (2002) **124**, 6818.

[16] Chr. Plützer, E. Nir, M. S. de Vries, and K. Kleinermanns *Phys. Chem. Chem. Phys.* (2001) **3**, 5466.

[17] J.-M. L. Pecourt, J. Peon, and B. Kohler. *J. Am. Chem. Soc.* (2001) **123**, 10370.

[18] C. E. Crespo-Hernández, B. Cohen, P. M. Hare, and B. Kohler *Chem. Rev.* (2004) **104**, 1977.

[19] J. W. Longworth, R. O. Rahn and R. G. Shulman *J. Chem. Phys.* (1966) **45**, 2930.

[20] M. Daniels and W. Hauswirth *Science* (1971) **171**, 675.

[21] P. R. Callis *Ann. Rev. Phys. Chem.* (1983) **34**, 329.

[22] V. M. Belyakova and V. L. Rapoport *J. Photochem. Photobiol. B: Biol.* (1993) **19**, 105.

[23] Z. Guan, N. L. Kelleher, P. B. O'Connor, D. J. Aaserud, D. P. Little, and F. W. McLafferty *Int. J. Mass Spectrom. Ion Processes* (1996) **157/158**, 357.

[24] J. C. Marcum, A. Halevi and J. M. Weber *Phys. Chem. Chem. Phys.* (2009) **11**, 1740.

[25] J. B. Fenn, M. Mann, C. K. Meng, S. F. Wong, and C. M. Whitehouse *Mass Spectrom. Rev.* (1990) **9**, 37.

[26] R. Wester *J. Phys. B: At. Mol. Opt. Phys.* (2009) **42**, 154001.

[27] G. Aravind, R. Antoine, B. Klaerke, J. Lemoine, A. Racaud, D. B. Rahbek, J. Rajput, P. Dugourd, and L.H. Andersen *Phys. Chem. Chem. Phys.* (2010) **12**, 3486

[28] H. Bluhme, M. J. Jensen, S. B. Nielsen, U. V. Pedersen, K. Seiersen, A. Svendsen and L. H. Andersen *Phys. Rev. A.* (2004) **70**, 020701.

[29] S. B. Nielsen, J. U. Andersen, J. S. Forster, P. Hvelplund, B. Liu, U. V. Pedersen, and S. Tomita *Phys. Rev. Lett* (2003) **91**, 048302-1.

[30] G. Aravind, L. Lammich, and L. H. Andersen *Phys. Rev. E.* (2009) **79**, 011908.

[31] H. B. Pedersen, M. J. Jensen, C. P. Safvan, X. Urbain, and L. H. Andersen *Rev. Sci. Instrum.* (1999) **70**, 3289.

[32] K. Støchkel, U. Kadhane, J. U. Andersen, A. I. S. Holm, P. Hvelplund, and M-B. S. Kirketerp, M. K. Larsen, M. K. Lykkegaard, S. B. Nielsen, S. Panja, S. and H. Zettergren *Rev. Sci. Instrum.* (2008) **79**, 023107.

[33] V. Larraillet, R. Antoine, P. Dugourd and J. Lemoine *Anal. Chem.* (2009) **81**, 8410.

[34] R. L. Cerny, M. L. Gross, and L. Grotjahn *Anal. Biochem.* (1986) **156**, 424.

[35] D. R. Phillips and J. A. McCloskey *Int. J. Mass Spectrom. Ion processes* (1993) **128**, 61.

[36] K. H. Kraemer *Proc. Natl. Acad. Sci. U.S.A.* (1997) **94**, 11.

A Route to Bose-Einstein Condensate and Current Status

Yeshpal Singh[*] **and Kai Bongs**

*Department of Physics and Astronomy, University of Birmingham
Edgbaston Road, B15 2TT Birmingham (UK)*
e-mail: y.singh.1@bham.ac.uk[*]

1. INTRODUCTION

In 1924, de Broglie established that there is a wave associated with every kind of matter. If p is momentum of a particle then the wavelength λ of the associated wave is given by the following relation-

$$\lambda = \frac{h}{p} \tag{1}$$

Where h is Planck's constant. In fact, the wave nature of matter is a key aspect to the physics of ultra cold atoms. For bosonic particles if λ is equal to or more than roughly the inter particle distance, a phase transition to a so called Bose-Einstein Condensate (BEC) takes place, where a macroscopic number of particles occupy the same quantum state and consequently form a coherent giant matter wave. This phase transition was predicted by S. N. Bose and A. Einstein in 1924-1925 but realized only in 1995 by E. Cornell and C. Wieman (JILA), and W. Ketterle (MIT), for which the latter three were awarded with the Nobel Prize in physics in 2001.

If n is the density of a gas then $n^{-1/3}$ is a measure of the inter particle separation. Therefore the BEC transition takes place when

$$\lambda \sim n^{-1/3}$$

Or more precisely if

$$n\lambda^3 \geq 2.612 \tag{2}$$

$n\lambda^3$ refers to the phase-space density of the gas. The thermal deBroglie wavelength, λ depends on the temperature of the gas in the following way-

$$\lambda = \frac{h}{\sqrt{2\pi m k_B T}} \tag{3}$$

Here m is the mass of an atom, k_B is Maxwell-Boltzmann constant and T is the temperature of the gas. From Eqns (2) and (3), it is obvious that it is the density-temperature combination that is important in order to achieve a BEC. The temperature at which a BEC transition takes place is called critical temperature. This suggests that by choosing a high density, one can bring the critical temperature in a regime which is easily accessible in a lab. However, the basic obstacle to this approach is that before one reaches the critical temperature, molecule/cluster/solidification takes place and hence no BEC. For instance, though in case of water, the critical temperature is in K regime, at 0°C (273K) water turns into ice phase. Therefore, one has to work in a metastable regime at ultra low densities (10^{13}-10^{14} per cm^3), where three-body collisions are dynamically suppressed as compared to two-body collisions. Energy and momentum conservation require three-body collisions in order to turn a gas into a liquid or solid, while two-body collisions are sufficient to thermalize the energy distribution in the gas phase. The low densities imply ultra low temperatures in the range of μK-nK in order to achieve a BEC. In a typical experiment, these temperatures are achieved with a combination of laser cooling followed by evaporative cooling.

2.1. LASER COOLING

The ground breaking developments in laser cooling in 1980's inspired a hot pursuit of realizing a BEC in a lab. The basic idea of laser cooling is that you extract more energy out of an atom by the emission of a photon than that given to it in the absorption of a photon (Fig 1).

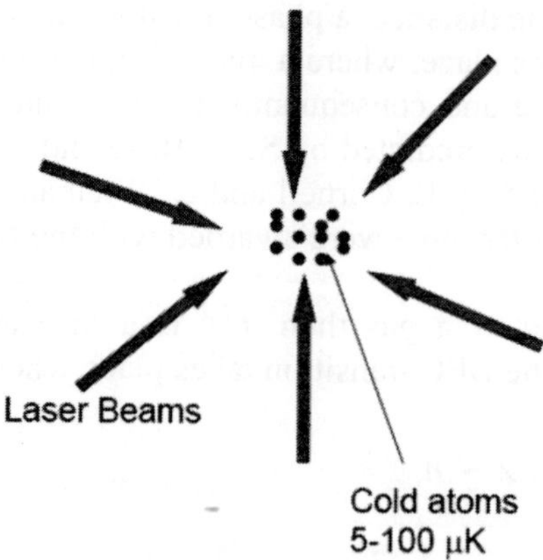

Fig. 1. Laser Cooling. Cloud of thermal atoms interacting with three mutually perpendicular pairs of laser beams.

If the absorbed energy ($\nabla\omega_a$) is less than the emitted energy ($\nabla\omega_e$), the atom loses energy equal to $\nabla(\omega_e - \omega_a)$ resulting in cooling of the gas. In the following, we discuss the interaction of an atom with light for two cases-1) single plane wave; 2) counter propagating planes waves which give rise to Doppler cooling. The approach we take is semi classical, where the centre of mass of the atom behaves like a classical particle.

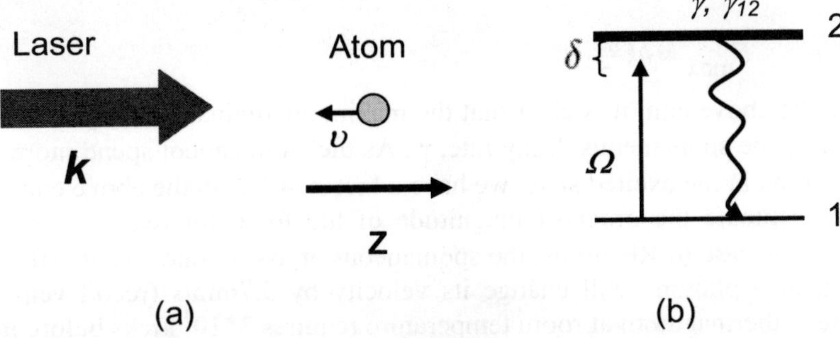

Fig. 2. Two Level Atom and Plane Wave. (a) An atom moving towards a laser beam. **(b)** Schematic level diagram of a two level atom interacting with a red detuned beam.

Case 1. Plane wave: Consider a two level atom interacting with a plane wave with a wave vector k. Suppose that v is the velocity of the atom, Ω ($= - \mu_e.E_0/V$; μ_e is electric dipole moment and E_0 is amplitude of the electric field) is the Rabi frequency, δ ($=\omega_L-\omega_a$; where ω_L and ω_a are laser and atomic resonance frequencies respectively) is the detuning from resonance, γ is the spontaneous emission rate and γ_{12} ($= \gamma_{21}$) is the generalized dephasing rate in the sense that this also includes the effects such as finite line width of an applied laser, atom-atom interaction etc (Fig 2). For a completely isolated atom we have $\gamma_{12} = \gamma/2$.

1) if $|\Omega| << \gamma$, the dissipative force F on the atom is given by [1]-

$$\vec{F} \approx \hbar\vec{k}\,\frac{2\gamma_{12}|\Omega|^2}{\gamma_{12}^2 + (\delta - kv)^2} = \hbar\vec{k}R_{opt} \qquad \text{for} \quad R_{opt} << \gamma \qquad (4)$$

Where $R_{opt} = \dfrac{2|\Omega|^2\gamma_{12}}{\gamma_{12}^2 + \delta^2}$ is called the optical pumping rate.

The physical interpretation of eqn (4) is that each time a photon is absorbed, the atom gets a momentum kick ∇k in the direction of the incoming photon. When the atom decays to the ground state, it randomly emits a photon into a solid angle 4π, such that the recoil momenta average to zero over many emission processes. So on average, motional state of the atom changes by ∇k during complete cycle of absorption and emission. However the absorption and re-emission takes place at the optical pumping rate R_{opt}, so the rate of change in momentum must be ∇kR_{opt} which is nothing but the force F on the atom.

2) if $|\,\Omega\,| >> \gamma$, the dissipative force F on the atom is given by-

$$\vec{F} \approx \hbar\vec{k}\,\frac{\gamma R_{opt}}{\gamma + 2R_{opt}} \tag{5}$$

If $R_{opt} \rightarrow \infty$, the force F is maximum on the atom. So

$$\vec{F}_{max} \approx \hbar\vec{k}\,\frac{\gamma}{2} \tag{6}$$

From the above eqn, it is clear that the maximum attainable dissipative force is limited by the spontaneous decay rate, γ. As the atom cannot spend more than half of its time in the excited state, we have a factor of $\gamma/2$ in the above eqn.

Let us estimate the order of magnitude of the force for real systems. For instance in the case of Rb atoms, the spontaneous emission rate γ is 36 MHz and each kick of a photon will change its velocity by 5.7mm/s (recoil velocity). Therefore, a thermal atom at room temperature requires $3*10^4$ kicks before it can be slowed down to rest. The time required for this slowing process is roughly 1 ms and the maximum deceleration experienced by the atom is approximately 10^5 m/s^2 which is approximately 10^4 times the acceleration due to Earth's gravity.

Case2. Counter Propagating Beams (Optical Molasses)- As we shall see laser cooling creates a   velocity dependent force on the atom, which around zero linearly increases with the velocity. In fact we are looking at an optical analogue of viscosity. To understand the origin of optical viscosity let us expand the force due to one beam in powers of velocity:

$$\vec{F}_{+} \approx \hbar\vec{k}\,\frac{2\gamma_{12}|\Omega|^2}{\gamma_{12}{}^2 + \delta^2}\left(1 + \frac{2\delta\vec{k}\bullet\vec{v}}{\gamma_{12}{}^2 + \delta^2}\right) \quad \text{for} \quad \vec{k}\bullet\vec{v} << \delta,\gamma_{12} \tag{7}$$

In eqn (7), the first term is velocity independent whereas the second term is velocity dependent. So the velocity dependent force originates from the second term. One should also notice that the first term depends on **k**, while the second term depends on k^2. Therefore, applying a counter propagating beam will only cancel the first term and consequently we will be able to retain the velocity dependent cooling term (Fig 3). The total force on the atom is-

$$\vec{F} = \vec{F}_{+} + \vec{F}_{-} = \eta_D m v_z \tag{8}$$

Where $\eta_D = \dfrac{8\hbar k^2}{m}\,\dfrac{\delta\gamma_{12}|\Omega|^2}{\left(\gamma^2{}_{12} + \delta^2\right)^2}$ is called friction coefficient for Doppler

cooling, m is the mass of the atom and v_z is the z-component of the atom's velocity. One should notice that η_D depends on the sign of detuning δ (=ω_L-ω_a). If $\delta<0$, the viscous force will cool the atom cloud while if $\delta>0$, it will heat the cloud. For $\delta<0$, due to the change in Doppler shifts, an atom absorbs a photon with lower energy than the consecutively emitted photon on average. In this way, the atom loses its kinetic energy with every cycle of absorption and emission. In case of $\delta>0$, the opposite is true. One can interpret the above eqn (8) in the following way. When $\delta<0$, atoms are shifted closer to the resonance with the laser beam opposing its direction of motion and absorb more photons from this beam

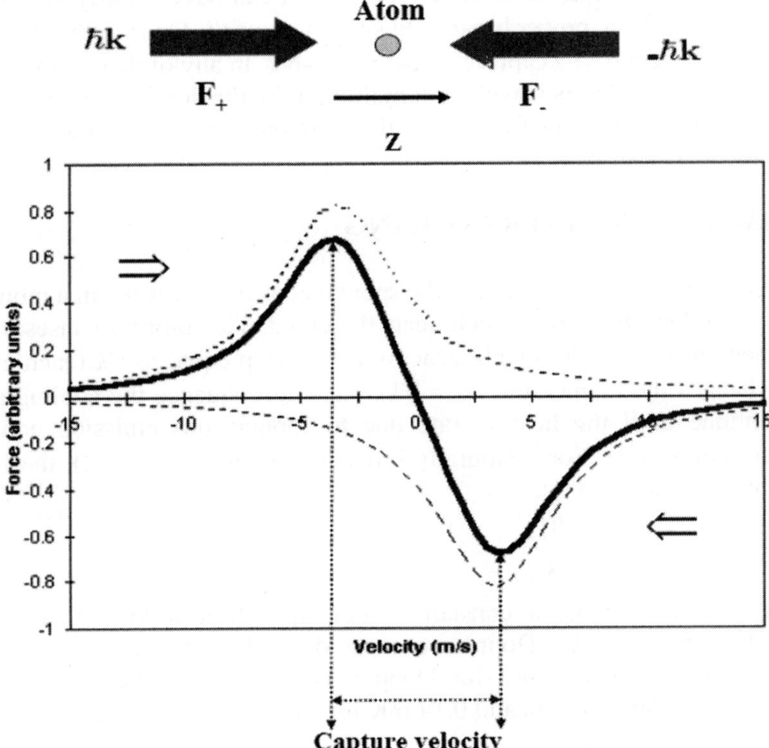

Fig 3. Doppler Cooling Force. A moving atom interacts with two red detuned and counter-propagating laser beams. The dotted/dashed line shows the force due to the right("-")/left ("+") beam and the solid line shows the net force on the atom. Note that for a typical alkali atom, Doppler cooling has a capture velocity of a few m/s and the force around zero velocity is linear in velocity.

than from the other one – thus reducing its velocity. For $\delta>0$ the opposite is true and the sample gets heated. It is worth mentioning here that in the above analysis, we have considered force only in one dimension Z; if we want to cool the atoms in all directions, we must apply 2 additional pairs of counter propagating beams in the other orthogonal directions. Such a viscous cooling in 3D is called optical molasses For small velocities (Figs 3, 10) the force varies linearly with the velocity of the atom, allowing to define a friction coefficient [2]. As the velocity increases, the friction coefficient becomes velocity dependent and one must consider the full expression for the Doppler force. In fact, for $\vec{k} \cdot \vec{\upsilon} \gg \delta, \gamma_{12}$ the force vanishes. The range of velocities over which the total force is linear, the atomic motion is most effectively damped by this cooling mechanism. However,   the critical velocity ($\pm\upsilon_c$) at which the force is maximum is often used as a simple criterion to estimate the fraction of atoms to be cooled by Doppler cooling. The velocity υ_c is called capture velocity and the range $-\upsilon_c \leq \upsilon \leq \upsilon_c$ is called capture range. Though, in general, the capture range depends on many parameters, in certain cases it can have a fairly simple form. For a strongly driven, power broadened system, $\upsilon_c \sim \Omega/k$. On the other hand, for a far detuned system $\delta \gg \gamma$ capture velocity, $\upsilon_c \sim \delta/k$. In any of these situations, the capture velocity scales as wavelength multiplied by the dominant rate. It must be mentioned that the viscous force cools the atom but can not trap it as the force is not position dependent.

2.2. LIMIT OF DOPPLER COOLING

In each absorption-emission cycle, the spontaneous decay results in a momentum kick in a random direction, which heats the cloud. This process causes a spread in momentum in a random walk process with a step equal to ∇k (each event of decay changes the momentum $\delta p = \nabla k$). As a consequence the cooling process will continue until the heating rate due to spontaneous emission equals the Doppler cooling rate. For a naturally broadened system ($\gamma_{12} = \gamma/2$), the limiting temperature T_D is given by:

$$T_D = \frac{\hbar\gamma}{2k_B} \tag{9}$$

Here k_B is the Boltzmann constant. From eqn (9) it follows that the final temperature achieved by Doppler cooling depends only on the spontaneous emission rate. Typical values for Doppler cooling are in the mK range. For example it is 0.25mK for Na and 0.14 mK for Rb.

2.3. SUB-DOPPLER COOLING

It was assumed that the Doppler cooling limit is the fundamental limit for laser cooling. However, in the early 1990's, experiments on alkali atoms (Na, Rb, Cs) showed much lower temperatures than the Doppler limit. This was a big puzzle as the Doppler cooling limit seems quite intuitive and perfectly reasonable. Dopper cooling is based on the classical motion of an atom in the radiation which is quite valid as at T_D, where the deBroglie wavelength of a typical atom is about 10 nm which much smaller than the wavelength of light. So where have we gone wrong? We have gone wrong in the assumption that an atom has only two levels, while in reality atoms possesses a multi-level structure.

The explanation for the lower temperatures came one year later [3]. It was motivated by two experimental facts- one that the temperature depended sensitively on the polarizations of the applied lasers; second that the detuning of the applied lasers was also an important parameter. It was found that for $\delta \gg \gamma$, it was easier to achieve sub-Doppler temperatures. As a far detuned laser induces significant AC stark shifts of the energy levels of the atom this experimental result suggested that AC Stark shifts should also be included in the theory. There are mainly two ways in which one can obtain sub-doppler cooling which are as follows:

1) "lin $\perp$ lin" Configuration (Sisyphus Cooling)- We consider two counter propagating linearly polarized beams such that their polarizations are orthogonal to each other. Suppose ε refers to the polarization of a wave, E_0 is amplitude of the wave and if the waves are traveling along Z-axis. Then the net field along the Z-axis is given by-

$$\vec{E} = E_0 \vec{\varepsilon}_x e^{i(kz + \omega t)} + iE_0 \vec{\varepsilon}_y e^{i(-kz + \omega t)} + cc$$

or

$$\vec{E} = E_0 e^{i\omega t}\left\{\left(\vec{\varepsilon}_x + i\vec{\varepsilon}_y\right)\cos kz + i\left(\vec{\varepsilon}_x - i\vec{\varepsilon}_y\right)\sin kz\right\} + cc$$

$$\qquad\qquad\qquad\qquad \sigma_+ \qquad\qquad\qquad\qquad \sigma_-$$

(10)

The electric field is the sum of two cross-circularly polarized standing waves which are shifted by 90^0 from one another. Eqn (10) says that the superposition of the two fields yields a polarization gradient such that after every $\lambda/8$ distance, the polarization changes from linear to circular. It is important to notice that at every $\lambda/4$ interval, circular polarization changes from σ_- to σ_+ or vice-versa (Fig 4).

In order to properly understand the mechanism of Sisyphus cooling, let us consider an atom with a specific level structure and a transition from $\mathbf{J} = 1/2 \rightarrow \mathbf{J'} = 3/2$ (see Fig. 5). When such an atom is subjected to circularly polarized light σ_+ with Rabi frequency Ω and far detuned such that $\gamma \gg \Omega$, AC Stark shifts have an important effect on it. Since Clebsch-Gordon coefficients for the two allowed transitions m $= -1/2 \rightarrow$ m $=+1/2$, m $= +1/2 \rightarrow$ m $= +3/2$ are different, the AC Stark shift is also different for the two ground states-

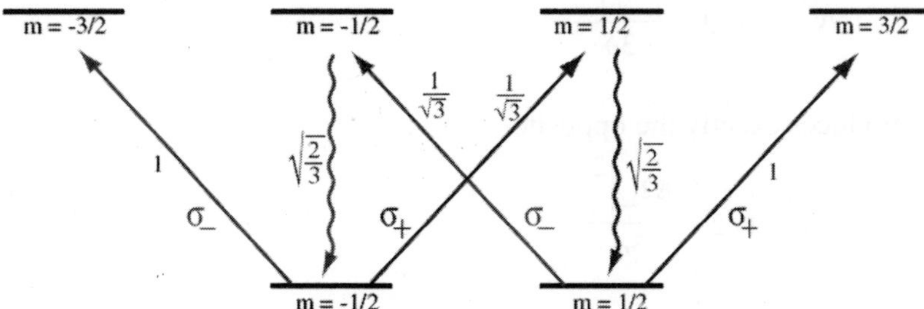

Fig. 5. Multilevel Atom Scheme for Sub-Doppler Cooling. For each transition, the Clebsch-Gorden coefficients are written. Notice that the m $= \pm 1/2 \rightarrow$ m $= \pm 1/2$ transitions are coupled only by spontaneous emission.

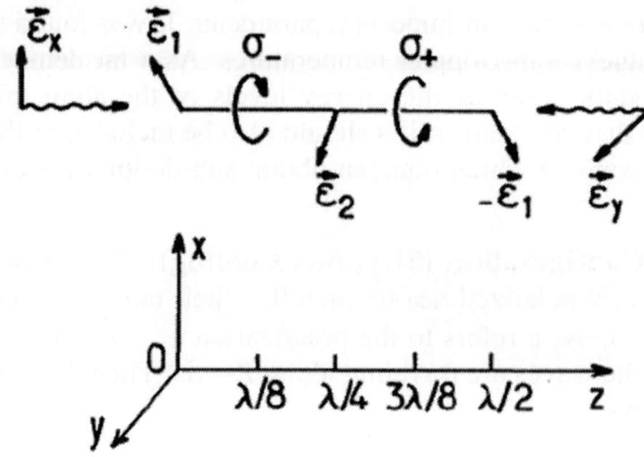

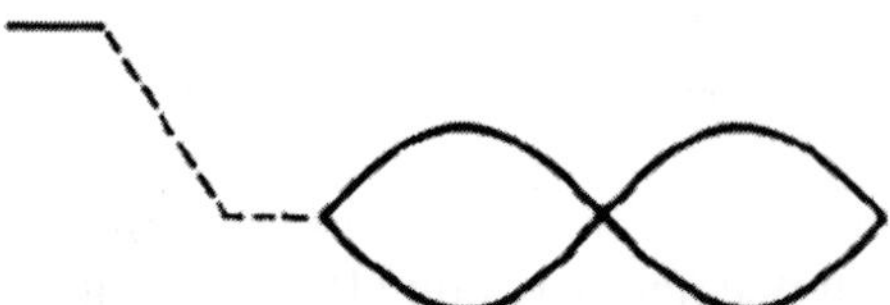

Fig. 4. Polarization Gradients in "lin ⊥ lin" Case. **(Above)** The two counter propagating waves have orthogonal linear polarizations which result in a polarization gradient in such a way that antinodes of $\sigma_\pm$ polarization are separated by $\lambda/2$ in space. **(Below)** The light shifted energies of different ground state m_F levels oscillate in space with a period $\lambda/2$. The figure has been adapted from the paper [3].

$$\Delta^{\sigma_+}{}_{1/2} = \frac{|\Omega_+|^2}{\delta} \tag{11}$$

$$\Delta^{\sigma_+}{}_{-1/2} = \frac{|\Omega_+|^2}{3\delta} \tag{12}$$

σ_- light produces exactly the opposite-

$$\Delta^{\sigma_-}{}_{1/2} = \frac{|\Omega_-|^2}{3\delta} \tag{13}$$

$$\Delta^{\sigma_-}{}_{-1/2} = \frac{|\Omega_-|^2}{\delta} \tag{14}$$

We know from eqn (10) that the lin⊥lin configuration results in two cross-circularly polarized standing waves, therefore we now have spatially dependent Rabi frequencies-

$$\Omega_+ = \Omega \cos kz \tag{15}$$

$$\Omega_- = \Omega \cos kz \tag{16}$$

For an atom moving through a far detuned laser field, the laser-atom interaction can be written as-

$$\hat{H}_{AL}^{eff} = \hbar \sum_{m = \pm\frac{1}{2}} \left(\Delta_m^{\sigma_+} + \Delta_m^{\sigma_-} \right) |m\rangle\langle m| \tag{17}$$

$$= \hbar \frac{|\Omega|^2}{\delta}\left(1-\frac{2}{3}\sin^2 kz\right)\left|\frac{1}{2}\right\rangle\left\langle\frac{1}{2}\right| + \hbar\frac{|\Omega|^2}{\delta}\left(1-\frac{2}{3}\cos^2 kz\right)\left|-\frac{1}{2}\right\rangle\left\langle-\frac{1}{2}\right|$$

If one moves along z axis, the ground state energies of an atom vary sinusoidally in space (see Fig. 4). As the energies vary in space and we are also far detuned, one obtains a force which is conservative in nature. However, a conservative force does not lead to cooling. So then how do we get cooling in lin⊥lin case? There is one more thing that is to be taken into account, which is optical pumping. From Fig. 5, it is evident that σ_+ light optically pumps atoms into m = +1/2 state while σ_- light optically pumps the atoms in m = -1/2 state. The point is that the optical pumping between the two sublevels of the ground state takes a finite time, say τ_P.

Suppose that the atom is initially at the bottom of a potential valley, say on the $E_{-1/2}$ curve (Fig 6) and suppose that the atom is moving to the right. If the velocity of the atom is such that it moves a distance of roughly $\lambda/4$ during the optical pumping time τ_P, it remains on the same energy level and climbs to the top of the potential hill. At this point it sees that the dominant polarization is σ_+ which optically pumps it into m = +1/2 state. In this process it loses some part of its kinetic energy. Now the atom starts to move on $E_{1/2}$ curve and climbs the potential hill. But this time it sees σ_- as the dominant light which optically pumps it into m = 1/2 and therefore again the atom loses some part of its energy. This process repeats and the atom continues to lose its kinetic energy as a consequence.

2) σ_+-σ_- Configuration: In case of two counter propagating beams with respective polarizations as σ_+ and σ_-, the superposition of the fields lead to rotation of polarization in space (see Fig. 7). In this case, energy of the ground state does not vary in space. Therefore the cooling mechanism is different from Sisyphus cooling. For a more detailed explanation, the reader is advised to refer to [3].

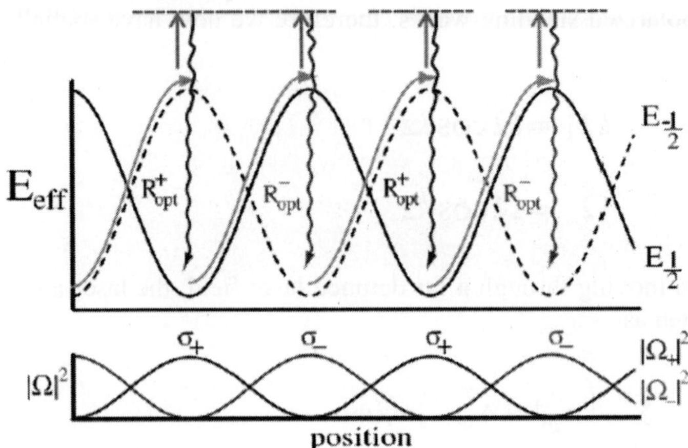

Fig 6. Sisyphus Cooling Mechanism. As an atom moves through the polarization gradient of the laser field, its internal energy level (green) follows the Stark shifted ground state. In regions where $\sigma_{\pm}$ light dominates, the atom is optically pumped into the m = ±1/2 state. For a red detuned laser, the atom is always optically pumped into the lower energy state, so it must lose kinetic energy with each cycle.

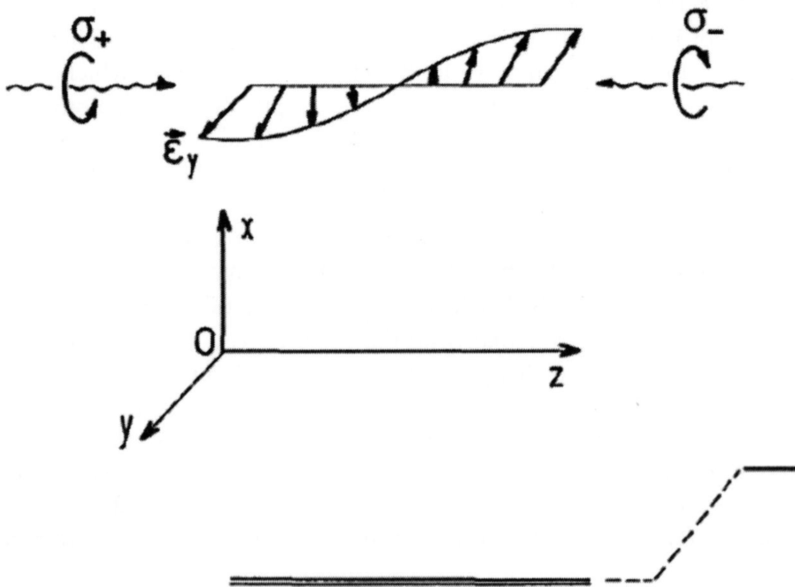

Fig. 7. Polarization gradient in σ_+-σ_-. Configuration. (Above) Two counter-propagating waves with σ_+ and σ_- polarizations respectively produce a linear polarization that rotates in space. **(Below)** For the "σ_+ and σ_-" case, the Stark shifted ground state energies do not vary in space. The figure has been adapted from the paper [3].

2.4. COMPARASION OF DOPPLER AND SISYPHUS

1) Viscosity-

In order to compare the two cooling schemes, it is instructive to look at their friction coefficients. Suppose that η_D and η_S are frictions coefficients for Doppler and Sisyphus cooling respectively, then in the limit of small powers and large detunings :

$$\eta_D = \frac{\hbar k^2}{m}\frac{\gamma_{12}|\Omega|^2}{\delta^3} \tag{18}$$

$$\eta_S = 3\frac{\hbar k^2}{m}\frac{\delta}{\gamma} \tag{19}$$

η_D depends on both laser power and detuning while η_S is totally independent of the power. η_D vanishes for weaker intensity and large detuning. On the other hand, η_S increases as the detuning increases. Therefore in the regime of a far detuned laser field, Sisyphus cooling is more efficient as $\eta_S \gg \eta_D$.

2) Capture Velocity: We have seen above that for weaker fields, the Doppler capture velocity $\upsilon_C{}^D \sim \Omega/k$ or $\upsilon_C{}^D \sim \delta/k$. However, calculations show that the Sisyphus capture velocity $\upsilon_C{}^S$ scales with AC Stark shift such that $\upsilon_C{}^S \sim \lambda|\Omega|^2/(18\pi\delta)$. As $\delta \to \infty$, the friction coefficient, η_S grows but unfortunately $\upsilon_C{}^S$ vanishes. So the higher friction in case of Sisyphus cooling comes at a price that the capture velocity is very small. Hence, Sisyphus cooling is very efficient only for velocities close to zero (see Fig. 8).

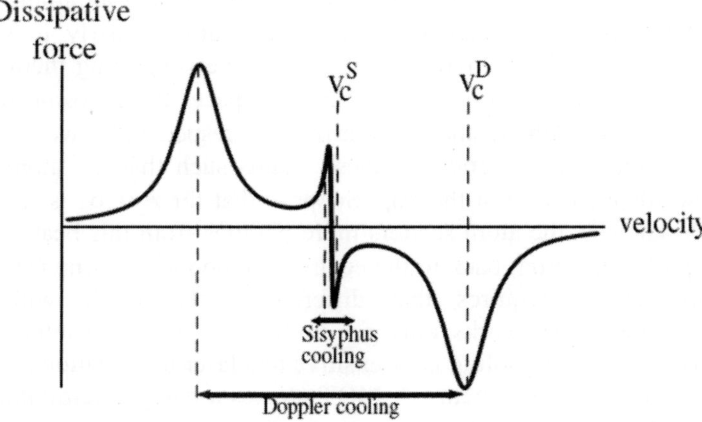

Fig 8: Doppler Cooling vs Sisyphus Cooling. The Doppler cooling has much larger capture velocity and hence can cool relatively hotter atoms to its limit. On the other hand, Sisyphus cooling provides much larger viscosity (slope at zero velocity) resulting in much lower temperatures in the end.

2.5. LIMIT OF SISYPHUS COOLING

From our above discussion, we know that Sisyphus cooling works when an atom has to climb the potential. In this process each time, the atom loses some part of its kinetic energy. Once the kinetic energy is so low that it can no longer make up to the top of the potential hill, Sisyphus cooling ceases to exist. So this kinetic energy sets the limit on Sisyphus cooling and the atom can be considered to be confined essentially to one potential well, the momentum spread can be given by $\delta p \approx 2\pi \nabla/\lambda$. But we know that $\delta p = \nabla k$ (recoil momentum) with the corresponding recoil energy $\nabla \omega_R = (\nabla k)^2/2m$. So the minimum kinetic energy when Sisyphus cooling is no more effective is given by-

$$E_{min}^{Kinetic} \sim \nabla \omega_R \tag{19}$$

For Na and Rb the corresponding temperatures T_R are 2.4μK and 0.37μK respectively. These temperatures are much lower than Doppler temperatures.

3. MAGNETO-OPTICAL TRAP (MOT)

So far we have seen that using lasers alone we are able to cool atoms down to recoil velocities, but we can not trap them. Therefore, we require something more. If we combine the laser field with a magnetic quadrupole field we can cool as well as trap the atoms. This is precisely the reason why it is called Magneto-Optical trap (MOT). We know that the magnetic field induces a Zeeman shift in the atom and the Doppler force depends on the detuning. The basic idea of the MOT is to combine optical molasses with a magnetic field gradient such that the detuning depends on space as well on velocity (see Figs 9 & 10). One of the simplest ways to create the gradient is to use two circular coils in an anti-Helmholtz configuration (Fig 9). Let us consider an atom with a **J=0** →**J'=1** transition and a magnetic field which varies linearly in space. The magnetic field lifts the degeneracy of the **J'=1** levels splitting them into **m=0, ±1** states whose energy also varies linearly in space. By choosing the counter propagating beams with σ_+ and σ_- polarizations respectively, we can cause an imbalance in the forces exerted by these beams such that the atom is always pushed towards the centre of the trap. Suppose that for z<0, σ_+ is resonant with the m=+1 state (i.e. the atom scatters more photons from this beam), then F_+ > F_-, which pushes the atom back to the centre. The opposite is true for σ_- for z>0. Since this scheme requires that different levels couple with different polarizations the MOT works only for multi-level atoms. Another important thing is that as Doppler cooling is insensitive to a laser polarization a MOT cools the atoms as well as traps them. A MOT creates a damped harmonic potential well for the atoms.

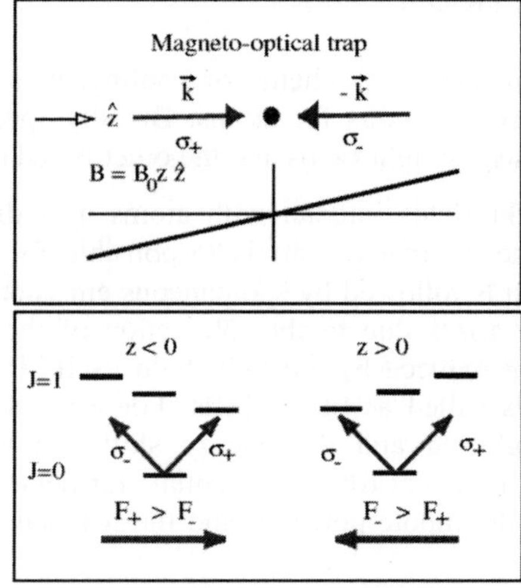

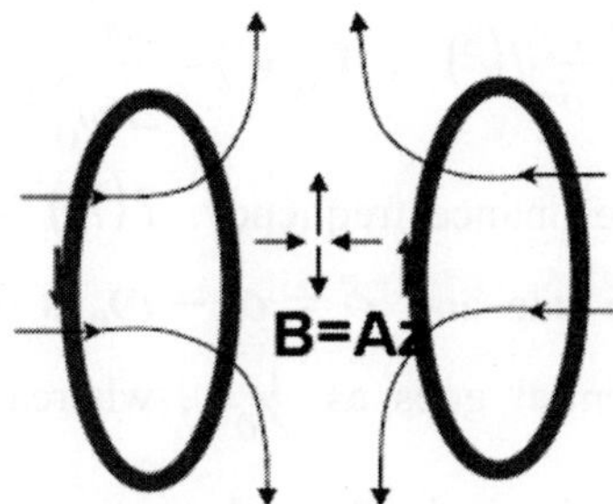

Fig. 9. Magneto-Optical Trap (MOT). (Left) Quadrupole field with two circular coils. **(Right)** Typical configuration of red detuned laser beams and magnetic field **(Top)**. Trapping forces for a J=0 ---> J=1 transition **(Below)**.

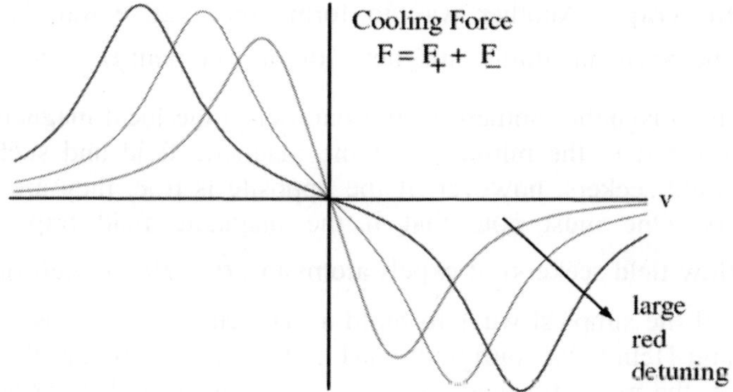

Fig. 10. Doppler Cooling Force. For red detuned beams, atoms with positive velocities experience a negative force, whereas the opposite is true for atoms with negative velocities. Notice the effect of red detuning on the slope around the centre as well as the shift in the maximum force.

4. CONSERVATIVE TRAPS

Before, we move onto our next scheme of cooling, it is quite appropriate to briefly discuss conservative traps for atoms. Broadly speaking, there are two types of traps: one based on light fields and the other based on magnetic fields.

1) Dipole Trap: Light fields interact with atoms in a dissipative as well as conservative way. The dissipative part is responsible for laser cooling where absorption of a photon is followed by spontaneous emission. On the other hand, the conservative part arises due to the interaction of the light field with the atomic dipole moment induced by the light field itself. This interaction causes potential energy shifts called ac-Stark shifts. For large detuning, spontaneous emission can be neglected and the energy shift can be used to create a conservative trapping potential (dipole potential) for neutral atoms. Within this limit, the strength of the dipole potential and the residual of scattering rate are given by-

$$V_{dip}(\vec{r}) \approx \frac{3\pi c^2}{2\omega_0^3}\left(\frac{\gamma}{\delta}\right)I(\vec{r}) \quad ; \quad \Gamma_{sc}(\vec{r}) \approx \frac{3\pi c^2}{2\hbar\omega_0^3}\left(\frac{\gamma}{\delta}\right)^2 I(\vec{r}) \tag{19}$$

Here ω_0 is the atomic resonance frequency, $I(\vec{r})$ is the laser intensity, γ is the spontaneous emission rate and $\delta = \omega - \omega_0$ is the detuning. One should note that the dipole potential goes as $1/\delta$, whereas scattering rate goes as $\left(1/\delta\right)^2$. Another thing to notice is the dependence on the sign of the detuning, δ .

If $\delta > 0$ (blue detuned), the dipole potential is repulsive while if $\delta < 0$ (red detuned), the dipole potential is attractive. So by having a red detuned laser, one can form a dipole trap even with a single laser beam.

2) Magnetic Trap: Another way to form conservative trap is to use the interaction between an atomic magnetic dipole moment $\vec{\mu}_m$ and a magnetic field $\vec{B}$. If the magnetic moment is anti-parallel to the local magnetic field, the atoms are trapped in the minimum of the magnetic field and such states are called low field seekers, however, if the opposite is true, they are called high field seekers. One must note that if the magnetic field traps spin states $|F, m_F\rangle$ (low field seekers), it expels atoms in $|F, -m_F\rangle$ (high field seekers) states. One of the simplest ways to build a magnetic trap is to use two circular coils in an anti-Helmholtz configuration (Fig 9). The limitation of this trap is that at the centre, the magnetic field vanishes, resulting in the loss of atoms due to spin flip (or Majorana) losses. The centre of the trap acts like a hole through which atoms leak out. This hole can be plugged by using either a blue detuned laser or an additional structure which produces additional magnetic field at the centre. In order to produce, a non-zero magnetic field at the centre, there are two main types of magnetic traps: the Ioffe-Pritchard trap and time-averaged orbiting potential (TOP) trap (Fig 11). In the Ioffe-Pritchard trap, four parallel rods with alternate current produce a transverse quadrupole field, while two circular coils with a distance much larger than in Helmholtz configuration produce longitudinal confinement. The resulting trap is cigar shaped. In the case of a TOP trap, a rotating magnetic bias field $\vec{B}_0$ is added to the spherical quadrupole field. The frequency of the rotating field is such that it is much higher than the orbiting frequency of the atoms but much lower than the Larmor frequency. The resulting time averaged trap is harmonic but much tighter than the one obtained by dc magnets of the same size. The rotating field moves the zero of the magnetic field along a circle called "circle of death".

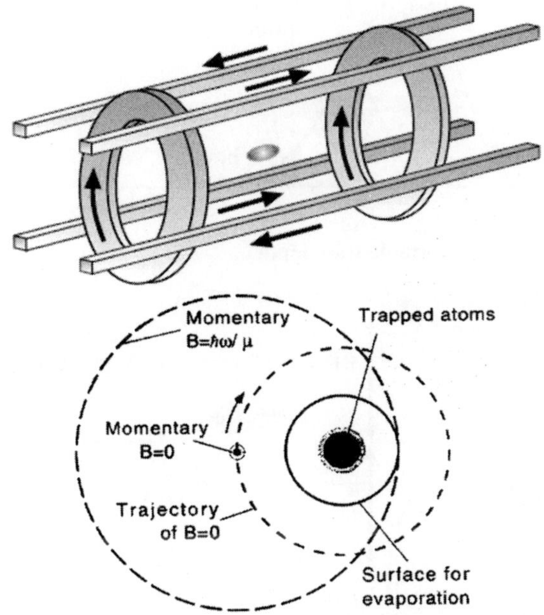

Fig. 11. Plugging the Zero Field in the Centre. (Left) Ioffe-Pritchard trap; Circular coils with a distance much larger than in Helmholtz configuration remove the zero in the centre and create a harmonic field minimum. **(Right)** Time-averaged orbiting potential trap (TOP). The additional field makes sure that the locus of zero magnetic field is a circle.

5. EVAPORATIVE COOLING

The basic limitation of laser cooling techniques is the recoil temperature. So in our pursuit of realizing a BEC, the phase-space density falls short off by many orders of magnitude. This gap can be bridged by using evaporative cooling [4]. The basic idea behind evaporative cooling is the following (Fig 12). Like in a cup of coffee when most energetic particles leave the surface, the coffee cools down. In the same way, if we continue to remove the high energy tail of the thermal distribution, the evaporated atoms carry more energy than the average kinetic energy of the distribution. However, in order to continue the process, we need to re-populate the tail for each cycle. This is done by the elastic collisions through which the sample re-thermalizes. After each cycle, the distribution shifts towards lower as well as narrower values in velocity. However, the rate of cooling will decrease with time, so we need to have an efficient mechanism in order to keep the cooling rate sufficiently large.

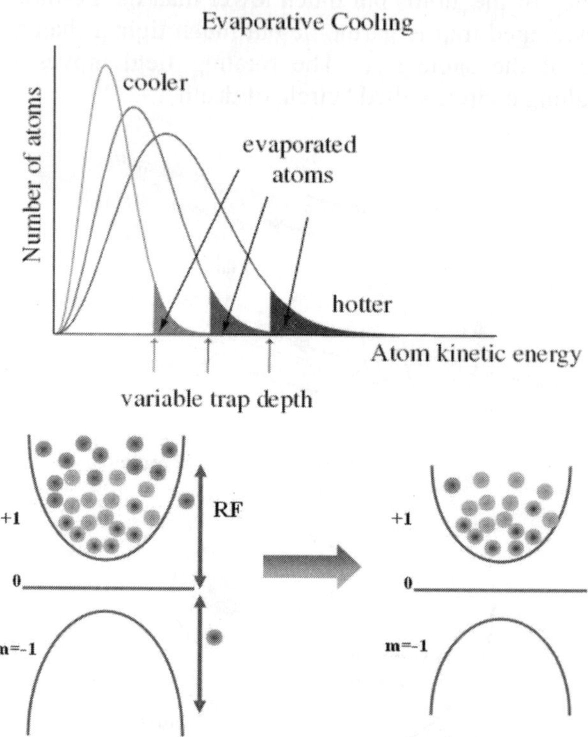

Fig. 12. Evaporative Cooling. (Left) For a given confining trap, only the hottest atoms have enough kinetic energy to escape the trap. The remaining atoms collide with each other to re-thermalize and consequently attain a lower temperature. By successively trimming the trap, colder temperatures can be reached. **(Right)** In a magnetic trap evaporation is performed using RF-radiation. RF-radiation flips the spin of the atoms from a trapped state to an untrapped state. Reducing the frequency of the RF is equivalent to going from the edge of the trap towards the centre of the trap.

The evaporative cooling is carried out by slowly lowering the potential barrier or resonantly exciting the atoms into an untrapped state. The former technique is employed by lowering the intensities of laser beams in a dipole trap. The later technique requires a magnetic trap where one can spin flip atoms from a trapped state to an untrapped state by using an RF-field. In a magnetic trap, only the weak field seeking state (m = +1 state in Fig 12) is trapped; on the other hand, the high field seeking state (m = -1 state in Fig 11) is untrapped. By resonantly coupling these two states, one transfers atoms from the m = +1 state to the m = -1 state. Although all the atoms have the same hyperfine structure irrespective of their energies, the magnetic trap induces spatially dependent Zeeman splittings. As a result, one can sweep the RF-frequencies such that the resonance moves from the edge of the trap to the centre of the trap. Consequently we can slowly move from hotter atoms to colder atoms which reside at the centre of the trap.

In addition to the removal of the most energetic atoms, it is equally important that the atoms thermalize through elastic collisions before each evaporation cycle. As the atoms have a finite life time, the elastic collision rate has to be much higher than the loss rate. The limitation in trap life time mainly comes from the loss processes such as three body recombination, inelastic scattering and collision with back ground gases. Three body recombination requires three particles to interact simultaneously such that when two particles relax to lower energy states, the excess energy and momentum is carried away by the third particle which leaves the trap. The probability for three particles to collide together is given by third order correlation function, $g^3(0)$ which is proportional to n^3 (n is density). Therefore, it limits the usable densities to ~10^{15} per cubic cm as at higher densities 3 body losses become important. If we are using a magnetic trap, it traps only one spin state and if collisions between two atoms result in the spin change, we lose the atom. Another important loss mechanism is collision with back ground hot atoms. In order to limit this loss, the experimental chamber is evacuated to ~10^{-11} mbar.

Therefore, evaporative cooling must be performed in such a way that on one hand, it is fast enough with respect to the life time of the trap but on the other hand, it is slow enough to give enough time to atoms to re-thermalize.

6. BOSE-EINSTEIN CONDENSATION (BEC)

A BEC is typically realized by combining laser cooling with evaporative cooling. At high temperatures (room temperature is also high for our considerations), atoms in a gas behave like classical particles with a very well defined position and momentum. But as we go down in temperature, the atoms are no more classical objects and they should be described by their associated deBroglie wave packets. In this regime, quantum statistics starts to play an important role. In order to understand what happens in the regime called quantum degeneracy, let us recall some of the basic ideas of statistical mechanics. Let us also recall that there are two types of particles, bosons and fermions. Bosons are full integer spin particles whereas fermions are half integer spin particles. There is no upper limit to how many bosons can simultaneously occupy the same quantum state, on the other hand, due to the Pauli exclusion principle, only one fermion can occupy a quantum state.

In the following discussion, we are going to follow the grand canonical ensemble approach. The grand canonical partition function is of the following form [5]:

For bosons
$$\tau(z,V,T) = \prod_K \left(1 - ze^{-\beta\varepsilon_K}\right)^{-1} \tag{20}$$

For fermions
$$\tau(z,V,T) = \prod_K \left(1 + ze^{-\beta\varepsilon_K}\right) $$

Where $z = e^{\beta\mu}$, $\beta = 1/k_BT$; z is called fugacity, μ is called chemical potential, k_B is called Boltzmann constant, ε_K are the possible energy values and K are the possible states (e.g. momentum states in free space). We have used K instead of P in order to avoid any confusion arising from the fact that we have reserved P for pressure. One can obtain thermodynamics of the system from the equation of state-

$$\frac{PV}{k_BT} = Ln\,\tau(z,V,T) \tag{21}$$

Where P is pressure and V is volume. The normalization is given by-

$$N = z\frac{\partial}{\partial z}Ln\,\tau(z,V,T) = \sum_P \langle n_K \rangle \tag{22}$$

$\langle n_K \rangle$ is the mean occupation of the Kth level, which is given by-

$$\langle n_K \rangle = \frac{1}{e^{\beta(\varepsilon_K - \mu)} \mp 1} \tag{23}$$

Here the -/+ sign is for bosons and fermions respectively. This leads to a very different physics. However, in the following discussion we shall focus only on bosons and make the assumptions that the gas is ideal (no interactions between particles) and it is in 3D free space. Let us make an important remark here that in free space, the single particle ground state is the zero momentum (**K=0**) state which means that $\varepsilon_K = K^2/2m \to \varepsilon_0 = 0$. For bosons, combing the eqns (20) and (21) gives us-

$$\frac{PV}{k_BT} = -\sum_K Ln\left[1 - ze^{-\beta\varepsilon_K}\right] \tag{24}$$

So for the ground state (**K=0**), if z =1, something remarkable happens, which we will know very soon that it is the onset of BEC. Since **K=0** term is something special, we separate the sum in the above eqn (24) into two parts, **K=0** term and the rest. Let us also convert, summation into integrals, one obtains-

$$\frac{PV}{k_BT} = -\frac{1}{V}Ln(1-z) - \frac{4\pi}{h^3}\int_0^\infty dk\,k^2 Ln\left[1 - ze^{-\beta\frac{k^2}{2m}}\right] \tag{25}$$

$$\frac{N}{V} = \frac{1}{V}\frac{z}{1-z} + \frac{4\pi}{h^3}\int_0^\infty dk k^2 \frac{1}{\left[e^{-\beta\frac{k^2}{2m}}z^{-1} - 1\right]}$$

Integration of the above eqn gives-

$$\frac{P}{k_B T} = -\frac{1}{V}Ln(1-z) + \frac{1}{\lambda^3}g_{5/2}(z) \qquad (26)$$

$$\frac{N}{V} = \frac{1}{V}\frac{z}{1-z} + \frac{1}{\lambda^3}g_{3/2}(z)$$

Where $\lambda = \left(\frac{2\pi\hbar^2}{mk_B T}\right)^{\frac{1}{2}}$ is the thermal deBroglie wavelength which we have

already seen before, and the functions $g_s(z)$ are defined as $g_s(z) = \sum_{l=1}^{\infty}\frac{z^l}{l^s}$. The

function $g_{3/2}(z)$ is particularly important for us. In the interval $0 \leq z \leq 1$, it is a positive and monotonously growing function of z such that $g_{3/2}(1) = 2.612$.

The fact that it reaches its maximum at z = 1 is crucial. Let us call $N_0 = \frac{z}{1-z}$ as

the population of the single particle ground state, then eqn (26) gives us-

$$\frac{N_0}{V} = \frac{N}{V} - \frac{1}{\lambda^3}g_{3/2}(z)$$

Or

$$\frac{N_0}{V}\lambda^3 = n\lambda^3 - g_{3/2}(z) \qquad (27)$$

$n = \frac{N}{V}$ is the density and $n\lambda^3$ is phase-space density. As we know that $g_{3/2}(1) = 2.612$, for low temperatures T or/and large enough densities, $n\lambda^3 > 2.612$ and when this happens we can not put any more particles in states different from the **P = 0** state. This might become clearer from the following argument-

$$N = N_0 + \frac{V}{\lambda^3} g_{3/2}(z)$$

$$\underbrace{}$$

$$\left. \begin{array}{c} \text{No. of particle in } \mathbf{K \neq 0} \text{ state} \\ \end{array} \right\} \sum_{K \neq 0} N_K < \frac{V}{\lambda^3} g_{3/2}(1) \qquad (1)$$

This suggests that all the extra particles must go to the $\mathbf{K = 0}$ state. This phenomenon is called Bose-Einstein condensation. Now let us look at the critical temperature for the onset of BEC. This is the temperature at which the total number of particles is equal to the maximal number of $\mathbf{K \neq 0}$ particles equals. Therefore:

$$N = \frac{V}{\lambda(T_c)^3} g_{3/2}(1) \rightarrow n\lambda(T_C)^3 = 2.612 \qquad \text{(Critical phase space density)}$$

Or $\quad k_B T_C = \frac{2\pi\hbar^2}{m}\left(\frac{n}{2.612}\right)^{2/3} \qquad$ (Critical temperature)

For $T < T_C$ and $z = 1$, the non-condensed number of particles is given by-

$$N_T(T) = \frac{N\lambda(T_C)^3}{\lambda(T)^3} = N\left(\frac{T}{T_C}\right)^{3/2}$$

Hence for $T < T_C$, the condensate fraction is :

$$\frac{N_0(T)}{N} = 1 - \left(\frac{T}{T_C}\right)^{3/2} \qquad (28)$$

At $T = T_C$, one has a phase transition into a condensed state. It is quite important to note that this transition does not require any interactions between particles. In that sense, it is a completely quantum statistical phenomenon.

It must be emphasized that the above treatment was carried out for an untrapped gas. However, in the case of a trapped case, the above eqn gets modified in the following way:

$$\frac{N_0(T)}{N} = 1 - \left(\frac{T}{T_C}\right)^3 \qquad (29)$$

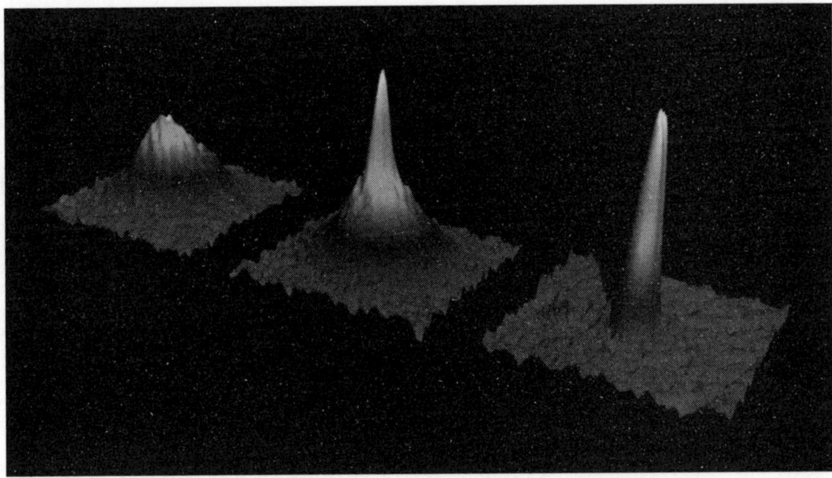

Fig. 13. Observation of Bose-Einstein condensation by Absorption Imaging. The "sharp peak" is the Bose-Einstein condensate (BEC), characterized by its slow expansion observed after 6 msec time of flight. **(Left)** Thermal cloud cooled to just above the BEC transition temperature; **(Middle)** cooled to just after the appearance of a BEC; **(Right)** cooled to until almost a pure BEC. The figure has been adapted from the paper [6].

7. DETECTION OF A BEC

A typical Bose-Einstein condensate contains 10^5-10^7 atoms with a typical density of 10^{14}-10^{15} cm^{-3} and a temperature of a few 100 nK. For such a sample, a "contact probe" cannot be used due to the low particle numbers, as the sample would equilibrate with the probe and not the probe with the sample. In this situation, fortunately light comes to our rescue and we can take advantage of light scattering by the atoms. The interaction of light with atoms results in three main processes-absorption of photons, re-emission of photons and a shift in phase of the transmitted light. Based on these three processes are three different detection techniques namely absorptive, fluorescence and dispersive imaging. Among these, absorptive imaging is the most popular. It is done by illuminating the atom cloud with resonant laser light, resulting in a shadow cast onto a CCD chip. The imaging of the density distribution can be carried out either *in-situ* or after a time-of-flight. In either case, we obtain an image which reflects the density distribution of atoms either in trap or after a ballistic expansion. A typical BEC inside the trap is quite small in size and hence it is much easier to image a condensate after some time-of-flight. For non-interacting atoms the density profile after a time-of-flight is a measure of the momentum distribution inside the trap.

Almost all the properties of a BEC are inferred by measuring the density distributions. So far we have assumed that the atoms do not interact with each other. However, in reality the atoms of a BEC do interact with each other. At ultra low temperatures and for dilute gases, the interaction is dominated by two body elastic collisions. Though the true inter atomic potential is quite complicated to calculate, for large inter atomic separations (>0.5 nm), we can approximate the interaction to be van der Waals. Since in the Bose condensed state the deBroglie wave length of an atom is much larger than the spacing between atoms, the actual shape of the potential is not so important. At these ultra cold temperatures, the scattering is dominated by s-wave scattering and hence the only important parameter is the s-wave scattering length, a_s. Therefore the inter-atomic potential can be assumed to be a contact interaction in the following way-

$$V(r) = \frac{4\pi\hbar^2 a_s}{m} \bullet \delta(r) = g \bullet \delta(r) \tag{30}$$

Where r is the relative separation between the two atoms and $g = \dfrac{4\pi\hbar^2 a_s}{m}$ is

called interaction strength. Many qualitative as well as quantitative properties of a BEC can be described by the following Gross-Pitaevaskii eqn which is actually a non-linear Schrodinger eqn-

$$i\hbar \frac{\partial}{\partial t} \psi(\vec{r},t) = \left(-\frac{\hbar^2 \nabla^2}{2m} + V_{ext}(\vec{r}) + g|\psi(\vec{r},t)|^2 \right) \psi(\vec{r},t) \tag{31}$$

$\psi(\vec{r},t) = \psi(\vec{r})e^{-i\mu.t}$ is the wave function for the macroscopic ground state, and is called order parameter, μ is the chemical potential and $V_{ext}(\vec{r})$ is external trapping potential which for our purpose we will assume to be harmonic . The solution of this equation also shows that one gets stable condensate if the interactions are repulsive ($a_s>0$) such that the trapping force is compensated by the repulsive force between atoms. In the opposite case ($a_s<0$), the condensate collapses unless the atom number is very small and the zero point kinetic energy compensates both the trapping force and the interaction energy.

Case 1:   weak interactions ($gn << \hbar\omega_{x,y,z}$)- In this case one can neglect the interaction term in eqn (31). The ground state wave function is simply the ground state of the harmonic potential, which is a Gaussian. For N particles, the ground state is given by-

$$n_c(\vec{r}) = \frac{N}{\pi^{3/2}} \prod_{i=1}^{3} \frac{1}{x_{i,o}} e^{-x^2_i / x_{i,o}^2} \tag{32}$$

Where c stands for condensate, $x_{i,0} = x_{i,HO} = \sqrt{\hbar/m\omega_i}$ is the rms width of the condensate wave function in the ith direction. During the free expansion of the condensate, the Gaussian wave function remains Gaussian except for a change in the phase factor. Thus after a time t, the condensate is rescaled as $x_{i,t}^2 = x_{i,0}^2 + v_{i,HO}^2 t^2 = x_{i,HO}^2\left(1+\omega_i^2 t^2\right)$; where $v_{i,HO} = \sqrt{\hbar\omega_i/m}$ is the rms velocity width of the trapped condensate. Therefore for $t >> 1/\omega_i$, the initial size of the condensate is not very important any more.

Case 2:   Strong interactions ($gn >> \hbar\omega_{x,y,z}$): Thomas-Fermi Limit- In this case one can ignore the kinetic term (first term on RHS) in eqn (31). The density is given by-

$$n_c(\vec{r}) = \max\left(\frac{\mu - V_{ext}(\vec{r})}{gn}, 0\right) \tag{33}$$

For a harmonic trap, the condensate density has a parabolic profile.

$$n_c(\vec{r}) = \frac{15}{8\pi}\frac{N}{\Pi x_{i,c,0}}\max\left(1 - \sum_{i=1}^{3}\frac{x_i^2}{x_{i,c,0}^2}, 0\right) \tag{34}$$

Where $x_{i,c,0} = \sqrt{2\mu/m\omega_i^2}$ is the half width of the condensate where density goes to zero. If a condensate is released from a time independent trap, it retains its parabolic shape except the rescaling in time. For instance, suppose a release from a cigar shaped trap with radial frequency ω_ρ and the aspect ratio $\omega_\rho/\omega_z = \varepsilon^{-1}$, the half lengths of the condensate evolve (to the lowest order in ε) [7] in the following way-

$$\rho_0(t) = \rho_0(0)\sqrt{1+\tau^2} \tag{35}$$

$$z_0(t) = \varepsilon^{-1}\rho_0(0)\left(1 + \varepsilon^2\left[\tau.\arctan\tau - Ln\sqrt{1+\tau^2}\right]\right) \tag{36}$$

Where $\tau = \omega_\rho t$. Eqns (35) and (36) describe three different stages during time-of-flight expansion: 1) For $\tau<1$, all the interaction energy is converted to kinetic energy, which results in an acceleration in the radial direction. 2) For $1<\tau<\varepsilon^{-2}$, the expansion takes place mainly in radial direction but with little in the axial direction. 3) For $\tau > \varepsilon^{-2}$, the system has reached the asymptotic regime and the ratio between the axial and radial direction is $z_0(t)/\rho_0(t) = \pi\varepsilon^2/2$. For typical frequencies ($\omega_\rho = 200 \times 2\pi$ Hz, $\omega_z = 30 \times 2\pi$ Hz), the interaction energy (called mean field energy) takes roughly 1-2 ms to convert into kinetic energy during the expansion.

7.1 COMBINATION OF A BEC AND A THERMAL CLOUD

Quite often one gets a BEC residing on a thermal pedestal. In this situation, the density distribution is fit with a bimodal curve where a Gaussian is used for thermal part and an inverted parabola is used for the condensed part. In a typical experiment, an image is taken after 20-30 ms time-of-flight.

8. OPTICAL LATTICES

Above we have seen that a light field can be used to create conservative potentials for the atoms. The basic idea behind optical lattices is to use the light field in order to realize periodic light traps for neutral atoms. In its simplest form, lattice potentials can be created by interfering two counter propagating laser beams. This idea can be extended to the two other orthogonal directions and in this way we can choose to have 1D, 2D or 3D lattices. If the laser is red detuned, the atoms are attracted to high intensity regions (anti nodes) whereas if the laser is blue detuned, the atoms are pushed away from high intensity regions to no intensity regions (nodes). In either case the atoms are trapped. The depth of the trapping potential can simply be varied by increasing or decreasing the intensity of the laser. The system of neutral atoms in optical lattices is very analogous to a system of electrons in a solid crystal. Here the role of electrons and ions is played by the neutral atoms and light traps respectively. Due to the periodicity of the potential, there are Bloch bands and atoms can be described by Bloch waves. For instance, like electrons in a solid, under constant accelerations, we can observe Bloch oscillations in optical lattices. In order to load a BEC at the centre of the first Brillion zone, the lattice is adiabatically ramped up. As long as a lattice is not very deep, the wavefunction for an atom is delocalized over many lattice sites and if the lattice is suddenly switched off, a time-of-flight measurement results in momentum peaks separated by the ∇/d (d is lattice constant), which is the result of interference between matter wave contributions from different lattice sites. Time-of-flight measurements are good for coherence measurements. However, if the lattice is ramped down adiabatically, one can map the Brillion zones in time-of-flight measurements [8].

Degenerate gases in optical lattices provide a very useful system to study quantum phase transitions. The unique advantage of cold atom systems is perfect control over parameters. Atoms in an optical lattice can hop from one lattice site to other, as well as interact with each other. The interplay between hoping and mutual interaction is the root cause for quantum phase transitions and leads to highly enriched physics. The transition from a super fluid to a Mott insulator is one such example [9]. In this example the system is tuned from a weakly interacting regime to a strongly correlated regime. For bosons (fermions) such a system is well described by the Bose-Hubbard (Fermi-Hubbard) model. In the following we will mainly focus on bosons but a similar treatment applies for fermions as well. The Bose-Hubbard model is written as-

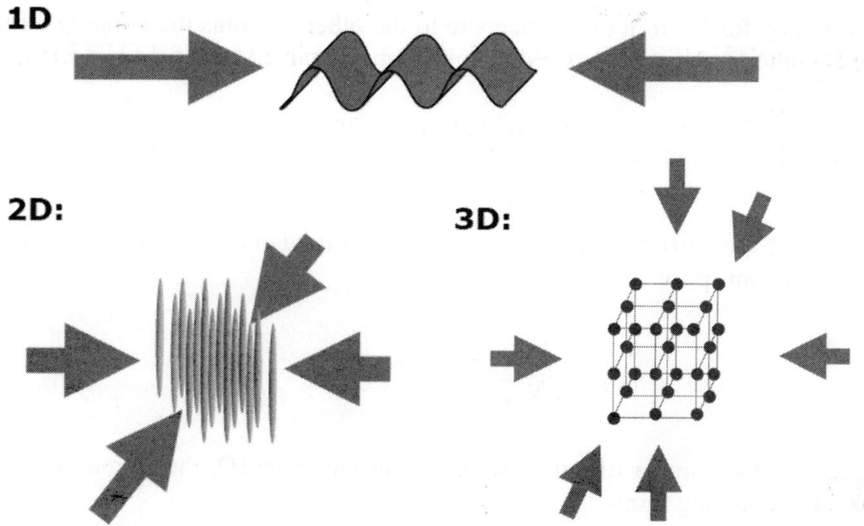

Fig. 14. Optical Lattice. Simple arrangements to realize optical lattices in 1D, 2D and 3D

$$\hat{H} = -J \sum_{\langle i, j \rangle} \hat{b}_i^+ \hat{b}_j + U_0 \sum_i \hat{n}_i \left(\hat{n}_i - 1 \right) + \sum_i \varepsilon_i \hat{b}_i^+ \hat{b}_i \qquad (37)$$

Where $\langle i, j \rangle$ refers to summation over all nearest neighbors, $\hat{b}_i^+ (\hat{b}_j)$ are single particle creation (anhilation) operators for the site i, $\hat{b}_i^+ \hat{b}_i = n_i$ refers to number operator for the site i, ε_i is the energy offset at the site i due to external trap. The first term describes that atoms can hop from one site to the other at a rate J. J basically determines the overlap of wavefunctions localized to adjacent lattice sites and it deceases exponentially with increasing the lattice depth. The second term describes that if any two atoms are on the same site, they repel each other and the onsite interaction energy is given by U_0 which is proportional to s-wave scattering length a_s for cold atoms. U_0 essentially tells us how much energy it costs us to put another atom to an already filled lattice site. The third term is the energy associated with putting a particle at the site i. One should notice a negative sign in front of the first term. It is due to the fact the effective potential for a lattice site is reduced from its bare value by the potentials for the neighbouring sites. In this Hamiltonian, the relative strength of tunneling J and U_0 is crucial. Below we discuss two extreme situations-

1) $J \gg U_0$: This regime is achieved by keeping the lattice shallow. Atoms are able to hop freely from one lattice site to the other and thus the hopping term in the Hamiltonian dominates the dynamics. Ignoring interactions we can write-

$$\hat{H}_{kin} = -J \sum_{\langle i, j \rangle} \hat{b}_i^+ \hat{b}_j \tag{38}$$

One can diagonalize the kinetic part of the Hamiltonian by defining the operators in momentum space.

$$\hat{b}_i = \frac{1}{\sqrt{N}} \sum_{\vec{k}} e^{i \vec{k} \cdot \vec{x}_i} \hat{b}_{\vec{k}} \tag{39}$$

Here N is the number of lattice sites. For instance, in 1D, the Hamiltonian will take the following form-

$$\hat{H}_{kin} = -2J \sum_{\langle i,j \rangle} \hat{b}_k^+ \hat{b}_k \cos 2ka \tag{40}$$

Where a is the lattice constant. Clearly the ground state of this Hamiltonian is the state where all the atoms are in the $k = 0$ state i.e.

$$\left| \psi_0 \right\rangle = \frac{1}{\sqrt{N!}} \left(\frac{1}{\sqrt{N}} \sum_i \hat{b}_i^+ \right)^N \left| vac \right\rangle \tag{41}$$

In the limit of large N, this state is a coherent state. As the atoms can freely tunnel from one lattice site to the other, this state is superfluid. As a coherent state follows a binomial atom number distribution, the number of atoms per lattice is not very well defined however the relative phase between two lattice sites is very well defined. One more point that we should make here is that the excitation spectrum for this state is gapless as it does not cost any energy to keep more than one atom at a lattice site.

2) $J \ll U_0$: In principle this regime can be achieved in three different ways: 1) increasing the lattice depth, which has been applied most often, but in this process both J & U_0 are simultaneously affected; 2) increasing the s-wave scattering length and thereby increasing U_0; for instance using a Feshbach resonance; 3) modulating the lattice which results in the re-normalization of the tunneling rate J such that U_0 remains unchanged [10]. In this situation the Hamiltonian is dominated by the interaction term-

$$\hat{H}_{int} = U_0 \sum_i \hat{n}_i \left(\hat{n}_i - 1 \right) \tag{42}$$

Suppose that there are N atoms and exactly N lattice sites. The lowest energy state will require that each lattice site is occupied by exactly one atom. Such a state is given by-

$$\left| \psi_0 \right\rangle = \prod_i \hat{b}_i^+ \left| vac \right\rangle \tag{43}$$

This ground state is known as Mott-insulator as the particles are frozen to their sites. There is no mobility in the system. Contrary to the superfluid state, this state is a Fock state which means that the number of atoms per lattice site is very well defined. However, now the phase is completely undefined. As it costs extra energy to put two atoms at a single lattice site, the excitation of this state shows a gap. Another important property of this state is its incompressibility.

8.1 SUPER FLUID TO MOTT INSULATOR TRANSITION

We have seen above that the relative strength of U_0 and J holds the key to tune beween the superfluid and Mott insulator regime. The beauty of this phase transition is that it is completely driven by quantum fluctuations and not by thermal fluctuations and therefore can take place even at $T = 0$. This is precisely the reason why it is called a quantum phase transition. In the following we make the assumption that the system is homogenous which means that we disregard the external trap. We also assume that the total number of atoms is equal to the total number of sites or in general we assume that the total number of atoms is an integer multiple of the total number of lattice sites (commensurate filling). In fact, with incommensurate filling there is strictly speaking no phase transition as the extra atoms or holes will delocalize to minimize their kinetic energy. The calculations show that for a single atom per lattice site, the phase transition occurs when $U_0/J = 5.8 \times z$, where z is the coordination number, for a 3D cubic lattice $z = 6$. As we seen above that the superfluid excitation spectrum is gapless while that of the Mott insulator is gapped with $\Delta\varepsilon \approx U_0 - J$. A vanishing excitation gap can be used a distinguishing feature for this phase transition.

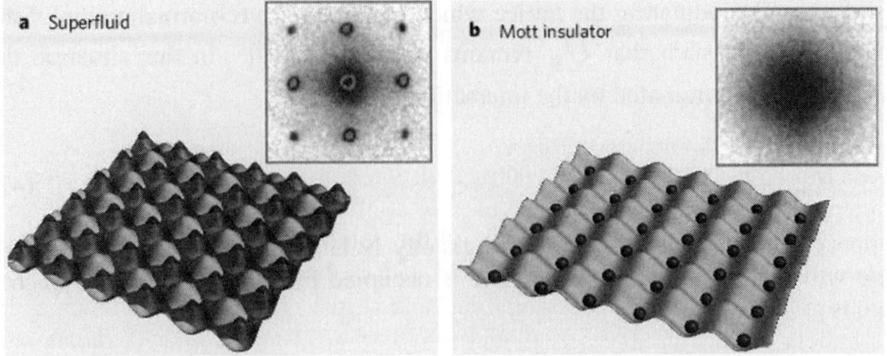

Fig. 15. Super Fluid to Mott Insulator Transition. (a) For weak interaction strength, the ground state of the system is a superfluid. The atoms are delocalized throughout the available space (purple color) and have a common macroscopic wave function. The time-of-flight measurement shows distinct diffraction peaks (inset). **(b)** For strong interaction strength, the ground state is a Mott insulator state. Now the atoms are completely localized to a specific lattice site. In this case, the diffraction pattern vanishes as the phase coherence as well as the super fluidity disappears (see inset). The figure is adapted from the reference, [11].

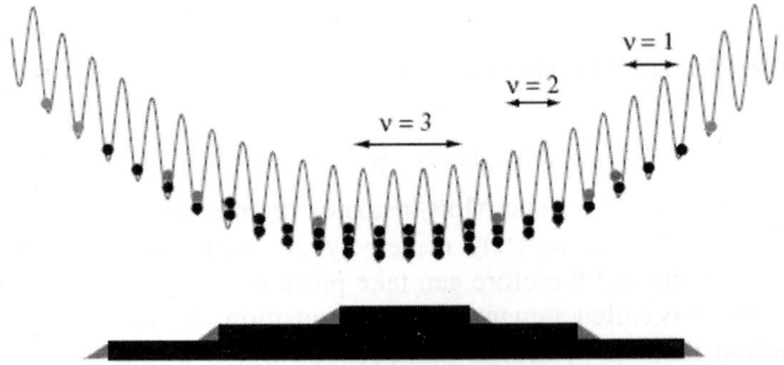

Fig. 16. Wedding Cake Structure. Optical lattice potential superimposed on an additional harmonic potential (**above**). In a strongly interacting regime, Mott insulator phases (with integer fillings, v) are separated by superfluid regions (non-integer fillings) This results in a wedding cake like structure (**below**).

However, in a real experiment, there is typically an additional harmonic trap confining the atoms in the optical lattice. The centre of the trap has higher density that the edges, which means that number of atoms per lattice site in the centre is higher than that of the edges. In fact this is visible in a Mott insulator phase which finally takes the shape of a wedding cake. Therefore, in a real situation, the filling factor increases in steps as one moves from the edges towards the centre. In an experiment, first inference is obtained by looking at the diffraction pattern for a relatively shallow lattice. By increasing the lattice depth one can then go from the superfluid regime to the Mott insulator regime and back, meaning that this process is reversible. In order to confirm the Mott insulator phase one can look at the excitation spectrum.

9. ATOM CHIPS

An atom chip [12] is a micro fabricated, integrated device, in which electric, magnetic and optical fields can be used to confine, control and manipulate cold atoms. Atom chips take advantage of the interaction between the magnetic moment of an atom and a magnetic field. If F is the total angular momentum of the atom, mF is its projection along the field B, μB is the Bohr magneton and g is Lande g-factor, the interaction energy is $E = g\mu_m B m_F$. If $gm_F > 0$, the atom is attracted towards the minimum of the magnetic field (weak field seekers) and if $gm_F < 0$, the atom is attracted towards the maximum of the magnetic field (high field seekers). However, Maxwell's eqns do not allow a maximum of the magnetic field in free space, only weak field seekers are trapped by atom chips.

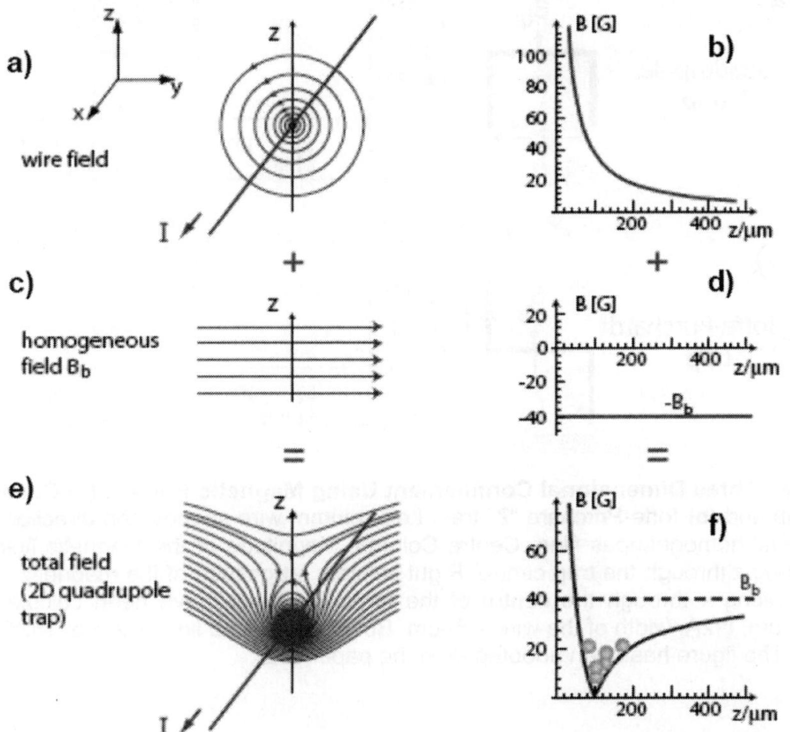

Fig. 17. (a) Thin and straight wire along x-axis with current I. **(b)** Magnetic field along z-axis. **(c)** Additional homogeneous field along y-axis. **(d)** Graphical representation of the same homogeneous field. **(e)** The two magnetic fields give rise to a 2D quadrupole field at z=z0 above the wire for I=2 Amp and Bb=40 G. **(f)** Graphical representation of the situation in **(e)**. This is clear that the atoms are in 2D but are free to move along the wire. The figure has been adopted from the paper, [12].

If we have a thin and infinitely long wire along x with current I through it, the curling magnetic field B scales as 1/z from the wire (see Fig17). If we add a constant, homogeneous magnetic field B_b perpendicular to the wire (say along y), it cancels the wire magnetic field along a line suspended above the wire at a distance, $z_0 = \dfrac{\mu_0 I}{2\pi B_b}$. Close to this zero field line, the resultant magnetic field is a 2D quadrupole field in the plane perpendicular to the wire. Atoms in weak field seeking states are attracted towards this field minimum and trapped in 2D but they are still free to propagate along the wire and hence this arrangement can be called a wire guide. The gradient of this 2D quadrupole field is $\nabla B = -\dfrac{\mu_0 I}{\left(2\pi z_0^2\right)}$. It is evident that for a given I, the trap is tighter if z_0 is smaller.

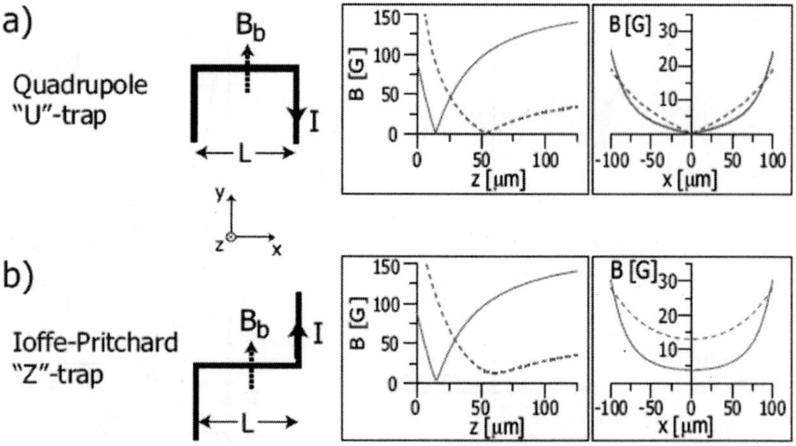

Fig. 18. Three Dimensional Confinement Using Magnetic Forces. (a) Quadrupole "U"-trap and **(b)** Ioffe-Pritchard "Z"-trap. Left column: wire shapes and direction of the additional homogeneous filed. Centre Column: Magnitude of the magnetic field on a line along z through the trap centre. Right column: Magnitude of the magnetic field on a line along x through the centre of the trap. The fields have been calculated for L=250 um, I=2A, width of the wire = 50um, Bb=54G (dashed line) and Bb=162G (solid lines). The figure has been adopted from the paper, [12].

3D traps can also be formed by bending the wire in "U" or "Z" shape (see Fig. 18). In either case the central part along with the additional field, still forms a 2D quadrupole trap in the transverse direction, whereas the axial confinement comes from the bent part. In case of "U" shape, the current flows in opposite directions and cancel each other. The resulting potential is a 3D quadrupole trap with field zero at x=0, y>0 and z≈z0. The shift along y is due to the field component B_z of the bent wires. However, in case of a "Z" wire, the current in the bent wires flows in the same direction and hence the fields along x add at the centre of the wire. The result is a 3D Ioffe-Pritchard trap with trap centre at x=0, y=0 and z≈z0. The curvature $\partial^2 B_x / \partial x^2$ provides the confinement in the axial direction. As the z-component of the field vanishes for this shape, the trap center remains unshifted.

10. BEC IN MICROGRAVITY

So far a typical experiment on cold atoms has been confined to earth bound laboratories. The success of these laboratories in exploring the physics at extremely low energy scales has been the motivation to continue the path towards lower energy scales by lifting Earth-bound laboratory restrictions. There are three main reasons why a gravity free environment is desirable. 1) It can provide an unperturbed evolution for long durations which is crucial for atom interferometers and atomic clocks. This emanates from the fact that the sensitivity of a clock or an interferometer increases with the observation time. In a microgravity environment, the precision of these sensors can be enhanced up to three orders of magnitude. 2) It provides a mass independent confining potential which is very important for the research on a mixture of quantum gases such as degenerate Bose-Fermi gases. 3) In a microgravity environment, it is possible to adiabatically lower the trapping potential resulting in ultra-large condensates. The effect of ultra-weak long range forces becomes important in these condensates, which is expected to lead to new kinds of low energy phase transitions. In addition, microgravity is a prerequisite for an experiment on the equivalence principle in quantum domain which is at the heart of general theory of relativity.

(a)

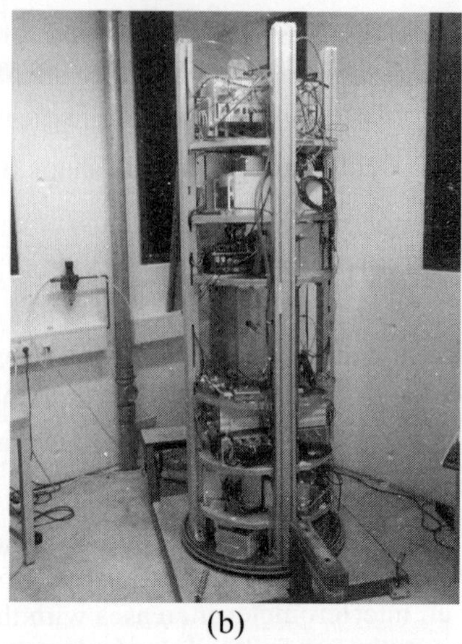

(b)

(c)

Fig 19. (**a**) Drop tower at Bremen. (**b**) Capsule containing a complete mobile BEC experiment. (**c**) Atom chip on its mount.

There are many platforms such as parabola flights, international space station (ISS), space carriers and ballistic rockets etc where one can access microgravity and a drop tower at ZARM in Bremen, Germany is one of them. This facility is completely automated and provides an excellent microgravity environment for approximately 4.7 seconds in a free fall mode. However, another mode which is catapult mode, gives microgravity time up to 9 seconds.

The facility permits either to drop a compact capsule of maximum height 173 cm and maximum diameter 60 cm or to launch smaller capsules with a catapult mode simulating parabolic flights. The capsule consists of all the components such as laser systems, power supplies, batteries, CCD cameras, computer control, vacuum chamber etc. required to realize and observe BECs. The experiment is fully remote controlled from a control room. Using this facility, the very first BEC in microgravity with ^{87}Rb atoms in the F = m$_F$ =2 state has been realised. In order to realise the BEC, an atom chip has been used. Atom chips are excellent devices in order to realize a compact and a robust setup owing to their small energy consumption, the possibility to create high magnetic field gradients necessary for a fast evaporation process. Radio frequency (RF) induced fast evaporation has enabled us to realize our BECs within 1.2 seconds implying that there are more than 3 seconds of free fall left for performing an experiment on the BEC. Simply by the virtue of a microgravity environment, it has been possible to work with much shallower traps than the one of a typical ground based experiment. Consequently it is possible to adiabatically open the trap and hence reduce the mean field energy resulting in a very cold gas of atoms. It is the first time that a free evolution of a BEC has been observed for a time as long as 1 second. After 1 second, the size of this giant, coherent matter wave is approximately 1.5 mm as compared to a typical size of 50 microns after 20 ms time of flight in a typical earth bound laboratory. For more details refer to our paper [13].

These activities are being carried under the project QUANTUS (QUANTen Gase Unter Schwerelosigkeit) which is a collaboration of the University of Hamburg, University of Ulm, Humboldt University of Berlin, MPQ Munich, ZARM at the University of Bremen and the Leibniz University of Hannover. It is supported by the German Space Agency DLR with funds provided by the Federal Ministry of Economics and Technology (BMWi) under grant number DLR 50 WM 0346.

11. OPTICAL CLOCKS

Atomic clocks are essential tools in modern society. Clocks operate in computers, data networks, are ubiquitous in scientific applications and are even operated in satellites, especially in navigation systems such as GPS, GLONASS and the currently developed European Galileo system. Time, or more precisely, time interval, is the most precisely measurable quantity. Since the speed of electromagnetic waves in vacuum is invariable, distance measurements can be performed by propagation time measurements. Thus, ultimately the precision of distance measurements is limited by the available precision in time measurements. In the International System of Units, the unit of time, the second, is based on an atomic hyperfine transition in neutral cesium (Cs) atoms. Laboratory clocks using cold Cs atoms have an inaccuracy of several parts in 10^{16} today. Although this is already by far the lowest inaccuracy of any physical unit, a major scientific development over the last decade, namely clocks based on optical rather than microwave transitions, has opened a new era in time/frequency metrology. In optical clocks the (laser) electromagnetic wave beats 10^{15} times per second instead of 10^{10} as in microwave clocks. Therefore, one can detect (and also correct) a minute change of the period much faster, allowing an enhanced stability. In addition, several perturbing effects on the energy levels of the employed atoms are smaller in relative terms for optical transitions compared to microwave transitions, and a large gain is therefore also possible in the accuracy. Optical clocks have now achieved a performance significantly beyond that of the best microwave clocks, at a fractional frequency inaccuracy of $8.6 * 10^{-18}$ [14]. It is therefore expected that in the mid-future the unit of time will be redefined via an optical transition. The essential techniques used in optical clocks are the confinement of the atoms to regions significantly smaller than the wavelength of light, provision of an environment as free of disturbing influences (magnetic and electric fields, residual gas, black-body fields) as possible, choice of adequate atomic species, and the narrowing of the spectral width of the clock laser to relative levels of 10^{-15} and less.

With the rapidly improving performance of optical clocks, in the future, most applications requiring the highest accuracy will require optical clocks. They cover the fields of fundamental physics (tests of General Relativity and its foundations), time and frequency metrology (comparison of distant terrestrial clocks, operation of a master clock in space), geophysics (mapping of the gravitational potential of the Earth), and potential applications in astronomy (local oscillators for radio ranging and interferometry in space).

The authors are also involved in setting up an optical clock based on Sr, which has turned into a work horse for these systems. In our experiment, we are using strontium (Sr) in a blue detuned optical lattice. Sr is an alkaline-earth element and has two electrons in its outer shell. This gives it many unique properties not existing in conventional alkali systems such as Rb or Na. The most prominent feature of Sr is that its two valence electrons can form spin singlet (electron spins antiparallel) or spin triplet (electron spins parallel) states. The ground state is a singlet, and transitions to the triplet states are nearly forbidden, resulting in long meta-stable lifetimes and narrow linewidths. These properties are used to build optical clocks and are at the heart of recent proposals in fields as diverse as the variation of fundamental constants, mHz linewidth lasers, and quantum simulation and computation.

Conclusion

The breakthroughs to create ultracold atoms, marked by the Nobel prizes in 1997 and 2001 , have initiated a major modern field of research. Degenerate quantum gases are an excellent tool to study and simulate the rich physics of fundamental interest including super fluidity, quantum phase transitions, super conductivity, coherence and correlations etc. These are excellent candidates for quantum information and quantum technology. In nutshell, ultrscold atoms have emerged as a new field of research in itself and connect to different fields of physics such as condensed matter, atomic and molecular, quantum optics and atom optics. In fact, with the recent realization of ultra cold molecules, one can envisage to control and engineer chemical reactions at ultimate quantum level. The unprecedented control over cold atoms has resulted in extremely precise sensors such as atomic clocks and atom interferometers. For example state-of the-art techniques have been applied to measure inertial forces and fundamental constants with high precision, better than or compatible with the best classical instruments. The gravitational constant G has been measured with 10^{-3} precision, [15], while sensitivities of $8*10^{-8}$ m/s^2*Hz$^{-1/2}$ for gravitational acceleration [16], $3*10^{-7}$ s^{-2}*Hz$^{-1/2}$ for gravity gradients [15] and 0.6 nrad*Hz [17] for rotations, better than or compatible with the best classical instruments. Optical clocks have already reached a fractional frequency inaccuracy of 8.6 $*10^{-18}$ [14] which is the highest precision ever. Cold atom gravity sensors could provide a new quality of global geodetic data allowing e.g. to monitor water circulation at an unprecedented level and thus providing crucial tests to current climate models in addition to laying the foundations for broadening of its technology platform to small scale optical frequency standards, which could provide ultra-precise timing in future high-capacity communication networks. More over cold atoms offer a novel approach to quantum ICT research such as secure private quantum communication, ultrafast databases and pattern recognition systems with even larger social impacts in the longer term.

It is must also be pointed out that so far all these developments have been made on the ground but now space has also come up as a new venue for the ultracold atom based activities. In short, given the perfect control over these systems, we not only simulate systems, but also create new systems.

Acknowledgement: Yeshpal Singh would like to acknowledge Marie-Curie Fellowship for financial assistance.

References:

[1]   Balykin et al. Rep. Prog. Phys. **63**, 1429 (2000).

[2]   Claude N. Cohen Tannoudji and William D. Phillips, Physics Today **43**, 33, (1990).

[3]   J. Dalibard and C. Cohen-Tannoudji, J. Opt. Soc. Am. B **6**, 2027 (1989).

[4]   Harald F. Hess, Phys. Rev. B **5**, 3476 (1986).

[5]   Kerson Huang, Book on Statistical Mechanics

[6]   Wolfgang Ketterle, Rev Mod Phys **74**, 1131 (2002).

[7]   W. Ketterle et al. Varenna Letures (1999); Y. Castin and R. Dum, Phys. Rev. Lett. **77**, 5315–5319 (1996); Yu. Kagan, E. L. Surkov, and G. V. Shlyapnikov, Phys. Rev. A **54**, R1753 (1996); Yu. Kagan, E. L. Surkov, and G. V. Shlyapnikov, Phys. Rev. A **55**, R18–R21 (1997); Pippa Storey and Maxim Olshanii, Phys. Rev. A **62**, 033604 (2000).

[8]   Markus Greiner et al., Rev. Lett. **87**, 160405 (2001).

[9]   D. Jaksch et al., Phys. Rev. Lett. **81**, 3108 (1998); Markus Greiner et al., Nature **415**, 39-44 (2002).

[10] Alessandro Zenesini et al., Phys. Rev. Lett. **102**, 100403 (2009), H. Lignieret al., Phys. Rev. Lett. **99**, 220403 (2007).

[11] Markus Greiner and Simon Fölling, Nature **453**, 736 (2008).

[12] J. Reichel, Appl. Phys. B **75**, 469–487 (2002).

[13] Tim van Zoest et al. Science 1540-1543 (2000).

[14] C.W. Chou et al., Phys. Rev. Lett. **104**, 070802 (2010).

[15] G. Lamporesi, A. Bertoldi, L. Cacciapuoti, M. Prevedelli, and G.M. Tino, arXiv:0801.1580v1 [physics.atom-ph] 10 Jan 2008.

[16] Holger Mueller, Sheng-wey Chiow, Sven Herrmann, and Steven Chu, arXiv:0710.3768v2 [gr-qc] 4 Dec 2007.

[17] T L Gustavson, A Landragin and M A Kasevich, Class. Quantum Grav. **17**, 2385–2398 (2000).

Bose-Einstein Condensates in a Harmonic Trap and Optical Lattice

Prasanta K Panigrahi[1], Rajneesh Atre[2], S Sree Ranjani[3], Priyam Das[1] and Kumar Abhinav[1]

[1] *Indian Institute of Science Education and Research (IISER)- Kolkata, Mohanpur, Nadia 741252, India*
[2] *Department of Physics, Jaypee Institute of Engineering and Technology, Guna 473226, India*
[3] *School of Physics, University of Hyderabad, Hyderabad 500 046, India*

e-mail:pprasanta@iiserkol.ac.in1

1 INTRODUCTION

The fact that, Bose-Einstein distribution

$$f(E) = \frac{1}{e^{\frac{E-\mu}{kT}} - 1},$$

(1.1)

can be singular at low temperatures, leads to the possibility of macroscopic occupation of of a quantum state [1, 2]. This condensation of bosons in to a low-energy state, below a critical temperature was first seen in liquid helium, below the so-called λ-point. Due to the fact that, helium was in a liquid phase and finite temperature effects were substantial, many novel features of this macroscopic quantum state remained illusive, till 1995, when BEC was achieved in a trap at nano-Kelvin temperatures [3, 4, 5]. The Gross-Pitaevskii equation (GPE) was found to capture the mean-field dynamics quite well. As mentioned earlier, the weakly interacting nature of this low density gas and almost complete absence of a normal component at $10^{-9}K$ temperature, were responsible for the success of GPE in capturing BEC dynamics. Subsequent demonstration of Feshbach resonance technique [6] for controlling the strength of the two-body interaction, as also the possibility of realizing this system in one and two dimensions, opened the possibility of observing novel phases of matter and nonlinear excitation in this quantum system.

In the following, we explicate the procedure to obtain the dynamical equations describing BEC in lower dimensional systems in the presence of harmonic trap, having tight confinement in certain directions. We concentrate here on one-dimensional cigar-shaped geometry, wherein, the harmonic confinement is taken to be strong in the transverse $x - y$ plane. As long as the interaction in the longitudinal direction is weak, so as not to excite the transverse degrees of freedom, the mean-field equation becomes effectively one-dimensional. We then establish the exact solutions of this one dimensional, weakly interacting system in the presence of a confinement along z-direction, as well as the optical lattice. The nonlinear mean-field equation in one dimension, being closely related to the nonlinear Schrödinger equation, admits soliton solutions [7, 8, 9, 10], which can be dark [11, 12], bright [13, 14, 15] or grey [16, 17]. All these experimentally observed solutions are then obtained in the presence of the trap. It was found that, their behavior can be effectively controlled through the two-body interaction, as well as trap parameters.

We find soutions of BEC in optical lattice [19], which reveal the experimentally observed dynamical phase transition between superfluid and insulator phases. The mean-field equation is then studied in the strong coupling regime [20]. The non-linear excitations are shown to be quite different from the weakly coupled systems.

2 THE CIGAR-SHAPPED GEOMETRY OF BECS

The GP equation, aptly capturing the dynamics of BECs, with harmonic trap is given by,

$$i\hbar\frac{\partial \Psi}{\partial t} = -\frac{\hbar^2}{2m}\nabla^2\Psi + V_{ext}\psi + U|\psi|^2\psi - \mu\psi. \tag{2.1}$$

In the case under consideration, we identify $V_{ext} = V_{HO}(x, y) + V_1(z, t)$ to be the external trapping potential, $U = \frac{4\pi\hbar^2 a}{m_a}$ as the atom-atom interaction strength and μ to be the chemical potential. When the trap is absent, the time independent equation

$$\frac{\partial V}{\partial \psi^*} = (g|\psi|^2 - \mu)\psi = 0, \tag{2.2}$$

is satisfied by a uniform constant solution, wherein, $|\psi|^2 = \sigma_0 = \frac{\mu}{g}$. The other phase is the normal one, where $\psi = 0$. Clearly, the repulsive, hardcore two-body interaction:

$$\begin{aligned} H_{int} &= \int\int \psi^*(x')\psi^*(x)V(x - x')\psi(x')\psi(x)d^3x d^3x' \\ &= \int\int \psi^*(x')\psi^*(x)g\delta^3(x - x')\psi(x')\psi(x)d^3x d^3x' \\ &= \int g|\psi^*(x)\psi(x)|^2 d^3x, \end{aligned} \tag{2.3}$$

along with a negative chemical potential, make it possible to obtain this macroscopic condensed phase with a uniform non-zero density σ_0. It should be pointed out here, that for the condensed alkali atoms, with only one valence electron, the structure of the particles is decided by the number of neutrons. As these atoms have equal number of protons and electrons, from the rule of angular momentum addition, the net angular momentum of these two species is integral. Thus, whether the atom is bosonic or fermionic, is entirely determined by whether the number of neutrons is even or odd, respectively. Hence, different isotopes of the same atom will have different statistics. Different groups have experimentally obtained condensation of Na, Rb, Li, Cs etc. Yb have shown both fermionic and bosonic behavior, for different isotopes, and have been observed in superfluid phases, when cooled.

2.1 GP equation in one dimension

The cigar shaped BEC is obtained through application of a strong oscillatory trapping potential with frequency $\omega_\perp$, $V_{HO}(x,y) = \frac{1}{2}m\omega_\perp(x^2+y^2)$, in the transverse direction, considered here to be the $x-y$ plane. Assuming tight transverse confinement, we consider a trial wavefunction, $\Psi(\mathbf{r},t) = \psi(z,t)G(x,y,\sigma)$ [8, 21, 22], where $G(x,y,\sigma)$ is the normalized transverse equilibrium ansatz wave function:

$$G(x,y;\sigma) = \frac{e^{-(x^2+y^2)/2\sigma^2}}{\pi^{1/2}\sigma}, \tag{2.4}$$

and,

$$\sigma(z) = \int dxdy|\Psi(x,y,z)|^2 = |f(z,t)|^2, \tag{2.5}$$

the local particle density. It is assumed that $G(x,y;\sigma)$ varies negligibly in z-direction [21]. After eliminating $G(x,y,\sigma)$ from Eq.2.1, we end up with,

$$i\hbar\frac{\partial f}{\partial t} = \left[-\frac{\hbar^2}{2m}\frac{\partial^2}{\partial z^2} + V + gN\frac{|f|^2}{2\pi\sigma^2} - \frac{\hbar^2}{2m\sigma^2} - \frac{1}{2}m\omega_\perp^2\sigma^2\right] \tag{2.6}$$

whereas, σ satisfies,

$$\frac{\hbar^2}{2m\sigma^3} - \frac{1}{2}m\omega_\perp^2\sigma + \frac{1}{2}gN\frac{|f|^2}{2\pi\sigma^3} = 0. \tag{2.7}$$

The above two equations yield the relation $\sigma^2 = a_\perp^2\sqrt{1+2a_sN|f|^2}$, where $a_\perp = \sqrt{\hbar/m\omega_\perp}$, which finally leads us to the non-polynomial equation [22],

$$\begin{aligned}
i\hbar\frac{\partial}{\partial t}f &= \left[-\frac{\hbar^2}{2m}\frac{\partial^2}{\partial z^2} + \frac{U_0}{2\pi a_\perp^2}\frac{|f|^2}{\sqrt{1+2aN|f|^2}}\right.\\
&\quad + \left.\frac{\hbar\omega_\perp}{2}\left(\frac{1}{\sqrt{1+2aN|f|^2}} + \sqrt{1+2aN|f|^2}\right)\right]f.
\end{aligned} \tag{2.8}$$

In the weak coupling limit $2aN|f|^2 \ll 1$, after suitable normalization, we arrive at [23]:

$$i\hbar\frac{\partial}{\partial t}f = \left[-\frac{\hbar^2}{2m}\frac{\partial^2}{\partial z^2} + 2\hbar\omega_\perp a|f|^2 - \mu\right]f, \tag{2.9}$$

with a cubic non-linear coupling. where $\mu = 2\hbar\omega_\perp a\sigma_0$ is identified to be the chemical potential of the system, a being the scattering length.

We consider two ansatz solutions to understand the excitations of such systems.

Ansatz 1: For $f(z,t) = A\operatorname{sech}(kz)\exp(-i\omega t)$, with A, k and ω being constants, we arrive at terms only linear or cubic in $\operatorname{sech}(kz)$ on substitution. On equating the coefficients of cubic terms, the amplitude turns out to be $A = \sqrt{\frac{\hbar^2 k^2}{2\hbar\omega_\perp am}}$, while the linear term yields the dispersion relation,

$$\omega = -\frac{\hbar k^2}{2m} - \frac{\mu}{\hbar}. \tag{2.10}$$

The energy ω stays positive only for a finite magnitude of the wave-number k, as the chemical potential, for BECs, is always negative.

Ansatz 2: Considering $f(z,t) = A\tanh(kz)\exp(-i\omega t)$, upon substitution, by comparing cubic terms in $\tanh(kz)$, we arrive at the same amplitude as in the previous example. However, the linear terms in $\tanh(kz)$ result in the dispersion:

$$\omega = \frac{\hbar k^2}{2m} - \frac{\mu}{\hbar}, \tag{2.11}$$

which is positive for a large spectrum of the wave-number. It is worth noting that $\tanh(kx)$ is a solution for the repulsive case, whereas $\operatorname{sech}(kx)$ is a solution for the attractive case.

2.2   Generic solution with time-independent trap and interaction

In order to analyze more general scenario, we reconsider Eq.2.1, which is reduced to an effective one dimensional equation by assuming the interaction energy of atoms to be much less than the transverse kinetic energy of the condensate [22]:

$$\Psi(\mathbf{r}, t) = \frac{1}{\sqrt{2\pi a_B a_\perp}}\psi\left(\frac{z}{a_\perp}, \omega_\perp t\right)\exp\left(-i\omega_\perp t - \frac{x^2 + y^2}{2a_\perp^2}\right). \tag{2.12}$$

The GP equation then reduces to the following one dimensional nonlinear Schrödinger equation in dimensionless units:

$$i\partial_t\psi = -\frac{1}{2}\partial_{zz}\psi + \gamma(t)|\psi|^2\psi + \frac{1}{2}M(t)z^2\psi + i\frac{g(t)}{2}\psi. \tag{2.13}$$

Here, $\gamma = 2a_s(t)/a_B$, $M(t) = \omega_0^2(t)/\omega_\perp^2$, $g(t) = \eta(t)/\hbar\omega_\perp$, $a_\perp = (\hbar/m\omega_\perp)^{1/2}$ and a_B is the Bohr's radius. Generically, $M(t)$ time dependent. A constant $M(t)$ implies an oscillator potential which, depending on whether $M > 0$ or $M < 0$, can be confining or expulsive respectively. In order to explicate the cases of a number of experimentally realizable profiles of the time dependent trapping potential and to obtain corresponding analytical solutions, the following ansatz solution has been assumed:

$$\psi(z,t) = \sqrt{A(t)}F[A(t)\{z - l(t)\}]\exp[i\Phi(z,t) + G(t)/2], \tag{2.14}$$

where, $G(t) = \int_0^t g(t')dt'$, $l(t) = \int_0^t v(t')dt'$ and the phase is assumed to have a quadratic form:

$$\Phi(z,t) = a(t) + b(t)z - \frac{1}{2}c(t)z^2, \tag{2.15}$$

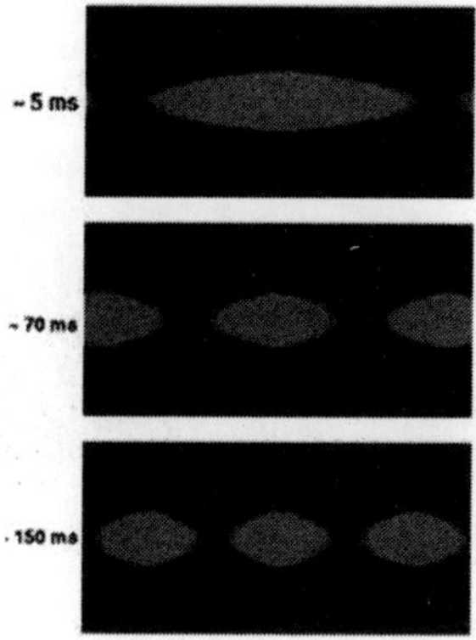

Figure 1: *Plots depicting the snapshots of soliton trains.*

where, $a(t) = a_0 + \frac{\lambda-1}{2}\int_0^t A^2(t')dt'$, and $c(t)$ obeys a Riccati type equation,

$$\frac{dc(t)}{dt} - c^2(t) = M(t). \tag{2.16}$$

Above equations show that $A(t), \gamma(t)$ and $l(t)$ are all non-trivially related to $c(t)$, the phase component. The following consistency conditions are also found:

$$\left.\begin{array}{c} \gamma(t) = \gamma_0 e^{-G(t)/2} A(t)/A_0, \quad b(t) = A(t) \\ A(t) = A_0 \exp\left\{\int_0^t c(t')dt'\right\}, \quad A_0 > 0 \\ \frac{dl(t)}{dt} - c(t)l(t) = b(t). \end{array}\right\} \tag{2.17}$$

Interestingly, via a change of variable,

$$c(t) = -\frac{d\ln[\phi(t)]}{dt}, \tag{2.18}$$

this equation can be expressed as a Schrödinger eigenvalue problem:

$$-\phi''(t) - M(t)\phi(t) = 0. \tag{2.19}$$

Utilizing this connection, below we show that, one can identify a soliton configuration, corresponding to each solvable quantum mechanical system. This gives us freedom to control the dynamics of BEC in a number of ways, which are analytically tractable. We will demonstrate only a few experimentally realizable examples in the text, although a host of time dependent oscillator frequencies can be addressed.

On substituting the ansatz Eq.2.14 in Eq.2.13 and from the consistency conditions, the differential equation for the function F is obtained, in terms of the new variable $T = A(t)[z - l(t)]$, as:

$$F''(T) - \lambda F(T) + 2\kappa F^3(T) = 0, \quad \text{where } \kappa = -\frac{\gamma_0}{A_0}. \tag{2.20}$$

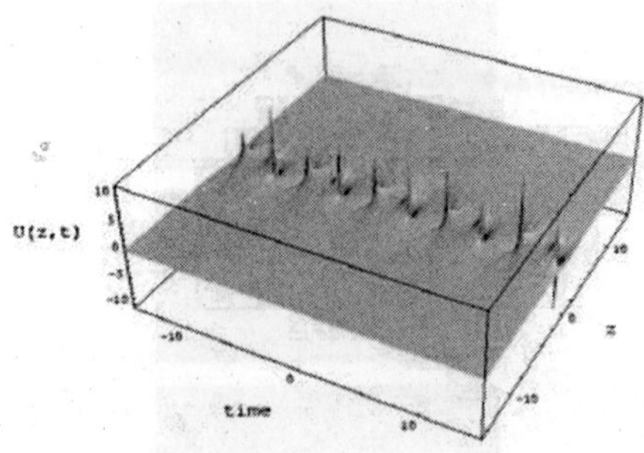

Figure 2: *Spatio-temporal condensate wave-function pattern in confining oscillator potential with constant attractive coupling.*

It is to be noted that Eq.2.20 possesses a number of different solutions, which can be identified as the twelve Jacobian elliptic functions *e.g.*, $\mathrm{cn}(T, m)$, $\mathrm{sn}(T, m)$ and $\mathrm{dn}(T, m)$ etc. Here m is the modulus parameter, with the range $0 \leq m \leq 1$. These functions becomes trigonometric and hyperbolic, in the limiting cases $m = 0$ and 1, respectively, and interpolate between the same for intermediate values of m. Appropriate parameter regimes give rise to a wide variety of solutions, including localized bright and dark solitons, and also soliton trains, akin to experimental observations of Strecker *et al.* [16].

Fig.1 depicts soliton trains at different times, for a particular set of parameter values, whereas Fig.2 shows the spatio-temporal behavior of the BEC wavefunction, subjected to transverse oscillator confinement under constant attractive coupling [24].

3   COMPLEX SOLITON SOLUTIONS IN A TRAP

We now explore more general complex profile solutions, known as Lieb mode in BEC [26]. For this purpose, the following ansatz solution is used,

$$\psi(z, t) = U(z, t)e^{i\Phi(z,t)}, \quad \text{with,} \quad U(z, t) = \sqrt{A(t)\sigma(T)} \exp[i\chi(T) + \frac{G(t)}{2}],$$

where, $T = A(t)(z - l(t))$ and $G(t) = \int_0^t g(t')dt'$. The- phase has a quadratic form exhibiting chirping: $\Phi(z, t) = a(t) + b(t)z - \frac{1}{2}c(t)z^2$, where $a(t) = a_0 - \frac{1-\bar{\mu}}{2}\int_0^t A^2(t')dt'$ with $\bar{\mu} = \mu + \lambda$, where, μ and λ, respectively, are the chemical potential and a constant parameter controlling the energy of the excitation [19]. Akin to the previous case, the chirped phase $c(t)$ can be determined from the following Riccati type equation, $\frac{\partial c(t)}{\partial t} - c^2(t) = M(t)$. The parameters are similar as before.

From current conservation, amounting to solving the imaginary part of the GP equation, one obtains, $\frac{\partial \chi}{\partial T} = u(1 - \frac{\sigma_0}{\sigma})$, where, σ_0 is the equilibrium density of the atoms in the center of mass frame T. Asymptotically, when $\sigma \to \sigma_0$,

$\frac{\partial \chi}{\partial T} \to 0$. In this frame, the density equation can be cast in the convenient form [?],

$$\left(\frac{\partial \sqrt{\sigma}}{\partial T}\right)^2 = (\kappa\sigma - u^2)\frac{(\sigma - \sigma_0)^2}{2\sigma}, \tag{3.1}$$

The solution takes the self-similar form,

$$\sigma(z,t) = \sigma_0 - \sigma_0 \cos^2\theta \, \text{sech}^2\left[\frac{A(t)(z - l(t))\cos\theta}{\zeta}\right], \tag{3.2}$$

where the Mach angle is given by, $\theta = \sin^{-1}\frac{V}{C_s} = \sin^{-1}\frac{u}{c_s}$.

The condensate wavefunction ψ can be explicitly written as,

$$\psi(z,t) = \sqrt{A(t)\sigma_0}\left[i\sin\theta + \cos\theta\tanh\left(\frac{\cos\theta}{\zeta}A(t)(z - l(t))\right)\right]e^{\frac{G(t)}{2}+i\Phi(z,t)}. \tag{3.3}$$

For the above solution, the phase step is a direct attribute of the complex envelope soliton.

4   CONTROLLING THE SOLITON DYNAMICS

For illuminating the control of soliton dynamics, we study the bright soliton trains of the form

$$\psi(z,t) = \sqrt{A(t)}\,\text{cn}\,(T/\tau_0, m)\exp[i\Phi(z,t)], \tag{4.1}$$

with $\tau_0^2 = -mA_0/\gamma_0$ and $\lambda = (2m - 1)/\tau_0^2$. For $m = 1$, equation Eq. (4.1) yields a bright soliton, where, for $\gamma_0 > 0$, one observes dark soliton trains of the form,

$$\psi(z,t) = \sqrt{A(t)}\,\text{sn}\,(T/\tau_0, m)\exp[i\Phi(z,t)], \tag{4.2}$$

with $\tau_0^2 = mA_0/\gamma_0$, $\lambda = -(m + 1)/\tau_0^2$. In the limiting case $m \to 1$, the last solution is a dark soliton.

4.1   Dynamics of the BEC with c(t) = ± B coth (t)

For the trap frequency of the confining trap being modulated using $c(t) = \pm B\coth(t)$, $B > 0$, the analytical solutions of the NLSE are obtained as given below.

Case (a)   For $c(t) = B\coth(t)$, $B > 0$, we have,

$$M(t) = -B^2 - B(B + 1)\text{cosech}^2(t). \tag{4.3}$$

$M_0 = -B^2$ signifies an expulsive oscillator scenario. It is to be noted that $V(x) = B(B + 1)\text{cosech}^2(x)$ is the supersymmetric Rosen-Morse potential, corresponding to superpotential $W = B\coth(x)$ [27]. $t = 0$ is a singularity of $V(t)$, causing a sudden change in the trap frequency. From the previous consistency conditions in Eqs2.17,

$$a(t) = a_0 - \frac{(\lambda - 1)A_0^2}{2}(\sinh(t) - 4t) \tag{4.4}$$

$$A(t) = A_0 \left(\frac{\sinh(t)}{\sinh(t_0)}\right)^B, \quad \gamma(t) = \gamma_0 \left(\frac{\sinh(t)}{\sinh(t_0)}\right)^B, \quad l(t) = l_0 \left(\frac{\sinh(t_0)}{\sinh(t)}\right)^B. \tag{4.5}$$

using these relations, bright and dark soliton trains are obtained:

$$\psi(z,t) = \sqrt{A_0 \sinh(t)}\, \text{cn}\left(A_0 \sinh(t)\frac{[z - l_0\text{cosech}(t)]}{\tau_0}, m\right)\exp(i\Phi(z,t)), \tag{4.6}$$

$$\psi(z,t) = \sqrt{A_0 \sinh(t)}\, \text{sn}\left(A_0 \sinh(t)\frac{[z - l_0\text{cosech}(t)]}{\tau_0}, m\right)\exp(i\Phi(z,t)) \tag{4.7}$$

The corresponding solution dynamics are plotted in fig 3(a) and 3(b), with matter wave density $|\psi(z,t)|^2$ is plotted as a function of t and z. The soliton trains get amplified with time away from the origin. For $m \to 1$, Eqs 4.6 and 4.7 yield bright and dark solitons, plotted in fig 3(c) and 3(d) respectively. They are highly localized and diverge from the $t = 0$ line in the $t - z$ plane. On re-labeling the initial time as t_0, instead of $t = 0$ in consistency conditions, and varying l_0, the center of mass location can be changed. The non-linearity $\gamma(t)$ is plotted as a function of t, for the attractive ($\gamma_0 < 0$) and the repulsive ($\gamma_0 > 0$) couplings in fig 3(e) and 3(f) respectively. It is observed that for a given γ_0, the sign of the non-linearity is unchanged, implying attractive or repulsive coupling regimes. The amplification and location of the solitons can be changed by changing the constants B, A_0, γ_0, l_0 and t_0.

Case (b)   For $c(t) = -B\coth(t)$, $B > 0$, $M(t) = -B^2 - B(B-1)\text{cosech}^2(t)$. We again are considering the expulsive oscillator scenario with the supersymmetric Posčhl-Teller potential $B(B-1)\text{cosech}^2(x)$ [27]. Interestingly, for $B = 1$, this potential maps onto the free particle problem, studied in [24]. On substituting $c(t)$ in Eqs 2.17,

$$A(t) = A_0 \left(\frac{\sinh(t_0)}{\sinh(t)}\right)^B, \quad \gamma(t) = \gamma_0 \left(\frac{\sinh(t_0)}{\sinh(t)}\right)^B, \quad l(t) = l_0 \left(\frac{\sinh(t)}{\sinh(t_0)}\right)^B. \tag{4.8}$$

We again obtain both bright and dark soliton train profiles with localized solitons, plotted in figure 4(a) - 4(f). As in case (a), though $V(t)$ is singular at $t = 0$, the soliton response is entirely different. In this case, the soliton train amplitude very high around $t = 0$ and they decay away from the origin, showing the spread of the soliton profile with time. By changing A_0, the amplitude and the decay time can be controlled . Figures 4(c) and 4(d) depict solitons which are highly localized, as before, with the exception of convergence towards $t = 0$ in the $t - z$ plane. Figures 4(e) and 4(f) show that, though for a given γ_0, the non-linearity does not change sign, it's magnitude rapidly increase about $t = 0$, whereas in the previous case, $\gamma(t)$ varies slowly with time.

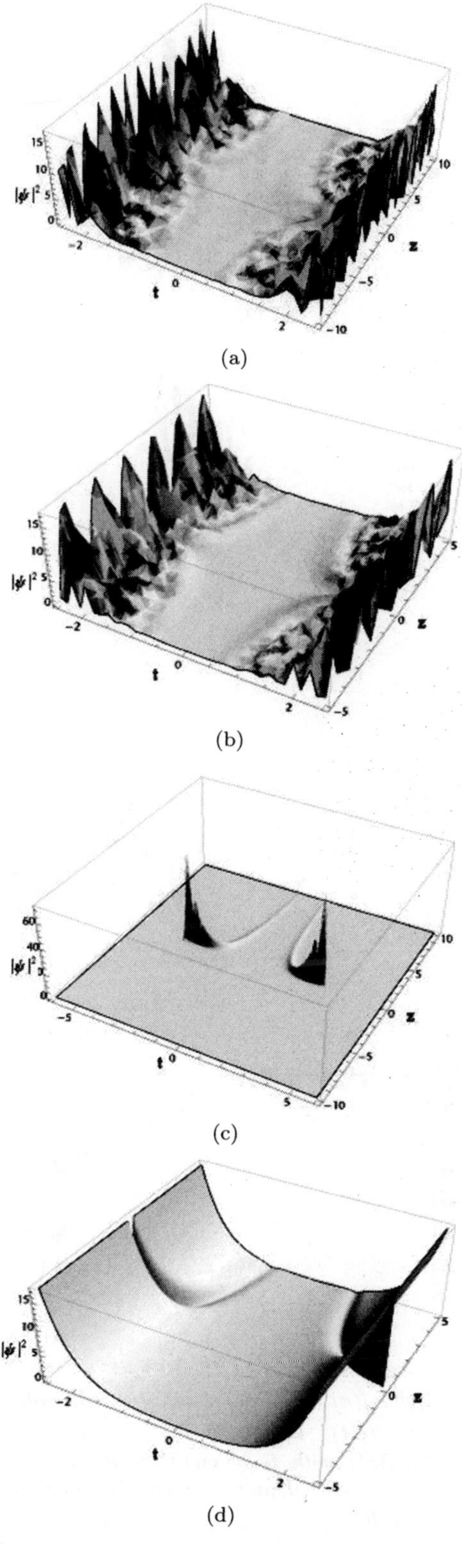

(a)

(b)

(c)

(d)

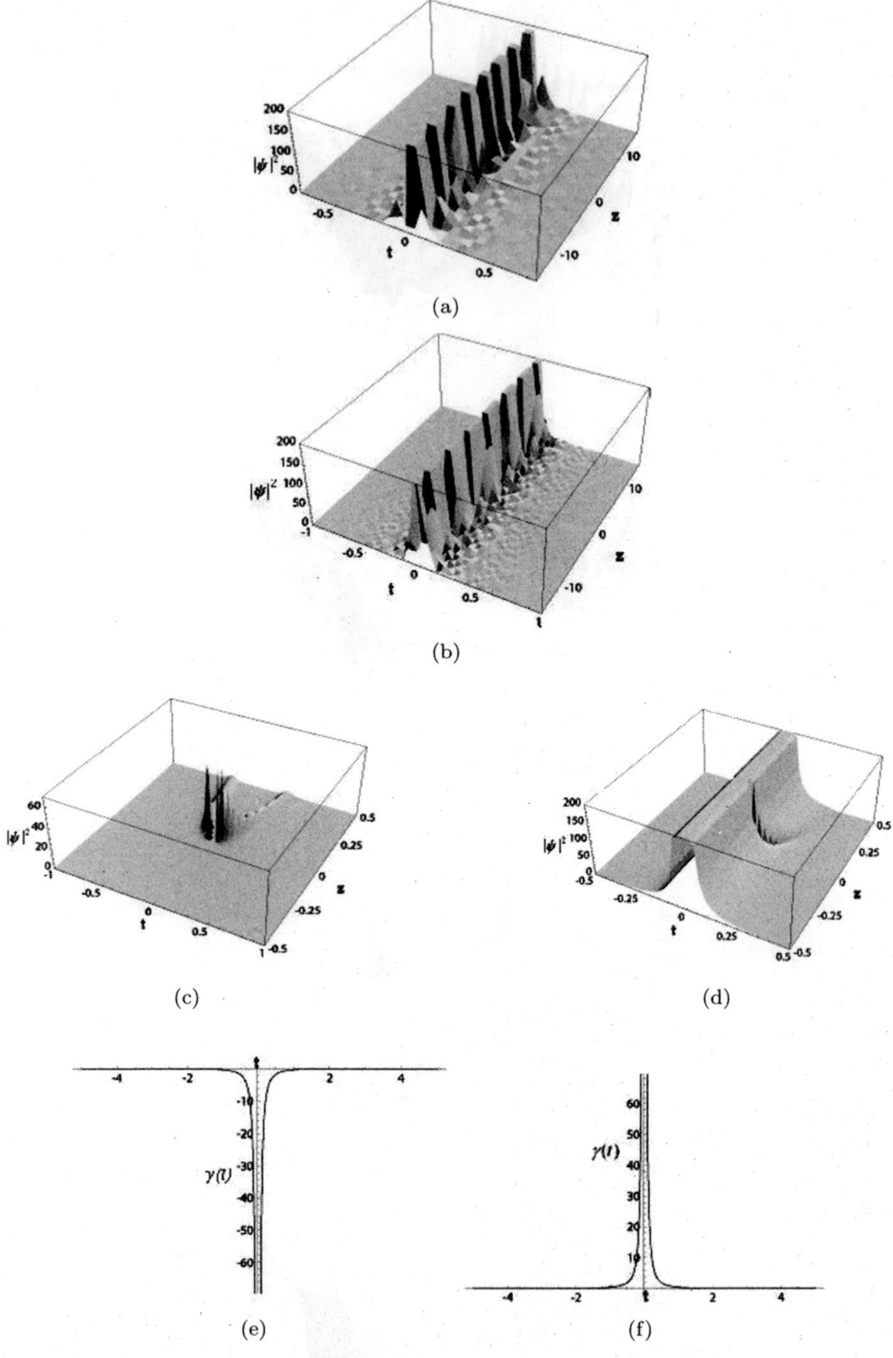

Figure 4: *Plots of bright $[(a),(c)]$ and dark $[(b),(d)]$ soliton solutions in the expulsive oscillator regime $(c(t) = -\coth(t))$, $\psi(z,t) = \sqrt{A(t)}F\{(A(t)[z - l(t)])/\tau_0, m\} \exp[\imath a(t) - \frac{\imath}{2}c(t)z^2]$ with $F = \mathrm{cn}\,(T/\tau_0, m)$ and $F = \mathrm{sn}(T/\tau_0, m)$ respectively, are shown. Plots (a)-(d) depict $|\psi(z,t)|^2$ versus t and z and (e) and (f) show the variation of $\gamma(t)$ with t.*

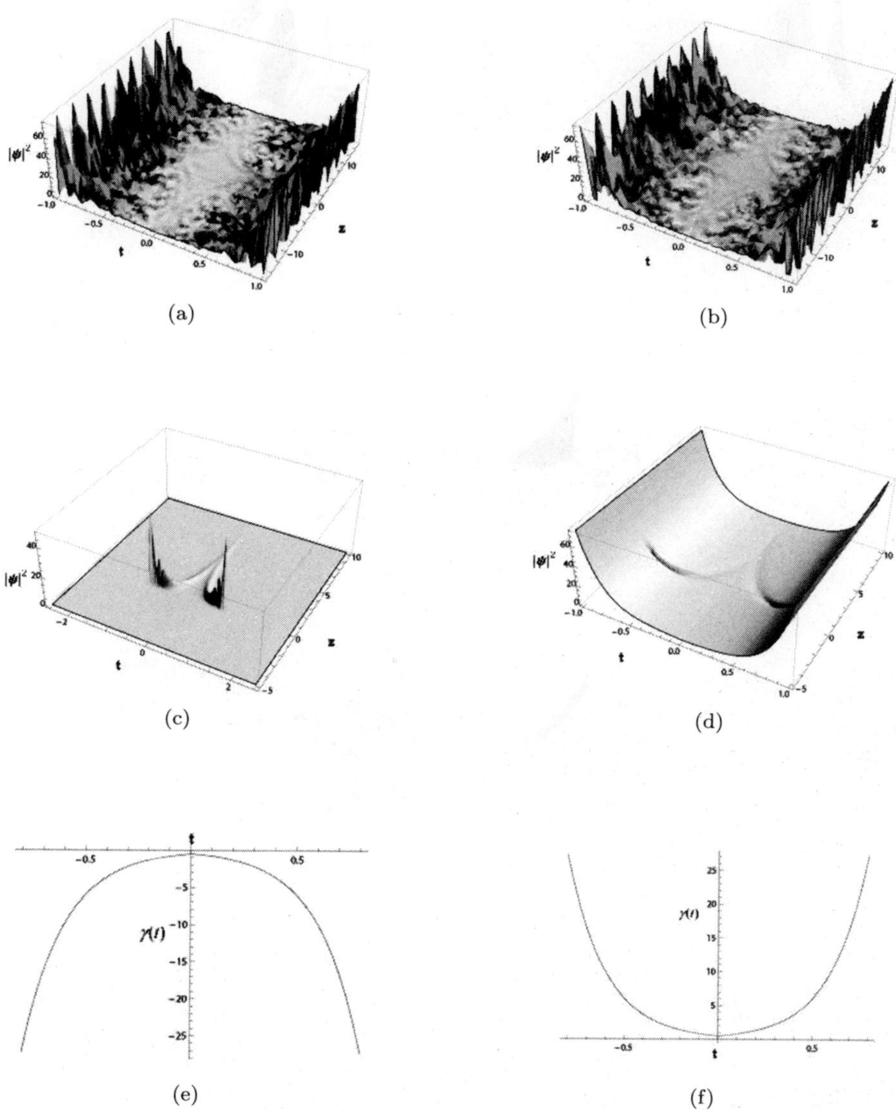

Figure 5: *Plots depicting attractive δ potential in the expulsive oscillator regime, $\phi(t) = \sqrt{V_0/2}\exp(-V_0|t|/2)$. Bright [(a),(c)] and dark [(b),(d)] soliton solutions, $\psi(z,t) = \sqrt{A(t)}F\{(A(t)[z-l(t)])/\tau_0, m\}\exp[\imath a(t) - \frac{\imath}{2}c(t)z^2]$ with $F = \mathrm{cn}\,(T/\tau_0, m)$ and $F = sn(T/\tau_0, m)$ respectively, are shown. Plots (a)-(d) depict $|\psi(z,t)|^2$ versus t and z, while (e) and (f) show the variation of $\gamma(t)$ with t.*

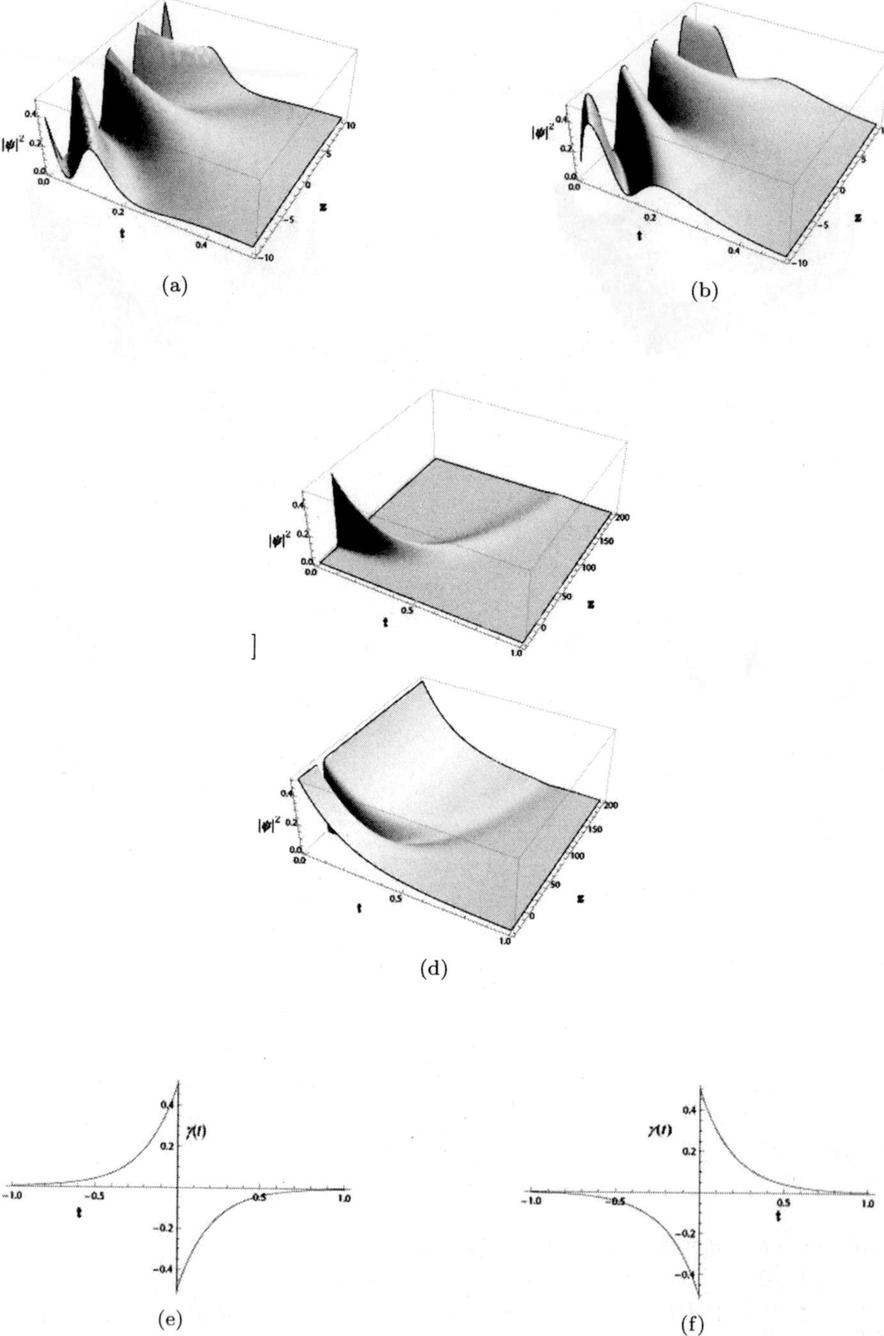

Figure 6: *Plots depicting attractive δ potential in the expulsive oscillator regime, $\phi(t) = \sqrt{2/V_0}\exp(V_0 t/2)$. Bright [(a),(c)] and dark [(b),(d)] soliton solutions, $\psi(z,t) = \sqrt{A(t)}F\{(A(t)[z - l(t)])/\tau_0, m\}\exp[\imath a(t) - \frac{\imath}{2}c(t)z^2]$ with $F = \mathrm{cn}\,(T/\tau_0, m)$ and $F = sn(T/\tau_0, m)$ respectively, are shown. (a)-(d) show $|\psi(z,t)|^2$ versus t and z, while (e) and (f) depict the variation of $\gamma(t)$ with t.*

4.2   Response of BEC to delta kicks in the trap frequency

The trap frequency, with an attractive delta potential at the origin, is varied and the response of the solitons to a sudden kick is analyzed. We consider,

$$V(t) = -V_0\,\delta(t), \quad V_0 > 0, \tag{4.9}$$

resulting in solutions of Eq.2.19:

$$\phi_1(t) = \sqrt{\frac{V_0}{2}}\exp(-\frac{V_0}{2}t) \quad (t > 0)$$

$$= \sqrt{\frac{V_0}{2}}\exp(\frac{V_0}{2}t) \quad (t < 0) \tag{4.10}$$

$$\phi_2(t) = \sqrt{\frac{V_0}{2}}\exp(\frac{V_0}{2}t) \quad (t > 0)$$

$$= -\sqrt{\frac{V_0}{2}}\exp(-\frac{V_0}{2}t) \quad (t < 0) \tag{4.11}$$

where $M_0 = -\frac{V_0^2}{2}$. That the second solution of a second order differential equation is obtainable from the first solution using the relation $\phi_2(t) = \phi_1(t)\int^t[dt'(\phi_1(t'))^{-2}]$, was used to obtain $\phi_2(t)$. $c(t)$ can be written in terms of $\phi(t)$ using Eq.2.18. The consistency conditions in terms of $\phi(t)$ are obtained as.

$$A(t) = A_0\frac{\phi(0)}{\phi(t)} \; ; \; \gamma(t) = \gamma_0\frac{\phi(0)}{\phi(t)} \, , \; l(t) = l_0\frac{\phi(t)}{\phi(0)}. \tag{4.12}$$

In the following, the soliton profiles are studied using the two solutions in Eqs 4.10 and 4.11 separately.

Case (a)   On considering Eqs 4.10 and 2.18, one obtains $c(t) = V_0/2$ for $t > 0$ and $c(t) = -V_0/2$ for $t < 0$. Eq.4.12 yields,

$$A(t) = A_0\exp(\frac{V_0}{2}|t|)\,, \; \gamma(t) = \gamma_0\exp(\frac{V_0}{2}|t|)\,, \; l(t) = l_0\exp(-\frac{V_0}{2}|t|) \tag{4.13}$$

and also,

$$a(t) = a_0 + \frac{(\lambda - 1)A_0^2}{2V_0}\left[\exp(V_0|t|) - 1\right]. \tag{4.14}$$

Substitution of $A(t), \gamma(t)$ and $l(t)$ in Eq.4.1 yields the solutions corresponding to bright and dark soliton profiles. These are plotted in figure 5, also showing the behavior of the nonlinearity. There, it is observed that rapid amplification of the soliton trains over time is caused by sudden temporal kick to the soliton profile. The solitons are highly localized and diverge from the origin, similar to the case (a) of the previous section. Variation V_0 results in the creation of a soliton with the required amplitude. For this case, the non-linearity $\gamma(t)$ is obtained to be an exponential function, making the scattering length $a_s(t)$ exponential in time [28]. It is known that an NLSE, with an exponentially varying scattering length, governs the BEC dynamics in an expulsive parabolic potential. The conditions for these systems being one dimensional and the parameter range corresponding to a stable condensate are discussed in [13].

Case (b)   The above analysis has been repeated considering Eq.4.11. In this case, the nonlinearity $\gamma(t)$ changes sign with the transition from negative time domain to positive time domain. The soliton dynamics has been studied for $t > 0$ yielding,

$$c(t) = -\frac{V_0}{2}\,,\ a(t) = a_0 - \frac{(\lambda - 1)A_0^2}{2V_0}\left[\exp(-V_0 t) - 1\right], \qquad (4.15)$$

$$A(t) = A_0 \exp(-\frac{V_0}{2}t)\,,\ \gamma(t) = \gamma_0 \exp(-\frac{V_0}{2}t)\,,\ l(t) = l_0 \exp(-\frac{V_0}{2}t). \qquad (4.16)$$

Figure 6 depicts the soliton profiles. The soliton train amplitude is observed to decays rapidly with increase in t, unlike the previous case. The trains are sustained for longer time for smaller values of V_0, enabling control over the decay of the solutions through variation of V_0.

5   BEC IN OPTICAL LATTICE WITH CUBIC NONLINEARITY

BEC in a periodic optical lattice is a subject of current research interest. We study the case of shallow lattice, wherein the mean field equation gives,

$$i\frac{\partial\psi(z,t)}{\partial t} = \left(-\frac{1}{2}\frac{\partial^2}{\partial z^2} + g|\psi(z,t)|^2 + V(z) - \mu\right)\psi(z,t), \qquad (5.1)$$

where, $V(z) = V_0\cos^2(z)$ is the optical lattice potential and the mass is scaled to unity. The pure two-body interactions $(g_1 = g)$ induce trigonometric solutions for the GP equation, of the form [19],

$$\psi(z,t) = \sqrt{\left(a - \frac{V_0}{g}\cos^2(z)\right)}\,e^{i\chi_2(z) - i\omega_2 t}, \qquad (5.2)$$

with background a being a free parameter. Here, $\chi_2(z) = \dfrac{c_2 \tan^{-1}[\sqrt{1 - \frac{V_0}{ag}}\tan(z)]}{\sqrt{a(a - \frac{V_0}{g})}}$ is the non-trivial phase, $\omega_2 = \frac{1}{2} - \mu + ga$ and $c_2^2 = \dfrac{(1 - 2\mu - 2\omega)(1 - 2\mu - 2\omega + 2V_0)}{4g^2}$ is the integration constant. The background a: $\nu_{SF} = a - \frac{V_0}{2g}$ is related with the average atom number density per lattice site. The filling fraction here is independent of n, the number of lattice sites. The presence of superfluid component is being indicated by the non-trivial phase $(c_2 \neq 0)$, along with a constant flow density $J_2 = c_2$. The ground state energy of the superfluid state is given by,

$$
\begin{aligned}
E_{SF} &= \frac{\pi g}{8}\left(\frac{3V_0^2}{g^2} - \frac{8aV_0}{g} + 8a^2\right) - \frac{\pi c_2^2}{\sqrt{a\left(a - \frac{V_0}{g}\right)}} \\
&\quad + \frac{\pi}{2}\left(2a - \frac{V_0}{g}\right) + \frac{\pi}{2}\sqrt{a\left(a - \frac{V_0}{g}\right)} \\
&\quad - \pi\mu\left(2a - \frac{V_0}{g}\right) + \pi V_0\left(a - \frac{3V_0}{4g}\right),
\end{aligned}
\qquad (5.3)
$$

indicating a branch cut at $a = a_1 = 0$ and $a_2 = \frac{V_0}{g}$, with a_{1T} and a_{2T} being the critical values of background. The super-current ($J_2 = c_2$) vanishes for $a_{1T} = 0$, $\omega_2 + \mu = \frac{1}{2}$ and at $a_{2T} = \frac{V_0}{g}$, $\omega_2 + \mu = \frac{1}{2} + ag$, causing the superfluid to transition to the insulating phase. Though the energy is continuous at the transition point (T), the first derivative is discontinuous. This indicates the phase transition to be of first order. In the insulating phase, the wave-function is, $\psi_{I_1}(z,t) = \sqrt{\left(-\frac{V_0}{g}\right)}\cos z\, e^{-i(\frac{1}{2}-\mu)t}$, with energy, $E_I^1 = \frac{\pi V_0}{8g}(8\mu - 3V_0 g - 4)$.

$\nu_I^1 = -\frac{V_0}{2g}$ is the corresponding number density, which indicates that in this insulating phase, the potential strength and interaction need to have opposite signatures. The second insulating phase at a_{2T} is characterized by $\psi_{I_2}(z,t) = \sqrt{\left(\frac{V_0}{g}\right)}\sin z\, e^{-i(\frac{1}{2}-\mu+ag)t}$, with energy $E_I^2 = \frac{\pi V_0}{8g}(5V_0 g - 8\mu + 4)$, and the average number density $\nu_I^2 = \frac{V_0}{2g}$.

Here, the interaction and potential strength should have same signature. These two phases, for a positive V_0, belong to attractive and repulsive regimes, respectively. It is noteworthy that both these insulating phases only exist for a given number density of atoms, for a fixed V_0 and coupling .

In one dimension, on considering the possibilities of charge density type of ground states, it is natural to ask whether the other type of solutions are allowed in this system, which still commensurates with lattice periodicity. Such an exact solution was obtained, exhibiting *rational* character, as

$$\psi_I(z,t) = \frac{a + b\cos^{\alpha}(z) + c\cos^{\delta}(z)}{1 + d\cos^{\beta}(z)} e^{-i\omega_3 t}, \tag{5.4}$$

with $\alpha = \beta = 1$ and $\delta = 2$. The superfluid phase is devoid of these solutions. This insulating phase corresponds to the parameters:

$$a = -\frac{3}{4}\sqrt{-\frac{V_0}{g}} \mp \frac{1}{4}\sqrt{\frac{12}{g} - \frac{9V_0}{g}}, b = \sqrt{-\frac{V_0}{g}},$$

$$c = \pm\frac{V_0}{2}\left(\sqrt{\frac{12}{g} - \frac{9V_0}{g}} - 3\sqrt{-\frac{V_0}{g}}\right) \text{ and}$$

$$d = \pm\frac{\sqrt{-V_0 g}}{2}\left(-\sqrt{\frac{12}{g} - \frac{9V_0}{g}} + 3\sqrt{-\frac{V_0}{g}}\right),$$

with, $\omega_3 = \frac{1}{2} - \mu$. It is observed here that the non-perturbative rational solutions with dual frequency character results from the competition of the lattice potential with the non-linearity, where both commensurate with the lattice potential. Interestingly, non-linearity and dispersion compensate each other in case of solitons, which leads to both stable localized solutions and periodic non-sinusoidal cnoidal waves. The dispersion affects the character of the solution, only marginally, at present. The non-singular solutions yields $V_0 \leq -\frac{1}{6}$ in repulsive interaction regime ($g > 0$), and $V_0 \geq \frac{4}{3}$ in attractive regime ($g < 0$).

Notably, the present density wave solution has a background, depending on both potential and the coupling, in contrast to the earlier found insulating phases independent of the background.

As $d \to 0, \pm 1$, the energy function is non-analytic. For $d = 0$, the Padé type of solutions, corresponding to the insulating phase, reduces to the insulating phase obtained earlier.

6   SOLITON SOLUTIONS IN PRESENCE OF STRONGLY COUPLED BEC IN A TRAP

In order to study the strong coupling scenario of BECs, we reconsider Eqs 2.1-2.9. Specifically in the strong coupling limit, $G(x, y; \sigma)$ varies negligibly in z direction, with corresponding error in numerical energy estimation being 10^{-7} [22]. Hence, Eq.2.9 is further justified, from which, in this regime, on considering $2aN|f|^2 \gg 1$ (but $N|\psi|^2 a \ll 1$ to satisfy the diluteness of BEC), one obtains the suitably normalized condensate equation [23]:

$$i\hbar \frac{\partial}{\partial t} f = \left[-\frac{\hbar^2}{2M} \frac{\partial^2}{\partial z^2} + 2\hbar\omega_\perp a^{1/2} \left(|f| - \sigma_0^{1/2} \right) \right] f, \qquad (6.1)$$

σ_0 being the equilibrium density of the atoms far away from the axis.

The the strongly interacting system with non-linear excitations can be probed through an ansatz solution,

$$f(z, t) = e^{i(kz - \omega t)} \rho(\xi), \qquad (6.2)$$

with a slowly varying envelope profile $\rho(\xi)$. Here, $\xi = \alpha(z - vt)$ and $v = \frac{\hbar}{M} k$, with $\rho(\xi)$ satisfying,

$$\alpha^2 \rho'' + g\rho^2 + \epsilon\rho = 0, \qquad (6.3)$$

where,

$$\begin{aligned}
g &= -4M\omega_\perp a^{1/2}/\hbar, \\
\epsilon &= 2M\omega/\hbar + 4M\omega_\perp (\sigma_0 a)^{1/2}/\hbar - k^2.
\end{aligned} \qquad (6.4)$$

It is straightforward to check that Eq. (6.3) is solved by the following ansatz solution,

$$\rho(\xi) = A + B\,cn^2(\xi, m), \qquad (6.5)$$

where A, B are constant parameters to be determined. A serves here as the background. Through substitution, we obtain,

$$\begin{aligned}
A &= \frac{1}{2g} \left[4\alpha^2(1 - 2m) - \epsilon \right], \\
B &= \frac{6}{g} \alpha^2 m,
\end{aligned} \qquad (6.6)$$

along with a relation between the width α and ϵ:

$$\epsilon^2 = 16\alpha^4 \left(m^2 - m + 1 \right). \qquad (6.7)$$

For BECs with profile as Eq. (6.2), the *effective* linear term ϵ (the effective chemical potential), given in Eq. (6.4), is a sum of three terms, only second one being positive definite. Hence, both the roots of the above equation are physically allowed.

The general periodic solution can then be written as,

$$\rho(\xi) = \frac{2\alpha^2}{g}\left[(1 - 2m) \mp \sqrt{m^2 - m + 1} + 3mcn^2(\xi, m)\right],\qquad (6.8)$$

representing a soliton train. The special case $m = 1$ yields localized solutions, corresponding to two specific values, $\epsilon = \pm 4\alpha^2$. The *positive* root requires presence of the background A, and the envelope takes the form,

$$\rho(\xi) = -\frac{\epsilon}{g}\left[1 - \frac{3}{2}sech^2(\xi)\right],\qquad (6.9)$$

which is a localized W-type soliton.

The Vakhitov-Kolokolov criterion [29] points out that the integral $N(\epsilon) = \int |\rho(\xi)|^2 d\xi$, when varied with respect to ϵ, indicates the stability of the solution $\rho(\xi)$. In the present case,

$$\frac{dN(\epsilon)}{d\epsilon} = -\frac{6\epsilon}{g^2},\qquad (6.10)$$

requiring that $\epsilon > 0$ for the stability of the solution, which is consistent with $\epsilon = 4\alpha^2$. For this case we find,

$$k^2 \geq 2\frac{M\omega}{\hbar} - |\epsilon|,\qquad (6.11)$$

setting a lower limit $\frac{\hbar|\epsilon|}{2M}$ for ω, if it is positive, in order to generate the W-type soliton. Such bound is not there for negative frequency. Physical, negative frequency is justified from the fact that the Schrödinger equation, having first derivative in time, admits both positive and negative values of energy $E = \hbar\omega$, forming two distinct sectors.

For the *negative* root of Eq. (6.7), the background vanishes for $m = 1$, and we get,

$$\rho(\xi) = -\frac{3\epsilon}{2g}sech^2(\xi),\qquad (6.12)$$

which, under the Vakhitov-Kolokolov criterion, yields,

$$\frac{dN(\epsilon)}{d\epsilon} = 6\frac{\epsilon}{g^2},\qquad (6.13)$$

and hence is stable again, as $\epsilon < 0$. In this case, $k^2 \geq |\epsilon| + 2\frac{M\omega}{\hbar}$, indicating that a finite wave number is needed to excite the solution, for the frequency being positive. There arise no such condition for negative frequency unless $|\omega| = \hbar|\epsilon|/2M$. This type of velocity restricted solitons have been identified in higher order non-linear Schrödinger equation, in the femtosecond domain [30], relevant for optical fiber pulses . The obtained analytical solutions exist only over the critical velocity, below which the soliton is numerically unstable.

We have numerically evolved the W-type solution, using the Crank-Nicolson finite difference method, which is unconditionally stable. The initial profile has been taken as $\psi(z, t = 0) = \psi(z, t = 0) + \tilde{\epsilon}$, with $\tilde{\epsilon}$ being a function, assuming random values at each point. Figure 7 shows that the W-type soliton remains unchanged with minor perturbation, where $\tilde{\epsilon}$ is taken to be 10 percent of the peak value of ψ. The minima positions and the width remained unaltered. The evolution was also checked to be unitary, conserving the number of particles, upto second order in dt.

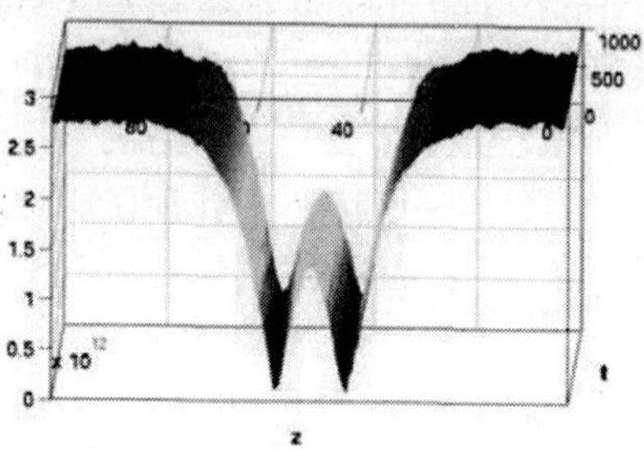

Figure 7: *Direct numerical evolution of W type soliton density profile. The temporal evolution indecates the stability of the W-soliton.*

CONCLUSION

In conclusion, the mean field Gross-Pitaevskii equation predicts a host of phenomena, which are directly verified by experiments. Here, we have shown, how dark, bright and so-called grey solitons can be obtained as exact solutions of the GP equation in the cigar-shaped geometry. In the presence of the trap, one can effectively control these excitations. Int is found that, the solutions of the elliptic function equation, together with a linear Schrödinger eigenvalue equation, suffice to describe the dynamics of the solitons, in the presence of time varying scattering length and trap frequency. Exact solutions are also found for the GP equation in an optical lattice, clearly showing the superfluid and the insulating phases, separated by a dynamical phase transition. The strongly couple regime is amenable for exact treatment, and yields solutions particularly different from the weak coupling sector. Hence, the analysis of the non-linear GP equation in other dimensions, with different confining potentials is a rich area of research. In particular, the dynamics of vortices in the plane is a subject of great interest. Another area of significant interest is the physics of deep optical lattice, wherein the Bose-Hubbard model provides the appropriate description. It can also be partly explored through discrete non-linear Schrödinger equation.

References

[1] S. N. Bose, Z. Phys. **26**, 178 (1924)

[2] A. Einstein, Sitzungsber. K. Preuss. Akad. Wiss., Phys. KI. **3**, (1925); A. Einstein, Sitzungsber. K. Preuss. Akad. Wiss., Phys. KI. **261**, (1924).

[3] M. H. J. Anderson, J. R. Ensher, M. R. Matthews, C. E. Wienman and E. A. Cornell, Science **269**, 198 (1995).

[4] K. B. Davis, M.-O. Mewes, M. R. Andrews, N. J. van Druten, D. S. Durfee, D. M. Kurn and W. Ketterle, Phys. Rev. Lett. **75**, 3969 (1995).

[5] C. C. Bradley, C. A. Sackett, J. J. Tollett and R. G. Hulet, Phys. Rev. Lett. **75**, 1687 (1995).

[6] J. Stenger, S. Inouye, M. R. Andrews, H.-J. Miesner, D. M. Stamper-Kurn, W. Ketterle, Phys. Rev. Lett. **82**, 2422 (1999); J. L. Roberts, N. R. Claussen, J. P. Burke Jr., C. H. Greene, E. A. Cornell, C. E. Wieman, Phys. KhaRev. Lett. **81**, 5109(1998); S. L. Cornish, N. R. Claussen, J. L. Roberts, E. A. Cornell, C. E. Wieman, Phys. Rev. Lett. **85**, 1795 (2000).

[7] V. E. Zakharov and A. B. Shabat, Sov. Phys. JETP **34**, 62 (1972).

[8] L. Salasnich, A. Parola, and L. Reatto, Phys. Rev. A **69**, 045601 (2004).

[9] A. Munoz Mateo, and V. Delgado, Phys. Rev. A **74**, 065602 (2006).

[10] V. E. Zakharov and A. B. Shabat, Zh. Eksp. Teor. Fiz. **64**, 1627 (1973) [Sov. Phys. JETP **37**, 823 (1973)]; J. R. Taylor, *Optical Solitons: Theory and Experiments* (Cambridge University Press, Cambridge, 1992).

[11] S. Burger, K. Bongs, S. Dettmer, W. Ertmer, K. Sengstock, A. Sanpera, G. V. Shlyapnikov and M. Lewenstein, Phys. Rev. Lett. **83**, 5198 (1999).

[12] J. Denschlag, J. E. Simsarian, D. L. Feder, Charles W. Clark, L. A. Collins, J. Cubizolles, L. Deng, E. W. Hagley, K. Helmerson, W. P. Reinhardt, S. L. Rolston, B. I. Schneider, and W. D. Phillips, Science **287**, 97 (2000).

[13] L. Khaykovich, F. Schreck, G. Ferrari, T. Bourdel, J. Cubizolles, L. D. Carr, Y. Castin, and C. Salomon, Science **296**, 1290 (2002).

[14] U. Al Khawaja, H. T. C. Stoof, R. G. Hulet, K. E. Strecker, and G. B. Partridge, Phys. Rev. Lett. **89**, 200404 (2002).

[15] S. L. Cornish, S. T. Thompson, and C. E. Wieman, Phys. Rev. Lett. **96**, 170401 (2006).

[16] K. E. Strecker, G. B. Patridge, A. G. Truscott, and R. G. Hulet, Nature **417**, 150 (2002); K. E. Strecker, G. B. Partridge, A. G. Truscott and R. G. Hulet, New J. Phys. **5**, 73 (2003).

[17] I. Shomroni, E. Lahoud, S. Levy and J. Steinhauer, *Nature* Phys. **5**, 193 (2009).

[18] S. Sree Ranjani, U. Roy, P. K. Panigrahi and A. K. Kapoor, J. Phys. B: At. Mol. Opt. Phys. **41** 235301 (2008).

[19] P. Das, M. Vyas and P. K. Panigrahi, J. Phys. B: At. Mol. Opt. Phys. **42** 245304 (2009) and citations therein.

[20] U. Roy, B. Shah, K. Abhinav and P. K. Panigrahi, J. Phys. B: At. Mol. Opt. Phys. **44** 035302 (2011).

[21] A. D. Jackson, G. M. Kavoulakis, and C. J. Pethick, Phys. Rev. A **58**, 2417 (1998).

[22] L. Salasnich, A. Parola, and L. Reatto, Phys. Rev. A **65**, 043614 (2002).

[23] A. D. Jackson, and G. M. Kavoulakis, Phys. Rev. Lett. **89**, 070403 (2002).

[24] R. Atre, P. K. Panigrahi and G. S. Agarwal, Phys. Rev. E **73** 056611 (2006).

[25] E. H. Lieb and W. Liniger, Phys. Rev. 130, 1605 (1963); E. H. Lieb, Phys. Rev. 130, 1616 (1963).

[26] E. H. Lieb and W. Liniger, Phys. Rev. 130, 1605 (1963); E. H. Lieb, Phys. Rev. 130, 1616 (1963).

[27] R. Dutt, A. Khare and U. Sukatme, Am. J. Phys. **56**, 35 (1988).

[28] Z. X. Liang, Z. D. Zhang, and W. M. Liu, Phys. Rev. Lett. **94**, 050402 (2005); L. Salasnich, Phys. Rev. A **70**, 053617 (2004); e-print cond-mat/0408165.

[29] M. G. Vakhitov and A. A. Kolokolov, Izv. Vyss. Uch. Zav. Radiofizika **16**, 1020 (1973) [English Transl. Radiophys. Quantum Electron **39**, 51]

[30] V. M. Vyas, P. Patel, P.K. Panigrahi, C.N. Kumar, W. Greiner, Phys. Rev. A **78**, 021803(R) (2008), and references therein.

Plasma and the Role of Atomic Physics in its Spectroscopic Diagnostic

Ram Prakash

Birla Institute of Technology (BIT), Mesra (Ranchi)
Jaipur Extension Centre,
27-Malviya Industrial Area, Jaipur-302017
email: ramprakash@bitmesra.ac.in and rplavania@yahoo.com

1. INTRODUCTION

The national and international scenario in Plasma Science and Technology in general [1] and Controlled Thermonuclear Fusion [2] in particular has undergone a sea change over the last few decades. Significant steps have been taken towards building nuclear fusion reactors for meeting the ever-growing energy requirements while maintaining harmony with the life sustaining natural processes. The study of magnetically confined high-temperature plasmas is central to the understanding of fusion research devices such as tokamaks [2] and stellarators [3]. The ultimate goal of this research is a future fusion reactor; the international tokamak experiment ITER in France being a critical milestone [4]. The experimental reactor ITER is a multibillion dollar joint international research and development project which aims to demonstrate the scientific and technical feasibility of fusion power and is an important pathway to develop nuclear fusion as a viable, long-term energy option. Recognizing the importance of nuclear fusion to meet the future national energy need, India has also joined the ITER project on 6th Dec, 2005 along with China, European Union, Japan, Republic of Korea, Russian Federation and United States of America. Spectroscopists have been involved in this research from the beginning by identifying unknown lines, making precision measurements, and continuously improving the knowledge base of atomic and molecular physics.

As the fusion plasma research has become more mission oriented in the last few decades, spectroscopists have worked hard to extract more quantitative information from the spectroscopic observations [5]. In fact, plasmas found in laboratory and fusion research devices consist of charged particles, neutral particles and fields. They also contain varying amount of impurity elements as a result of wall erosion, diagnostic purposes or for producing radiating coronal mantle. At the same time these impurities interact with most plasmas and produces atomic radiations. A fraction of the neutrals and ions diffuse into the hot plasma core where they are ionized to high ionization stages. The emitted radiation therefore ranges from visible to the x-ray region, and spectroscopic study requires experimental tools as well as large databases of pertinent atomic data, often from theory, to model the emission through a collisional-radiative calculation.

A worldwide effort is underway to develop new measurement techniques for plasma devices from the visible to the x-ray range [6]. Simultaneously there is continued development of new methods for spectral analysis by establishing and maintaining internationally recommended numerical databases on atomic and molecular collision and radiative processes [7], atomic and molecular structure characteristics [8], particle-solid surface interaction processes [8] and physico-chemical and thermo-mechanical material properties for use in fusion energy research and other plasma science and technology applications [7]. The radiative processes in plasmas are important in the design of fusion facilities and can be used to control conditions in fusion plasmas [9]. In turn, fusion scientists and facilities have played a central role in the development of plasma spectroscopy.

Interesting part with plasma radiations is that it contains information on the nature of the plasma and can be utilized to understand the plasma environment spectroscopically. Nevertheless, interpretation of spectroscopic measurements is not straightforward. The delicacy lies in the complex nature of the most of the plasmas and subsequently the manifold atomic processes. In order to infer the plasma parameters from the measured intensities of spectral lines, proper calculations of the population densities of the excited states are highly required [10].

Most plasmas are typically neither in local thermodynamic equilibrium (LTE) [11] nor in coronal equilibrium [12]. As a result, valid and credible modeling of such plasma involves solving the complete rate equations for all constituents of the plasma. This aim can only be achieved through the provision of large amount of input atomic data for all rates connecting energy levels within and among all ionization stages of importance in such plasmas. Very accurate data are also needed for specific processes in spectral studies of such plasmas. So, in order to evaluate the population of excited states quantitatively an appropriate theoretical model and a large atomic database are required. The Collisional Radiative (CR) model that has evolved from pioneering work of Bates *et al.* [13] provides the theoretical basis for the calculation of the population density. At present there are a few CR-model based codes and database -like ADAS [14], ALADDIN [15], and CHIANTI [16] etc used in the plasma physics community.

As it is evident, plasmas are complex systems of electrons, ions and atoms where many processes are happening simultaneously, it would be fascinating to understand plasma in general and explore basic understanding of many atomic processes in particular for actual use in plasma spectroscopic diagnostics. A brief description of the plasma is given in section 2. The section 3 is related with the general terms and consideration related to plasma spectroscopic diagnostics. Section 4 contains brief information about the plasma emission spectroscopic diagnostic technique. Section 5 deals with the radiation processes in plasma and in section 6 the plasma models including Corona, LTE, and Collisional-Radiative (CR) models that are used for quantitative plasma spectroscopic diagnostics have been discussed. In section 7 the actual use of spectra for plasma diagnostics is presented. The chapter is concluded in section 8.

2. WHAT IS PLASMA?

Plasma is known as fourth state of matter. Plasmas are conductive assemblies of charged particles, neutral particles and fields that exhibit collective effects [17]. The plasma state can exist in any solid, liquid or gas phases. However, it is to be noted that the plasma state differs with the above states due to its very nature of having additionally long-range electromagnetic interaction forces. To understand as a layman let us take a piece of ice, which is a solid. Now if we heat the ice, it will get converted into a liquid, and if we heat it further it will get converted into gas. What will happen if we heat the gas? The gas molecules break into atoms, and then atoms into electrons and ions. When this ionization becomes large enough, then the charged particles start behaving collectively. As there is a saying "either crowd decides or God decide", this collective approach of charge particles leads to an aggressive state of matter, which is the fourth state of matter, called plasma. So mere ionization in any gas, solid, liquid does not make plasma state. The charge particles must exhibit collective behavior for plasma state to exist. This is what happens due to large number of charge particles. The examples of plasmas in gases, liquids and solids are given in the table below.

Sr. No.	Phase	Examples
1.	Gas Phase Plasma	H_2, O_2, He, Ar, Cl_2, D+T, SF_6 Plasmas etc. Used in many technological applications
2.	Liquid Phase Plasma	Sodium in ammonia, electrolytes etc.
3.	Solid Phase Plasma	Free electrons & holes in semi-conductors, Highly correlated dusty plasmas etc.

Plasmas are described by many characteristics, such as temperature, degree of ionization, and density, the magnitude of which, and approximations of the model describing them, gives rise to plasmas that may be classified in different ways.

Ideal and non-ideal plasmas

At low densities, a low-temperature, partly ionized plasma can be regarded as a mixture of ideal gases of electrons, atoms and ions. The particles travel at thermal velocities, mainly along straight paths, and collide with each other only occasionally. With an increase in density, mean distances between the particles decrease and the particles start spending even more time interacting with each other, that is, in the fields of surrounding particles. So, an ideal plasma is one in which Coulomb collisions are negligible, otherwise the plasma is non-ideal. In fact, condition for gaseous plasma existence is that the average kinetic energy of an electron should substantially exceed the average Coulomb energy needed for an ion to bind the electron.

$$\frac{3}{2}kT_e \gg \frac{e^2}{4\pi\varepsilon_0 d^2}.$$

The terms used in this expression have their usual meanings. This expression even can be further used in terms of Debye shielding of the particles and ultimately to understand the collective behaviors of the plasmas [17]. When mean energy of interparticle interaction becomes comparable with the mean kinetic energy of thermal motion, the plasma becomes non-ideal [18].

Cold, warm and hot plasmas

A plasma composed of the same number of electrons and ions. In low-pressure gas discharge, the collision rate between electrons and gas molecules is not frequent enough for non-thermal equilibrium to exist between the energy of the electrons and the gas molecules. So the high-energy particles are mostly composed of electrons while the energy of the gas molecules is around room temperature. Under these circumstances we have $T_e \gg T_i \gg T_g$ where T_e, T_i and T_g are the temperatures of the electron, ion and gas molecules, respectively. This type of plasma is called as "cold plasma". "In cold plasma, the degree of ionization is below 10^{-4} [19].

In a high pressure gas discharge the collision between electrons and gas molecules occurs frequently. This causes thermal equilibrium between the electrons and gas molecules. We have $T_e \sim T_g$. We call this type of plasma "hot plasma".

Hot plasma (thermal plasma)

A hot plasma in one which approaches a state of local thermodynamic equilibrium (LTE). A hot plasma is also called as thermal plasma. [20]. Such plasmas can be produced by atmospheric arcs, sparks and flames.

Cold plasma (non-thermal plasma)

A cold plasma is one in which the thermal motion of the ions can be ignored. Consequently there is no pressure force, the magnetic force can be ignored, and only the electric force is considered to act on the particles [21]. Examples of cold plasmas include the Earth's ionopshere (about 1000 ^{0}K compared to the Earth's ring current temperature of about $10^{8\,0}$K), low discharge in a fluorescent tube etc.

Plasma ionization

The **degree of ionization** of a plasma is the proportion of charged particles to the total number of particles including neutrals and ions, and is defined as: $\alpha = n^+/(n + n^+)$ where n is the number of neutrals, and n^+ is the number of charged particles.

In a plasma where the degree of ionization is high, charged particle collisions dominate. In plasmas with a low degree of ionization, collisions between charged particles and neutrals dominate. The degree of ionization, which determines when a gas becomes a plasma, will vary between different types of plasmas. It may be as little as 10^{-6}:

Example: Let us take a low-temperature, low-density, weakly ionized plasma for industrial applications. By low temperature, we mean "cold" plasmas with a gas temperature normally ranging from 300 ^{0}K and 600 ^{0}K, by low density we mean plasmas with neutral gas number densities of approximately 10^{13} to 10^{16} molecules cm^{-3} (pressure between ~ 0.1 to 10^3 Pa), weakly ionized means degree of ionization lies between 10^{-6} to 10^{-1} [22]. At very low ionization level ($<10^{-3}$) neutral collision dominates. On the other hand for a few percent ionization the Coulomb collisions dominates over collisions with neutrals in any plasma [23].

When the input energy to the plasma increases gradually, the degree of ionization jumps suddenly from a fraction of 1 percent to full ionization. Under certain conditions, the border between a fully ionized and a weakly ionized plasma is very sharp [24].

There are many nomenclatures –like, Collisional plasmas, Non-collisional plasma, Neutral plasmas, Non-neutral plasma, High density plasma, Low density plasma, Magnetic plasma, Non-magnetic plasma, Collisional plasma, High Energy Density Plasmas, Dusty plasmas, Grain plasmas, Active and passive plasmas etc are quite popular. The classifications are based on dominance of the specific property of plasma, which is reflected by itself from the name.

Where Does Plasma Exist?

It may be surprising but plasma is by far the most common form of matter. The plasma state exists in natural form all over the cosmos and or also is created under unique conditions for specific purposes in laboratories. Plasma in Stars and in the tenuous space between them makes up over 99% of the visible universe and perhaps most of that which is not visible. The plasma found in nature and laboratories cover a very large range of electron densities and temperatures as shown in the figure 1.

3. GENERAL TERMS AND CONSIDERATION RELATED TO PLASMA SPECTROSCOPIC DIAGNOSTICS

Spectroscopic plasma diagnostic enables us to obtain simultaneously a large amount of information about the plasma without disturbing it. So it is a passive diagnostics. It can also be used in active ways. An active spectroscopy involves injection of a suitable beam in the plasma.

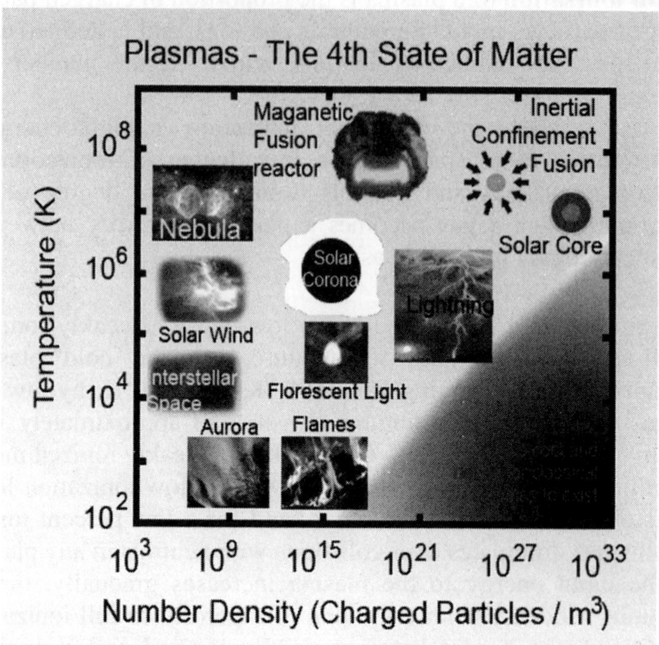

Figure 1.   Known ranges of different densities and temperature of the plasmas

(a) Common considerations

With increasing temperature of plasma the maximum of continuum radiation and the strong lines emitted occur at shorter wavelengths and typically plasma spectroscopy has to cover the spectral range from the infrared down to x-rays region. The physical concepts underlying the emission of radiation are more or less same over this spectral region but different experimental techniques have to be employed resulting in a corresponding partition into several spectral ranges such as *Infrared*, *Visible* (380-800 nm), *Ultraviolet* (200-380nm), *Vacuum Ultraviolet* (VUV) (105-200nm), *Extreme Ultraviolet* (EUV) (105-30 nm), *Soft Xrays* (30nm-0.15nm) and *Hard X-rays* etc.

In visible region, radiation is transmitted through air with practically no losses but this is not true for other regions. In addition, absorption by windows has also to be considered. Flint glass starts to absorb near 380 nm and for shorter wavelengths windows of quartz are advised, ordinary quartz cuts off at around 210nm, but best quartz is satisfactory down to just below 180 nm. The most commonly used window in the VUV is lithium fluoride (LiF), however, it looses it transmission gradually because of radiation damage. On the lower end of spectrum metallic or organic films or foils are used. At 5 nm, for example, 100 μm thick beryllium foil transmits about 50%.

An extremely useful data bank for atomic spectroscopy is that of National Institute of Standards and Technology (NIST) [25], it gives wavelengths of atoms, energy levels, ionization stages, configurations and transition probabilities. The energy E of the levels is given in the units of cm⁻¹ (1eV = 8065.541 cm⁻¹). A molecular database HITRAN [26] may be consulted for molecular spectra.

(b). Radiometric quantities

The radiant flux Φ (through the surface of the plasma) is the total energy emitted by a plasma as electromagnetic radiation per unit time and its unit is the watt (W). The SI quantity radiance L is defined as the radiant flux per unit projected area per solid angle Ω. Figure 2 illustrates the geometry, where θ is the angle between the normal of the surface element dA and the direction of the radiation.

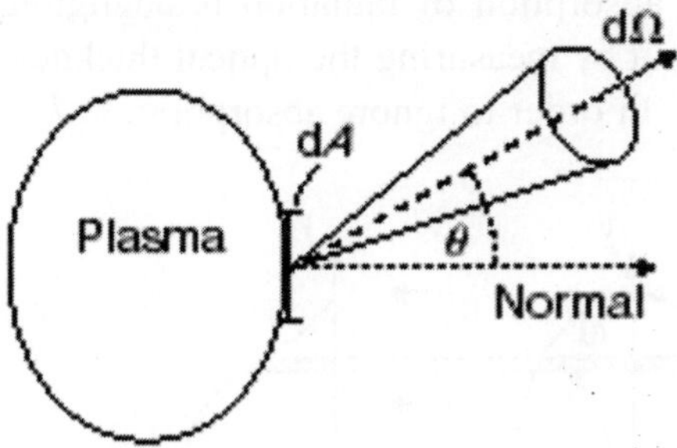

Figure 2. Definition of radiance

$$L = \frac{d^2\Phi}{dA\cos\theta d\Omega} \tag{1}$$

Its unit is Wm⁻² sr⁻¹. Although the radiance is the correct SI unit, the rather familiar word intensity is being commonly used in much of the plasma spectroscopy literature. The official SI quantity radiant intensity I is the flux per solid angle emitted from a source which makes it a property of a source rather than of a radiating surface: $I = d\Phi/d\Omega$, the unit being W/sr. The local emission at the position r in the plasma is characterized by the emission coefficient $\varepsilon(r)$:

$$\varepsilon(r) = \frac{d^2\Phi(r)}{dVd\Omega} \tag{2}$$

where dΦ (r) is the radiant flux from the volume element dV at r. The unit of the emission coefficient is Wm⁻³sr⁻¹. Usually measured spectral quantities are denoted by the pertinent subscript λ or v. They are derivative quantities and yield the total quantity when integrated over wavelength or frequency. For example

$$\varepsilon(r) = \int \varepsilon_\lambda(r,\lambda)d\lambda \tag{3}$$

The change in specific intensity $L\lambda$ in a layer with thickness dx is determined by the difference between the emitted and absorbed radiation (neglecting the scattering) using radiative transport equation:

$$dL_\lambda = \varepsilon_\lambda \, dx - k_\lambda \, L_\lambda \, dx \tag{4}$$

Here ε_λ is emissivity and k_λ is the absorption coefficient. The diagnostics methods based on the emission spectra can be applied directly to optically thin plasma objects where the absorption of radiation is negligible. Verification of absorption can be carried out by measuring the optical thickness $\tau_\lambda = k_\lambda \, l$, where l is the optical path length. In order to ignore absorption, $k_\lambda \, l \ll 1$.

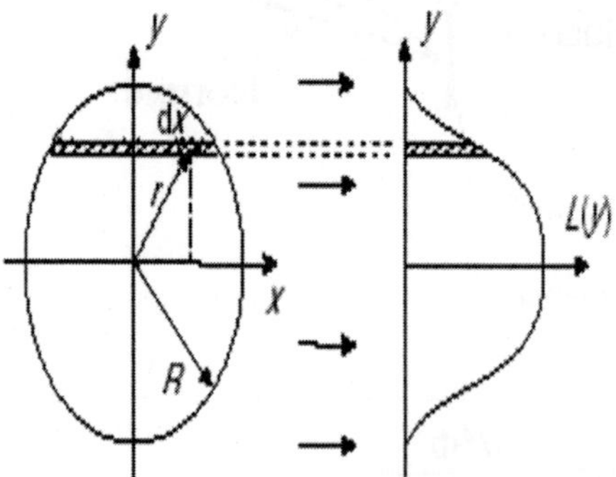

Figure 3. Radiance parallel to the y-axis of a cylindrically symmetric plasma column

From optically thin plasma, the quantity measured is always the local radiance at the surface of the plasma, which is given by integration of the emission coefficient along the line of sight. Generally, an average emission coefficient can be measured over the depth of plasma along line of sight. For a homogeneous plasma this corresponds to the true emission coefficient. In cases of axially symmetric plasma columns the derivation of local emission coefficients $\varepsilon(r)$ is possible if the radiance of the plasma is measured over the cross section of the column. Integration along a chord parallel to the x-axis yields after substituting $x^2 = r^2 - y^2$ (Fig. 3) the one-dimensional profile of the radiance,

$$L(y) = 2\int_0^{x_{max}} \varepsilon(r)dx = 2\int_0^{\sqrt{R^2-y^2}} \varepsilon(r)dx = 2\int_y^R \frac{\varepsilon(r)rdr}{\sqrt{r^2-y^2}} \qquad (5)$$

This equation is of the Abel type [27], and the local emission coefficient ε (r) is recovered by the Abel inversion which can be written analytically as

$$\varepsilon(r) = -\frac{1}{\pi}\int_r^R \frac{dL(y)}{dy}\frac{dy}{\sqrt{y^2-r^2}} \qquad (6)$$

In case of plasmas without symmetry, the mathematical problems associated with the general reconstruction of the local emission ε(r) throughout the plasma are essentially identical to those encountered in computer-aided tomography [27]. The radiance of the plasma has to be recorded in numerous directions, and the Radon transformation and its inverse then provides the mathematical basis for the reconstruction.

(c) Spectroscopic equipments

Depending on the application equipments such as spectrometer/ spectrograph, detector and imaging optics can be chosen. Details of various spectroscopic systems and their components are given in details in standard textbooks on plasma spectroscopy [28-31]. A typical spectrometers/ spectrograph consists of an entrance slit in the focal plane of a collimating spherical mirror, a grating as a dispersing element and a focusing mirror which forms an image of the entrance slit in the focal plane and a exit slit (see figure 4).

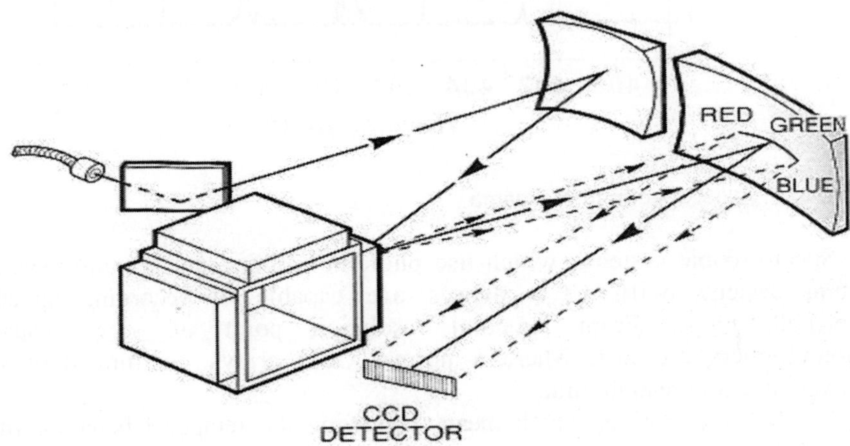

Figure 4. Schematic diagram of spectrometer/spectrograph

A detector is mounted on the exit slit. The plasma (emitting source) is either imaged by an optical arrangement on the entrance slit or coupled by fibers to the slit. The choice of grating (lines/mm) decides spectral resolution and blaze angle of grating determines the useful wavelength range. The focal length of the spectrometer influences the spectral resolution and together with grating defines the aperture and thus the throughput of light. The width of the entrance slit is also of importance for light throughput, which means a large entrance slit gives more intensity.

At the image plane of exit either a photomultiplier is placed behind the exit slit or a CCD array is mounted. The width of exit slit or the pixel size of CCD influences the spectral resolution of the system. The overall sensitivity of the system strongly depends on the detector: photomultipliers with different cathode coatings or CCD arrays with different sensor types (intensified, back-illuminated etc.) [32]. A typical spectra obtained by CCD based monocromator is shown in fig.5.

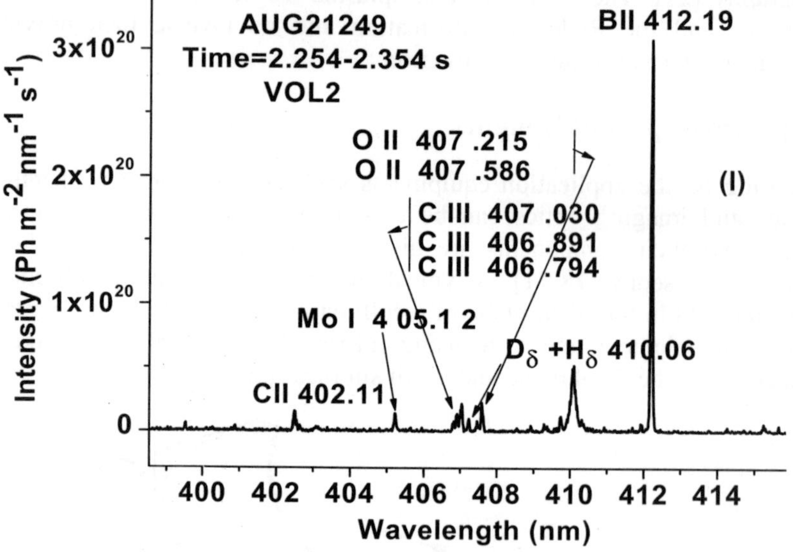

Figure 5.   Typical spectra from plasma

Spectroscopic systems, which use photomultipliers, are scanning system whereas systems with CCD arrays are capable of recording specific wavelength range. From temporal resolution point of view, usually photomultipliers are fast, where a normal CCD arrays are limited by the exposure time and readout time.

For line monitoring, which means following the temporal behavior of a particular emission line, pocket size spectrometers are suitable although they have a poor spectral resolution $\Delta\lambda \approx$ 1-2 nm. A spectrometer/ spectrograph with a focal length of 0.5-1 m ($\Delta\lambda \approx$ 40 pm) having a grating of 1200 lines/mm and a 2-D CCD array gives a good choice for spectral and temporal resolution. An Echelle spectrometer [33] provides an excellent spectral resolution ($\Delta\lambda \approx$ 1-2 pm) by making use of the higher orders of diffraction provided by the special

Echelle grating. They are an excellent tool for the measurement of line profiles and line shifts.

The calibration of spectroscopic system is also an important issue [32]. The calibration of wavelength can be done using low-pressure lamps (pencil sources, such as mercury-cadmium lamps). The calibration of intensity can be either relative or absolute calibration. A relative calibration takes in to account only the spectral sensitivity of the spectroscopic system along the wavelength axis. For absolute calibration, the light sources are required for which the spectral radiance is known. In visible spectral range, from 350nm to 800 nm, tungsten ribbon lamps are used and down in the UV range 200nm, the continuum radiation of deuterium lamps are used. In calibration procedure, one must make sure that solid angle is conserved [32]. Nevertheless, some basic principles can be applied even if only relative calibrated systems or systems without calibration are available.

4. PLASMA EMISSION SPECTROSCOPIC TECHNIQUES

In plasma emission spectroscopy, electromagnetic radiation emitted from plasma is recorded, spectrally resolved, analyzed and interpreted in terms of either parameters of the plasma or characteristic parameters of radiating atoms, ions or molecules. The implementation of these techniques are straightforward since only an optical port in plasma chamber is needed for the radiation to emanate and the apparatus required to resolve the radiation may also be least complicated. However, since the observer always samples radiation along the line of sight, measured values are generally line of sight average value. To get spatially resolved information, special arrangements have to be made or assumption regarding axial symmetry has to be taken. The optical emission spectroscopic methods are based on measuring the intensity of spectral lines, continuum spectrum or line profile measurements. The plasma parameters can be determined from spectral line widths.

Although spectra are generally easily obtained, their interpretation should be done carefully taking the appropriate plasma model depending on physical conditions. It should be emphasized that most diagnostic technique require an optically thin plasma that is very near homogeneous along the line of sight and remains in steady state for the duration of the observation. Very recently certain calculations of optical thick plasma have also been reported [34]. The accuracy of spectroscopic diagnostic techniques depends critically on the availability of accurate atomic data especially transition probabilities and broadening parameters.

(a). General identification and precursor studies

The emission spectroscopy is an easiest means to identify the particle species (atoms, ions, molecules) in the plasma, provided that particle emits radiation. Since the wavelength is a fingerprint of an element, it is sufficient to have a wavelength-calibrated spectrometer with imaging optics or a fibre. It may be quite useful in identifying the radicals, impurities etc and their wavelengths and intensities. As an example, a few identified species and their wavelengths are shown in the figure 6.

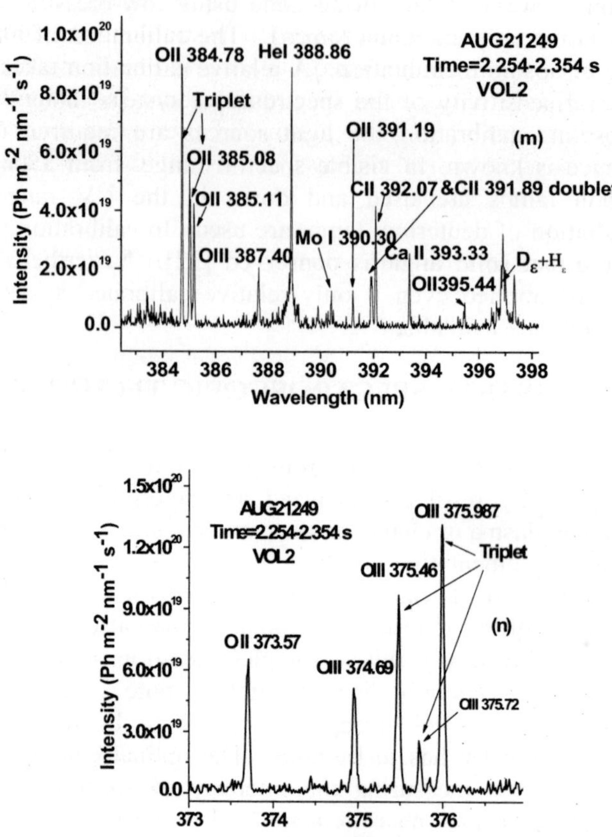

Figure 6.   Identification of species and wavelengths in typical spectra from plasma

5. RADIATION PROCESSES IN PLASMA

The plasma being an ionized state of matter at larger temperature it emits radiation over a wide range of electromagnetic spectrum. The radiations are mainly caused by the collisions taking place between particles leading to ionization or recombination and also due to acceleration or deceleration of charged particles. A study of radiation processes in plasma is important because it provides information about plasma parameters. The radiation study is also important from the point of view of energy balance in plasma.

The emitted radiation may exhibit either a continuum or line spectrum or both. Analysis of continuum intensities, line intensities and line profile can provide considerable information about plasma parameters such as electron temperature, electron density, ion temperature, plasma motion etc. The radiation frequency range may lie in microwaves, optical and X- rays regions depending on plasma parameters under consideration.

There are basically three types of radiation processes in un-magnetized plasmas, namely   (a) Bound- Bound or line radiation, (b) Free-Bound or continuum radiation, and (c) Free-Free or Bremsstrahlung radiations. Cyclotron radiation, which occurs in magnetized plasmas, is due to magnetic centripetal acceleration of charge particles as they spiral about the magnetic field lines. Blackbody radiation emitted from plasma in thermodynamic equilibrium is important only in astrophysical plasmas in view of the large size required for plasma to radiate as a black body.

(a). Bound-Bound or line radiation (excitation radiation)

The principal characteristic of bound-bound radiation is that the radiation is emitted at discrete frequencies. Whenever plasma contains atoms whose orbital electrons have not been completely stripped, line radiation will appear. Electrons changing their orbits remain bound to the nucleus, before and after the transitions, so this type of transition is called bound-bound transition. The corresponding wavelength extends from infrared to far ultraviolet. The radiation frequency is characteristic of both the atom or ion and the emitting levels. Bound-bound transition is a more dominant phenomenon in low-temperature or moderate temperature plasmas. At very high temperatures, plasma can be completely ionized, striping the atoms of all its orbital electrons. Due to their electronic excitation, atoms and ions emit a spectrum of lines such as

$$E_u - E_l = h\nu_{ul} = \frac{hc}{\lambda_{ul}} \tag{6}$$

where u and l are respectively, upper and lower excited levels between which the transition takes place. This is a spontaneous transition. Since the energy expected to be emitted in time dt is $h\,\nu_{ul}\,A_{ul}\,dt$, so the power radiated at ν_{ul} is $h\,\nu_{ul}\,A_{ul}$. The power density of bound- bound radiation is proportional to $P \propto Z^6$, thus even a very small fraction of impurities of high Z value can be detrimental for energy balance.

In plasma, spectral lines are broadened due to several factors, including collisions and pressure [30]. The Doppler and Stark broadening play important role in shaping the spectral lines. Consequently, the study of line profiles offers a powerful tool for plasma diagnostics. In processing plasmas, the spectral study of molecules and radicals may also be required, which is far more complex than those of single atom. The electronic excitation of diatomic molecules gives band systems in the UV, visible and IR regions. If the electronic state does not change during a transition, infrared bands called the vibration-rotation bands may be observed.

(b). Free-Bound or Continuum radiation (Recombination radiation)

Free electrons in plasma can recombine with ions or in some case be captured by neutral molecules. The energy lost by the electron in these processes may appear as radiation. Here the originally unbound charged particle is captured by another particle, and it emits the radiation, the process is called free-bound radiation

The excess energy of an electron with velocity radiation according to the relation ϑ_e is converted into radiation according to relation,

$$m_e \frac{\vartheta_e}{2} + E_{N^+}^I - E_{N_j} = h\nu \tag{7}$$

Where $E_{N^+}^I$ is the ionization energy corresponding to the reaction $N \rightarrow N^+ + e$ and E_{N_j} is the energy of the excited state j to which the electron is trapped. Since free electrons have any value of velocity ϑ_e , its recombination with an ion will results in continuum radiation. A threshold value exists for the wavelength resulting from the trapping of electrons with zero velocity in the excited state E_{N_j}

$$\Delta' E_N = h\nu_{min} \text{ or } \lambda_{max} = \frac{hc}{\Delta' E_N} \tag{8}$$

where $\Delta' E_N = E_{N_I}^I - E_{N_j}$ Because the electrons are distributed over a large spectrum of energies, their recombination into energy level j will give rise to a spectrum between ν_{min} and large value of ν. Electrons can be trapped into every available energy state of an atom, so the continuous spectrum coincides with the number of available energy levels.

Power density of free bound emission to the j th level is

$$P_j^{jb} \propto Z^4 \frac{N_e N_i}{\sqrt{T}} \tag{9}$$

So its value increases with number density and high Z value. Generally for processing plasmas, the continuum associated with free-bound transition will be dominant [30].

(c). Free-Free or Bremsstrahlung radiation

Accelerated charges radiate Electromagnetic (E.M.) Energy proportional to square of their acceleration. Bremsstrahlung or breaking radiation is the energy emitted when a free charged particle makes a transition between two states of a continuum in the field of an atom or ion. In this process a free electron makes a transition to another state of low energy with emission of photon. Since an electron is free before and after the coulomb interaction, Bremsstrahlung is refereed as "free- free" radiation. It is a continuous radiation.

At low energies, ions are massive to suffer the necessary acceleration, and Bremsstrahlung radiation arises almost entirely from electron radiation in electron-ion interaction. The wavelength (maximum) related to temperature is given by the expression,

$$\lambda = \frac{hc}{kT} \approx \frac{14000}{T_e(eV)} \overset{0}{A}$$

(10)

For high temperature, it usually occurs in X-ray or UV wavelength range.

The radiant energy W emitted by an electron per unit time is proportional to the square of their acceleration.

$$W = \frac{e^2 a^2}{6\pi\varepsilon_0 c^3}$$

(11)

where a is the acceleration of the particle and is given by

$$a = \frac{Ze^2}{4\pi\varepsilon_0 mr^2}$$

(12)

The power radiated per unit volume from plasma in electron-ion interaction with Maxwellian velocity distribution is given by

$$P_b \cong \frac{8\pi^2 Z^2}{3me^2c^3} \left(\frac{e^2}{4\pi\varepsilon_0}\right)^3 \frac{\sqrt{3mkT}}{h} N_e N_i \quad W/m^3$$

(13)

Where T_e is in ^0K and N_e and N_i are electron and ion densities in m^{-3}. Usually frequency of Bremsstrahlung radiation is higher than the plasma frequency, and thus Bremsstrahlung radiation propagates as if plasma were not just there. Since P_b scales on N_e, Bremsstrahluug becomes important in high-density plasma. Even more important is the dependence on Z^2, this shows that even a small percentage of high-Z ions in the plasma can cause a large increase in energy loss by radiation. This is the reason why in fusion plasma, a lot of attention is paid to the reduction of high-Z impurities.

(d). Cyclotron radiation

The radiation occurring when an electron or ion moves with in a magnetic field has the fundamental cyclotron frequency of rotation in the magnetic field in which it is trapped. There usually will be higher harmonics of this frequency present but of much weaker intensity. For low electron energies, it occurs as a line at the electron Larmor frequency, for higher electron energies, one finds radiation emitted at harmonics of the electron frequency in addition to the fundamental. The further details could be found elsewhere [30].

e). Black body radiation

Plasma is said to be optically thin when radiation trapping within the plasma can be neglected. Otherwise, it is optically thick. In optically thin plasma, the radiation represents a volume effect whereas in optically thick plasma the observed radiation is from surface. If the electromagnetic radiation generated within the plasma is absorbed and reemitted many times before reaching the boundaries of plasma, the radiation will come into equilibrium with itself and with plasma particles. In such plasma, concepts of radiations involving single or binary particle interaction can no longer be applied. Instead plasma is considered to be a medium in thermal equilibrium. In such a case, emission and absorption radiation are balanced and Kirchoff's law applies. A body, which absorbs completely all radiations at all temperatures, is termed as a black body. This equilibrium radiation is called "the black body radiation". The total radiation from the surface of a black body is given by Planck's law and depends only on the temperature of body

$$I = \sigma T^4 \ \mathrm{W/m^2} \tag{14}$$

Plasma having complete thermodynamical equilibrium exists only in stars or during the short interval of a strong explosion. Plasma size should be extremely large so that it can radiate as a black body.

6. PLASMA MODELS (CORONA, LTE, AND CR MODELS):

It is very well known that the impurity spectral lines one observes are the spontaneous emissions of excited impurity atoms/ions and obviously their intensities are proportional to the excited state population density N_u, which is generally determined by the local plasma parameters. The measured intensity is generally described by the simple relationship,

$$I_{ul} = \frac{1}{4\pi} \int_{x_1}^{x_2} N_u A_{ul} dx \tag{15}$$

For plasma the absolute intensity (or upper level population density N_u) depends not only on the properties of the isolated radiating species, but also on the properties of the plasma in the intermediate environment of the radiator. The population of excited states is described by a Boltzmann distribution provided that they are in thermal equilibrium. Since low-pressure plasmas are non-equilibrium plasmas, which means, in them population density does not necessarily follow a Boltzmann distribution. Thus, according to physical situations encountered, the plasma is said to be in Local Thermodynamic equilibrium (LTE) or Coronal Equilibrium (CE) or Collisional–radiative (CR) equilibrium. Before discussing these theoretical models, let us consider all the processes ultimately responsible for change in upper level population densities;

POPULATION KINETICS MODEL:

i). Collisional Processes
Bound-Bound transitions

Collisional de-excitation:
N_S (u) + e $\rightarrow$ N_S (l) + e **X_{ul}** de-excitation rate coefficient Collisional excitation:
N_S (l) + e $\rightarrow$ N_S (u) + e **X_{lu}** excitation rate coefficient

Free-bound transitions:

Collisional ionization:
N_S(u) + e $\rightarrow$ N_i + e + e **S(u)**Ionization rate coefficient 3-body recombination:
N_i + e + e $\rightarrow$ N_S (u) + e **β (u)** three body recombination rate coefficient

ii). Radiative processes
Bound–Bound transitions:

Spontaneous decay:
N_S (u) $\rightarrow$ N_S (l) + hν **A_{ul}** Spontaneous transition probability
Photo excitation:
N_S (l) + hν $\rightarrow$ N_S (u) + e **A_{lu}** Spontaneous absorption probability

Free-bound transitions:

Radiative recombination:
N_i+ e $\rightarrow$ N_S (u) + hν **α(u)** radiative recombination rate coefficient
Photo ionization:
N_S (u) + hν $\rightarrow$ N_i + e   **B(l,u) J** Photo ionization rate coefficient

For non-hydrgenic plasmas, additional processes must be included

Dielectronic recombination:
N_i + e $\rightarrow$ N_S^{**} $\rightarrow$ N_S (u) + hν α^D**(u)** Dielectronic recombination rate coefficient

and auto-ionization
N_S (u) + hν $\rightarrow$ N_S^{**} $\rightarrow$ N_i + e **B^D(l,u) J^D** auto-ionization rate coefficient

The set of equations describing the temporal relaxation of the density n(u) of level u and its collisional/radiative destruction and production is given by,

$$\frac{\partial n(u)}{\partial t} = P(u) - n(u)D(u) \tag{16}$$

Where P(u) is so-called production term and D(u) is destruction factor.

$$\frac{\partial n(u)}{\partial t} = \left(N_e \sum_{l \neq u} n(l) X_{lu} + \sum_{l < u} n(l) A_{lu} + N_e N_i \left\{ \alpha(u) + N_e \beta(u) \right\} \right)$$

Production term

$$- n(u) \left[N_e S(u) + N_e \sum_{l \neq u} X_{ul} + \sum_{l < u} A_{ul} \right] \qquad (17)$$

Destruction term

The solution of this equation is not straightforward. It is also not necessary to solve such equation always.   One can take certain valid assumptions and reduce the complexity. In view of this there are three theoretical models available in plasma spectroscopy for spectral analysis purpose.

a). Local Thermodynamical Equilibrium (LTE) model

b). Coronal model

c). Collisional-Radiative (CR) model, which is more generalized model

The following figure (7) can easily depict these models,

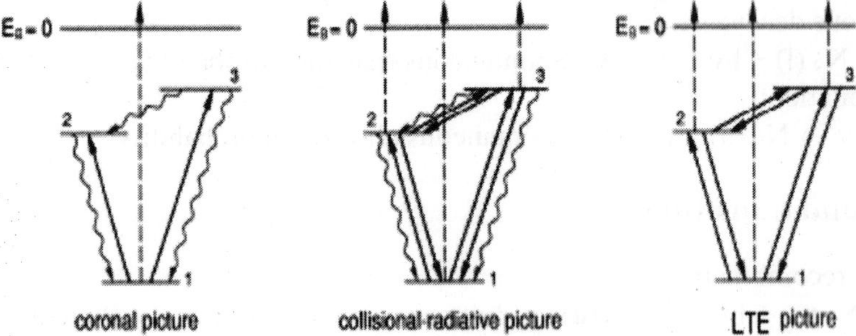

Figure 7. Comparison of coronal, CR-model and LTE model for three level atom (ground state and two excited states)

In coronal model an assumption is that all upward transitions are collisional and all downward transitions are radiative. In LTE model an assumption is that all most all processes are collisional, whereas CR-model is a general model, which involves solving complete coupled rate equation (17). Nevertheless, it is not always necessary to deal with all atomic processes and infinite energy levels simultaneously to solve equation (17) under CR-model consideration. To qualify these assumptions the probability for spontaneous decay of an excited atom and electron collision rate plays an important role. For example, Radiative decay rate for ground state Hydrogen $A_{ul} \approx 10^8 \ Z^{-4} \ s^{-1}$ and electron collision rate are given by $\nu_e = 2.91 \times 10^{-6} \ N_e \ (cm^{-3}) \ \ln \Lambda \ T_e^{-3/2}$ (eV) s^{-1} which is electron density dependent. The simplifications provided by these assumptions are described below.

a). LTE model:

If electron plasma density exceeds $N_e \sim 10^{15}$ cm^{-3}, the collisional rates will increase until radiative processes can be neglected. It means most excited states are depopulated by collisions mainly. In this case the excited states are in local thermodynamic equilibrium with the ground state.   Hence for high-density plasma region, if any spontaneous decay happens, an electron comes by that time and collides and takes it immediately to the higher energy state. So in the high-density case only collisional dominated transitions happen and radiative process becomes negligible. Thus, the distribution of population densities of the electron is determined exclusively by particle collision processes and collisions take place with sufficient rapidly so that the distribution responds instantaneously to any change in the plasma conditions. In such circumstances the each process is accomplished by its inverse process and these pairs of process occurs at equal rates by the principle of detailed balance [30]. Hence, the distribution of population densities of energy levels of electrons is same as it would be in a system in complete thermodynamical equilibrium. The population distribution is then determined by the statistical mechanical law of equilibrium among energy levels and does not require knowledge of atomic cross-sections for its calculation. Thus, one can describe the population distribution as,

Bound-bound case:

For bound-bound case, the Boltzmann relation describes the distribution of population levels within a given species,

$$\frac{N_S(u)}{N_S} = \frac{g_S(u)}{B_S(T_e)} \exp\left(-\frac{\chi_S(u)}{kT_e}\right), \tag{18}$$

$$\text{where } B_S(T_e) = g_S \exp\left[-\frac{\chi(l)}{kT_e}\right] \tag{19}$$

Where $N_S(u)$ is the population density of excited state u of an atom in the ground state, N_S is the ground state atom density of a specie, $g_S(u)$ is the statistical weight of the excited state u and $B_S(T_e)$ is the partition function corresponding to lower level ground state atom. T_e is the electron temperature and k is the Boltzmann's constant.

Free-bound case:

The relative total populations of successive ionization is given by Saha equation as,

$$\frac{N_e N_i}{N_S} = \frac{g_i(u)}{B_S(T_e)} . 2\left(\frac{2\pi mkT_e}{h^2}\right)^{3/2} \exp\left(-\frac{\chi_S(\infty)}{kT_e}\right) \tag{20}$$

Where N_e is the electron density and N_i is the ground state ion density and N_S is the ground state atom density.

Note:

1). Although the plasma temperature and density may vary in space and time, the distribution of population densities at any instant and point in space depends entirely on the local values of temperature, density and chemical composition of the plasma.

2). The condition of LTE is even valid if the transitions are coming from higher excited energy state u even though the electron density is low enough (below 10^{15} cm^{-3}) because the value of the atomic transition probability decreases as we go up in the excited states. Hence even for small N_e values the radiative processes can be easily neglected from the higher-level transitions. Furthermore, these transitions produces spectrum in the Infrared region of the spectrum, which is in general not under the scope of the common spectroscopic studies. Due to this fact the assumption of LTE is taken valid when $N_e \geq 10^{15}$ cm^{-3} for most plasma studies.

b). Corona-Model:

When the electron density is low, a rather simple model, which acquires its name from its applicability to the solar corona, may be used. This was a model basically proposed to explain the spectrum of solar corona. In corona model assumption is that all upward transitions are collisional and all downward transitions are radiative. In fact, in low-density plasma ($N_e \leq 10^{10}$ cm^{-3}), the probability for spontaneous decay of an excited atom is much higher than for any collisional depopulation process. Hence it can be easily assumed that,

(i) All upward transitions are collisional
(ii) All downward transitions are radiative

Further these assumptions are also valid even for $N_e \geq 10^{10}$ cm^{-3} provided that the transitions are happening very near to the ground state. Such transition produces spectrum in the Extreme Ultra-Violate (EUV) region of the spectrum, which is again not under the scope of the many spectroscopic studies. Hence the assumption for Corona model is taken valid only for low densities for most plasma studies.

We have already seen for low-density case the following two conditions satisfies,

i). All upward transitions are collisional,
$\sqrt{}$ Collisional excitation:

N_S (l) + e $\rightarrow$ N_S (u) + e   **X_{lu}** excitation rate coefficient

$\times$ Photo excitation:

N_S (l) + hv $\rightarrow$ N_S (u) + e   **A_{lu} (radiative negligible)**

Which is true when photon density in the system is very less. It means no photo excitation and absorption –all escapes –which is true for low-density optical thin plasma cases.

ii). *All downward transitions are radiative,*

$\sqrt{}$ Spontaneous decay:

$N_S (u) \rightarrow N_S (l) + h\nu$ **A_{ul}** Spontaneous transition probability

$\times$ Collisional de-excitation:

$N_S (u) + e \rightarrow N_S (l) + e$ **X_{ul}** de-excitation rate coefficient **(collisional negligible)**
Which is valid when the electron density is very small because for low-density cases X_{ul} becomes negligible.

The net result for bound-bound and free-bound transitions is expressed below.

Bound-bound transition:

Since for a detailed balance, the reverse process balances each and every process, the above assumption of corona model says that the spontaneous decay is not balanced by photo-excitation and the collisional excitation is not balanced by collisional de-excitation. But in steady state, the number of atoms being in a particular energy state is constant in time, which means for bound-bound transition in coronal steady state case, the collisional excitation is simply balanced by spontaneous decay.

$$N_e \, N_S \, X_{lu} = N_u \sum A_{ul} \tag{21}$$

Free-bound transitions:

We know that the processes of collisional ionization and three-body recombination are the inverse processes and must take place at equal rates in LTE,

$$N_S + e \Leftrightarrow N_i + e + e$$

However, the ionization rate $\propto N_e$ and three-body recombination rate $\propto N_e^2$. Positive ion may also recombine with electron by radiative process.

$N_i + e \rightarrow N_S (u) + h\nu$

The rate of this process $\propto N_e$. At a sufficiently low density it is more important than the three-body recombination, which is $\propto N_e^2$.

Furthermore, for the low-density optical thin plasma cases, the radiation density is very low and hence photo-ionization becomes negligible. Then the ionization equilibrium is balanced between radiative recombination and ionization by electron collisions.

$$N_e \, N_S \, S = N_e N_i \, \alpha \qquad\qquad (22)$$

These equations (21) and (22) are sufficient to describe the spectrum under coronal model case.

c). Collisional–Radiative (CR) model:

The collisional radiative (CR) models are 0-dimensional plasma model (& can be extended to 1-2D if impurity transports are important), which are used to calculate atomic state populations for one or more species, as a function of electron density and temperature. The model works under the condition that the two types of process cause an atom to change its excited state are collision and radiative processes. This is a general model because for intermediate density case (10^{10} cm^{-3} $\leq N_e \geq 10^{15}$ cm^{-3}), collisional frequency is higher, which leads to a competition between collisional processes and radiative decay processes. Thus for general model treatment one has to include all collisional and radiative processes and solve aforesaid equation (17) stepwise. How it is done is described below.

As one goes up in higher levels from the ground state the rate of spontaneous decay/ de-excitation decreases quite rapidly. So, in due course one will end up having certain level above which even in low densities before a spontaneous decay occurs, an electron comes and collides and excite that to a higher state, which is called as LTE. Then one comes to point that the levels above this certain level are in LTE. But how one decides these levels? Here comes the time information. If spontaneous decay time (A_{ul}) for a particular level is comparable to the electron collision frequency (which depends upon the electron density) then the level above that level will not be in so-called coronal equilibrium.

Then how one will treat these levels in reality? This treatment is mainly described in the CR-model. Most important thing to be noticed is that in the regimes where neither the LTE, nor coronal approximation is valid, the problem is essentially to be solving the complete coupled differential equation describing the population and de-population simultaneously as expressed in eq. (17). Considerable advances have been made in the last couple of year owing, partially, to the increasing knowledge of cross-sections necessary for the rate coefficients and, partially, to the use of high-speed computers for the calculation of such coupled equations. The CR model is the outcome of this advancement.

Simplifications in CR-model:

1). To avoid dealing with impossibly large summations and equally large number of differential equations, it is necessary to notice one of the most important properties of the CR-model. In fact, with increasing quantum number the level spacing becomes closer, the probability of collisional processes (within bound-bound) becomes grater while at the same time probability of radiative process becomes smaller. Thus to any desired accuracy there is always some level above which the effect of the radiative processes may be neglected. So simply one can assume that the bound-bound transition above this level is like free-bound transition. It is like above this level the atom goes into next ionizing level. Under these circumstances a modified form of Saha's equation may be used to evaluate the population densities of these upper levels u. In literature this is called as equilibrium population densities $\{n_E (u)\}$.

$$N_E(u) = N_S^B(u) = \frac{g_S(u)}{B_S(T_e)} \exp\left(-\frac{\chi_S(u)}{kT_e}\right) N_S$$

From same Boltzmann (23)

$$N_E(u) = N_{Si}^S(u) = \frac{B(T_e)}{2B_i(T_e)} \cdot \left(\frac{h^2}{2\pi mkT_e}\right)^{3/2} \exp\left(\frac{\chi_S(\infty) - \chi_S(u)}{kT_e}\right) N_e N_i$$

Modified Saha eq. (24)

Equation (24) is known as modified Saha equation. It applies in the restricted sense when $u = u_S$. If one normalizes the whole equation (17) with equation (24), one can compute the value of N (u) easily because the simulations in the rate equation (17) are no longer infinitely larger and the number of these equations that need to be considered is reduced to a manageable number.

Now how one will find the value of u_S? The value of $u = u_S$ (above which modified Saha equation may be used) may be found by changing it and checking that the result of the calculation has not changed by more than the required accuracy (within ± 10 %). For $p_S = g$ it will reduced to LTE limit.

Hence the first simplification in CR-model is that to a desired accuracy there is always some level above which the effect of the radiative processes may be neglected and we can treat these levels under LTE limit. The schematic of energy levels is shown below in figure 8 describing the LTE-and Coronal limits.

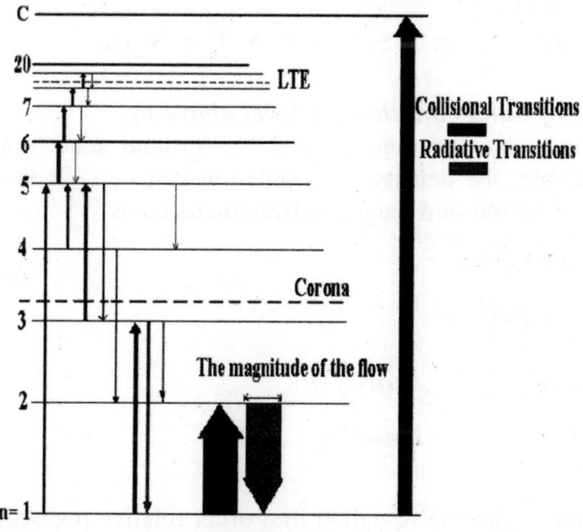

Figure 8. Schematic of energy levels of an atom

2). On examination of rates of various processes one can find that the relaxation time associated with the ground level atoms or ions are some orders of magnitude longer than those for transitions among the excited states and the continuum of free electrons. For the excited levels the relaxation times are the life times for spontaneous radiative decay or shorter. Compared with the relaxation time for the ground level population these others may be regarded as instantaneous. It is to be noted that for most of atomic plasmas, this condition holds well. For example, for the lower levels, radiation life times are typically much smaller than even with the transport times (10^{-7} s vs 10^{-4} s), and for higher excited levels, where radiation life times are larger, the more efficient collisional excitation/de-excitation by electrons usually satisfies the small life time requirement.

Hence the second simplification in CR-model is that the dominant populations are ground state atoms and ions (and also metastables if any), and their sum is constant. Compared with the relaxation times for the ground level population these others may be regarded as instantaneous.

$$N_S + N_i = Constant = N_{S,Total} \tag{25}$$

Thus a quasi-steady state solution of the set of equation (17) can be derived as,

$$\frac{dN_u}{dt} = 0 \quad \text{except for u} \neq 1 \tag{26}$$

Now the sets of eq (17) reduced to

1). A set of (u_S-1) simultaneous equations,

$$\sum_{u > 1} C_{eff} n(u) = N_i C_{iu} + N_S C_{Su} \tag{27}$$

Where C is the function of S, X, α, β, A, T_e & N_e etc.

2). Any number of Saha equation for level above u_S.

Since N_i (ground state ion density) and N_S (ground state atom density) are independent variable, the density of population can be split in atoms and ion ground level by using the said Saha and Boltzmann eq as

$$N_i(u) = r_i'(u) N_S^S(u)$$

$$N_S(u) = r_S'(u) N_S^B(u)$$

$$N(u) = r_i'(u) N_S^S(u) + r_S'(u) N_S^B(u)$$

$$\text{or } N(u) = r_i(u) N_i N_e + r_S(u) N_S \tag{28}$$

Where r'_i and r'_S are the so-called first order relative populations whereas r_i and r_S are the relative population including Saha-Boltzmann coefficient and these values can be found by solving eq (27) for N_i=0 and N_S =0, respectively.

***3). And equation (1) for ground level where we have to retain the time
derivative.***

These equations then give ionization and recombination of a system of ions
under consideration. These equations are described in terms of the effective rate
coefficients.

$$\frac{dN(1)}{dt} = \alpha_{CR} N_i N_e - S_{CR} N_S N_e \tag{29}$$

Where the collisional-radiative ionization and recombination rate
coefficients are expressed in terms of population coefficients for u>1 as

$$\alpha_{CR} = \beta(1)N_e + \alpha(1) + \sum_{u>1} r_S(u)[X_{u1}N_e + A_{u1}] \tag{30}$$

$$\alpha_{CR} = S(1) + \sum_{u>1} X_{1u} - \frac{1}{N_e}\sum_{u>1} r_i(u)[X_{u1}N_e + A_{u1}] \tag{31}$$

Here u>1 means all excited levels.

$$\text{When } \frac{dN(1)}{dt} = 0 \text{ or } \alpha_{CR} N_i N_e = S_{CR} N_S N_e \tag{32}$$

The plasma is called the ionizing equilibrium plasma.

It is important to note that the ionizing coefficient includes r_i and not r_S and
vise versa for recombining coefficient and there is no direct dependence of either
α_{CR} or S_{CR} on N_i and N_S. This makes it convenient to use these coefficients in
plasma transport model where only ground level atoms, electrons and ions are
considered. The ionization and recombination coefficients can then be used to
calculate the transformation flow from ground level atoms to ions and vise versa.

The solutions of these equations are time dependant solutions that allows
account to be taken of the finite time require for ionization and recombination
processes. Due to this fact this model is also quite useful even for transient
plasma cases. The steady-state value can also be deduced in the different regions
of models like corona & LTE for given N_e and T_e.

8. APPLICATIONS OF CR-MODEL

In CR model, the populations of various encrgy levels of atoms (or ions) in a
plasma are calculated assuming a quasi-steady state to exist among the excited
levels, which depends on the plasma parameters N_e and T_e. As an example we
have calculated the intensities of the B II lines and their ratios using CR-model
based data from ADAS code [14]. With an assumption that the average electron
density and temperature in an emission length x, the photon intensity $I(\lambda_{ul})$ of
a spectral line can be written from the quasi-steady-state approximation of the
CR-model as,

$$\tilde{I}(\lambda_{ul}) = \overline{PEC}_{recombining}\tilde{N}_e(\tilde{N}_i x) + \overline{PEC}_{excitation}\tilde{N}_e(\tilde{N}_S x) \tag{33}$$

where $\overline{PEC}_{recombining} = r_i\,A_{ul}$ and $\overline{PEC}_{excitation} = r_S\,A_{ul}$ represents the effective photon emission coefficients ($photons\ cm^3\ sec^{-1}$) for recombination and excitation processes respectively in an average measurement. Here $\widetilde{N}_S x$ and $\widetilde{N}_i x$ are the average column densities of the ground state atoms (here referring to BII) and ions (here referring to BIII). The ADAS package derives PEC values for a particular line λ_{ul} after calculating the population distribution of levels. This is done by solving a set of coupled rate equations for the number of levels of the ionization stage i. In each equation, one includes all the processes of populating and depopulating the level by excitation, deexcitation, spontaneous emission, ionization and recombination from adjacent ionization stages etc [10-16]. The proportionality factors –the effective photon emission coefficients (PEC's) in equation (1) –are functions of electron densities and temperatures.

The rates at which the ionization of ground state atoms into ions proceeds (and vice versa) are given by

$$\frac{dN_g}{dt} = -\frac{dN_i}{dt} = \alpha_{CR} N_i N_e - S_{CR} N_S N_e \tag{34}$$

Here α_{CR} and S_{CR} are effective collision-radiative (CR) recombination and ionization rate coefficients, respectively. These collisions-radiative rate coefficients are also function of N_e and T_e only, and are calculated from the code by considering all the processes of populating and depopulating the level u by excitation, deexcitation, spontaneous emission, ionization and recombination from the adjacent ionization states etc. In a plasma, the net population balance between N_i and N_S is determined, not by the plasma alone, but also the net influx/outflux of N_S (or B II) and N_i (or BIII).

Under steady-state approximation we can write,

$$\frac{S_{CR}}{\alpha_{CR}} = \frac{N_i}{N_S}. \tag{35}$$

Now, if $\dfrac{S_{CR}}{\alpha_{CR}} \gg 1$, the plasma is called "ionizing plasma" [35]. Under this condition, the effective ionization dominates the effective recombination in the plasma. In true sense the ionizing plasma condition holds well when $\dfrac{S_{CR}}{\alpha_{CR}} \gg \dfrac{N_i}{N_S}$ [36]. So a plasma in which $\alpha_{CR} N_i \ll S_{CR} N_S$, is said to be an "ionizing plasma" (and the reverse case is referred to as a "recombining plasma"). Fig 9 shows that unless the temperatures are very low (< 3 eV) $S_{CR} \gg \alpha_{CR}$. Consequently, in a wall conditioning plasma sustained by a supply of boron atom from the feed gas, the "ionizing plasma" condition applies. Since most of the tokamak plasmas have temperature above 3 eV in the edge and divertor regions, purely ionizing condition to derive the impurity line ratios is a good approximation. The analogy is also true for the terms available in equation (33). The population that is proportional to the ground state atom density is the recombining component of plasma and proportional to the ground state atom density is the ionizing component of plasma.

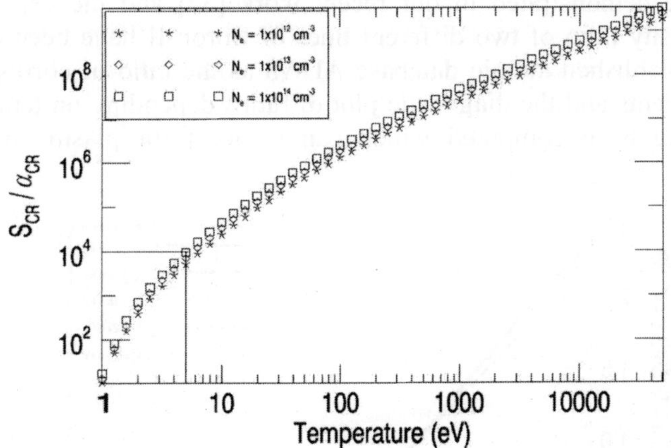

Figure 9. Plot for the ratio $\dfrac{S_{CR}}{\alpha_{CR}}$ at different N_e and T_e

(Star for N_e = 1x10^{12} cm^{-3}, Diamond for N_e = 1x10^{13} cm^{-3},

Square for N_e = 1x10^{14} cm^{-3}).

Impurity line ratios

Under purely ionizing condition, the first term in equation (33) can be taken to be negligibly small [17] and the line intensity $I(\lambda_{ul})$ for a transition from level u to level l is expressed as,

$$\widetilde{I}(\lambda_{ul}) = PE\overline{C}_{excitation}\widetilde{N}_e(\widetilde{N}_S x). \tag{36}$$

and the ratio of intensities of two lines becomes,

$$\frac{I_{ul1}}{I_{ul2}} = \frac{PE\overline{C}_{excitation1}}{PE\overline{C}_{excitation2}}, \tag{37}$$

where I_{ul1} and I_{ul2} are the absolute intensities of two spectral lines of a specific specie of interest. The terms $PE\overline{C}_{excitation1}$ and $PE\overline{C}_{excitation2}$ are corresponding PEC's which are functions of N_e and T_e only. The other terms will be cancelled in the ratio output. Thus, it is easy to calculate the expected ratio of two lines under different plasma conditions of N_e and T_e. The line ratios that are sensitive to one of the quantities (either temperature or density) and insensitive to other quantity have to be identified for this purpose.

It is somewhat repetitive procedure to identify suitable line-pairs because PEC's are complex functions of various atomic rate coefficients (which are again functions of N_e and T_e) and their nature is unpredictable till carefully explored.

This is what we demonstrated in our recent work [37] and the experimental observable intensity ratio of two different lines of Boron-II have been obtained from the well established atomic database ADAS as the ratio of corresponding emission coefficients and the diagnostic plot of ratios depending on temperature and density have been computed which can be used for plasma parameter measurements.

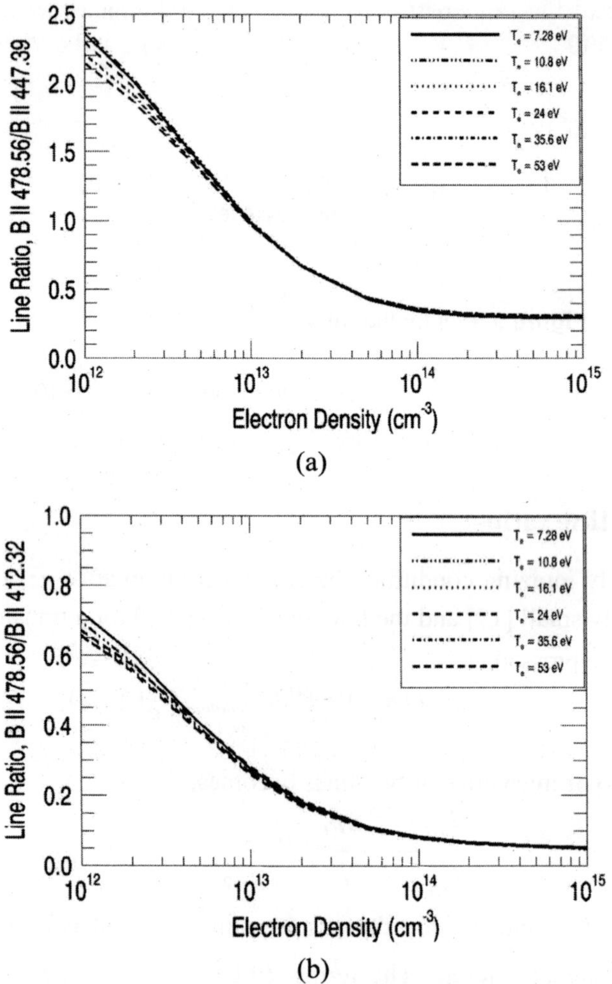

(a)

(b)

Figure 10. a) Calculated line intensity ratios at different b). Calculated line intensity ratios at different T_e for the ionizing plasma B II 478.56/ B II 412.32. T_e for the ionizing plasma B II 478.56/ B II 447.39.

The line ratio B II 478.56/ B II 412.32 and B II 478.56/ B II 447.39 are very useful for the electron density measurements between $1x10^{12}$-$5x10^{13}$ cm^{-3} as shown in figure 10(a) and figure 10(b) that is the typical densities in the edge regions of the most tokamaks. Many such line ratios B II 478.56 / B II 412.32, B II 478.56/ B II 447.39, B II 494.19/ B II 419.60, B II 628.77/ B II 419.60 and B II 494.19/ B II 628.77 etc have been reported [37].

The line pairs B II 419.60/ B II 703.28 and BII 412.32/ B II 628.77 [see figure 11(a) and figure 11(b)] are identified temperature sensitive under the ionizing plasma conditions, which could be utilized to derive the effective electron plasma temperatures in tokamak plasmas. The line ratio B II 419.60/ B II 703.28 as shown in figure 11(a) is very useful for the temperature estimation from 1 eV to 40 eV. The line ratio BII 412.32/ B II 628.77 as shown in figure 11(b) is useful for temperature estimation in the region 10 to 40 eV. These are the typical temperatures for most tokamaks in the edge and divertor regions.

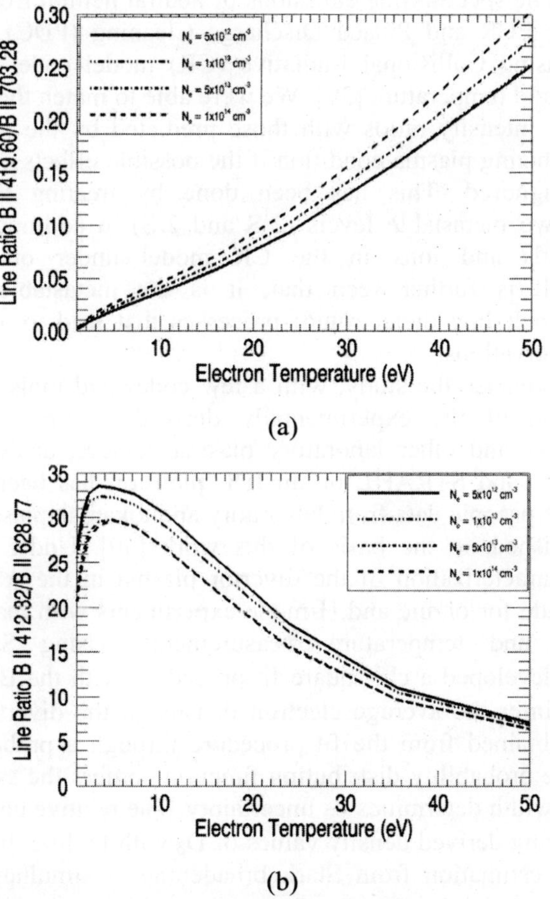

(a)

(b)

Figure 11. a) Calculated line intensity ratios at different. b). Calculated line intensity ratios at different N_e for the ionizing plasma B II 419.60/B II 703.28. N_e for the ionizing plasma B II 412.32/ B II 628.77.

These line pairs can be used to characterize plasmas containing boron, both during the wall-conditioning and in the edge regions of regular tokamak discharges followed by boronization.

In the similar ways, we utilized the CHIANTI code for generating synthetic spectra for a few impurity ions found in the ADITYA tokamak discharges [38]. The CHIANTI code is a freely available code and is generally used to interpret the spectrum of astrophysical plasmas. Though in its present form this code is tuned for the astrophysical situations, its CR kernel has been used for generating synthetic spectra for a few impurity ions found in the ADITYA tokamak plasmas.

We also applied CR-model for helium spectroscopy in ADITYA tokamak [39]. In fact, the helium atom is an attractive diagnostic specie spectroscopically because it has various advantages, for example, well-known atomic data, strong-visible lines and availability of many density or temperature sensitive line pairs for diagnostics. The spectral line emissions of neutral helium from two different plasmas namely, ECR and Pulsed Discharge Cleaning (PDC) plasmas –have been analyzed using Collisional Radiative (CR) model code, to estimate the electron density and temperature [39]. We were able to match the experimentally obtained relative intensity ratios with those predicted by the model under the assumption of ionizing plasma condition if the possible effects of the metastable states are not ignored. This has been done by treating the independent populations of two metastable levels (2^1S and 2^3S) in addition to the ground states of neutrals and ions in the CR model under quasi-steady state approximation. It is further seen that, it is the metastables and not the recombination (including dielectronic) processes that lead to a better fit with experimental observations.

To further advance the study, with a few codes and tools for the analysis and interpretation of the experimentally derived spectroscopic data from different tokamaks and other laboratory plasma devices, analytical procedure integrating ADAS and STRAHL on an IDL platform has been developed for interpreting spectroscopic data from laboratory and tokamak plasmas. A research report was published on the basis of this work [40]. Under these efforts a spectroscopic characterization of the divertor plasma in the ASDEX Upgrade tokamak was made for ohmic and H-mode experiments with main emphasis on plasma density and temperature measurements. Using Stark-broadening tabulations, we developed a chi-square fit procedure with the Balmer series D_δ and D_γ lines to infer the average electron density in the divertor plasma. The densities were obtained from the fit procedure through a probability analysis. The center of the probability distribution function confers the estimated density whereas its half-width determines its uncertainty. The relative uncertainties were tested by comparing derived density values of D_δ with D_γ line. In addition to the electron density estimation from Stark broadening, a simultaneous fit of the deuterium Balmer series D_β, D_γ, D_δ, D_ε line intensities with ADAS collisional-radiative (CR) model data allows in determining the effective electron temperature, the average ground state atom and ion densities [40].

On the basis of estimated basic plasma parameters and subsequently derived S/XB values from ADAS database, the impurity influx measurements for ohmic discharges were also carried out [40]. In comparison to the ion fluxes at the target, helium ground state and deuterium fluxes were increased during recombining phase whilst fluxes of boron, carbon and oxygen have reduced drastically. The particle fluxes obtained from spectroscopy lies within a factor of two than that from Langmuir probe, but the estimated plasma densities were around one order of magnitude larger. Detached plasmas were observed during ohmic discharges. The found degree of detachment from spectroscopic observations was ~4 which was comparable with other methods determination. During detached divertor conditions, the plasma temperature has dropped from 7-10 eV to ~2 eV in the outer divertor region.

In our recent efforts we have developed a novel and simple technique for VUV spectrometer-detector calibration [41]. In fact, a VUV spectrograph is always required to be absolutely calibrated in preparation for its use on tokamak plasmas. There a few complicated and costly methods already exist. In this work a simple and low cost method to calibrate such spectrometer is presented. A standard penning discharge lamp with helium background gas has been used for this purpose. The spectrum is recorded in the both visible and VUV region simultaneously from the penning plasma source at different values of the background pressures and discharge currents. From the many lines in the visible system (where absolute intensity calibration of a spectrometer was available), we characterized the discharge plasma of the lamp using Atomic Data and Analysis Structure (ADAS) code and database. We developed a minimization procedure using large number of experimentally observed spectral lines in the visible region and the predicted spectrum from the full collisional-radiative (CR) model of the ADAS database to determine simultaneously the important plasma parameters including electron density, electron temperature, ground state atom and ion densities and also metastable density for helium 2 ^{3}S state. With the proposed procedure, we are able to calculate the metastable populations of helium 2 ^{3}S state non-conventionally. There are limited efforts to calculate the metastable populations of helium using atomic absorption spectroscopy and a spectral line that terminates on the singlet metastable level. With the increase of discharge current at constant pressure in the penning plasma source the metastable population increases whereas with the increase of gas pressure at constant discharge current the metastable population decreases. Using the derived values of the basic plasma parameters, we computed the spectral intensities in the VUV region, and subsequently a comparison with the recorded spectrum in the VUV region yielded the desired sensitivity curve.

Conclusion

The spectral fingerprints of emitted radiation from different plasmas can provide information about many basic plasma parameters like electron temperature, electron density, ground state atom density, ground state ion density and identification of precursor species by measuring the line intensities or profiles etc. Emission spectroscopic diagnostics are non-intrusive and simple to set up. Basic configuration of Spectrometer / Spectograph have been discussed. In optically thin plasmas, usually these techniques measures average values of plasma parameters along the line of sight. The accuracy of spectroscopic diagnostic techniques depends critically on the availability of accurate atomic data. Some of the techniques used by us have been illustrated in this chapter.

References

[1]   J. R. Roth (2003) Industrial Plasma Engineering (IOP publishing) ISBN 0-7503-0317-4

[2]   J. Wesson (2004) Tokamaks Third edition (Oxford Science Publications) ISBN 978 0 19   850922 6

[3]   J. H. Harris, J. L. Cantrell, T. C. Hender, B. A. Carreras, and R. N. Morris (1985) "A flexible heliac configuration" *Nuc. Fusion* **25**: 623

[4]   http://www.iter.org/

[5]   Proceedings of IAEA International Workshop on "Challenges in Plasma Spectroscopy (CPS) for Future Fusion Research Machines" held at BIT Jaipur (20-22 Feb. 2008) Edited by Dr. Ram Prakash ISBN 978-81-903047-8-8

[6]   R. J. Hawryluk et al. Phys. Plasmas, Review Art. 5 (1998) 1577

[7]   K. Behringer, Plasma Phys. and Contr. Fusion 33 (1991) 997

[8]   K. Behringer, J. Nucl. Mater. 176 & 177 (1990) 606

[9]   R. Neu et al., Plasma Phys. Control Fusion, 49 (2007) B59-B70

[10] Fujimoto T., (2004) "Plasma Spectroscopy" (International Series of monograph on Physics) ISBN: 9780198530282

[11] A Hartgers *et al* J. Phys. D: Appl. Phys. 38 (2005) 3422

[12] T. Fujimoto, J. Phys. Soc. Jpn. 34 (1973) 216

[13] D R Bates, A E Kingston, *I.* R W P McWhirter (1962) *Optically thin films Proc. Roy. Soc.* A257 297

[14] H.P. Summers (1994) ADAS users manual, *JET –IR 06* (*Abingdon: JET Joint* undertaking)

[15] R.E.H. Clark and J A Stephens "The atomic and molecular data unit of international atomic energy agency" A/P conf. proc. 547 (2000) 167

[16] Dare K P, Landi E,Mason H E, Monsignori Fossi B and Young P R Astron. Astrophys. Suppl. Ser. 125 (1997) 149

[17] F.F. Chen (1984) "Introduction to Plasma Physics", 2nd edition, (Plenum Press)

[18] V. E. Fortov, Igor T. Iakubov, (2000) "The physics of non-ideal plasma" (World Scientific), ISBN 9810233051, ISBN 9789810233051

[19] Kiyotaka Wasa, Shigeru Hayakawa, (1992) *"Handbook of Sputter Deposition Technology: Principles, Technology and Applications"* (Materials Science and Process Technology Series), William Andrew Inc., 304 pages (page 95)

[20] Maher I. Boulos, Pierre Fauchais, Emil Pfender, (1994) *"Thermal Plasmas: Fundamentals and Applications"* (Springer) page 6.

[21] Marcel Goossens, (2003) *"An Introduction to Plasma Astrophysics and Magnetohydrodynamics"* (Springer), page 25.

[22] Loucas G. Christophorou, James Kenneth Olthoff, (2004) *"Fundamental Electron Interactions With Plasma Processing Gases"*, page 39

[23] Robert J. Goldston, (2003) Paul Harding Rutherford, "Introduction to Plasma Physics, Fully and Partially Ionized Plasmas" page 164

[24] Hannes Alfvén and Gustaf Arrhenius, (1976) *"Evolution of the Solar System"*, Part C, (Plasma and Condensation)

[25] http://www.physics.nist.gov/PhysRefData/contents-atomic.html

[26] http://www.hitran.com

[27] R. Bracewell, (1965) *"The Fourier Transform and its Applications"* (New York: McGraw-Hill) ISBN 0-07-007016-4. & F Natterer and F Wuebbeling (2001) "Mathematical methods in image reconstruction" (SLAM, Philadelphia)

[28] R H Huddlestone and S L Leonard (1965) Plasma diagnostic techniques (Academic Press, New York)

[29] W Lochte- Holtgreven (1968) "Plasma diagnostics" (North Holland, Amsterdam)

[30] I H Hutchinson (1987) "Principles of plasma diagnostics" (Cambridge University Press)

[31] A A Ovsyannikov, (2000) "Plasma diagnostics" (Cambridge International Scienc Publications, Cambridge)

[32] Hans-Joachim Kunze (2009), "Introduction to Plasma Spectroscopy" (Springer Heidelberg Dordrecht London New York) ISBN978-3-642-02232-6

[33] D. A. Skoog, F.J. Holler and T.A. Nieman, *(1998) "Principles of instrumental analysis" (Florida, Harcout Brace & Conpany)*

[34] Jalaj Jain, Gheesa Lal Vyas, Ravindra Kumar, R. Manchanda and Ram Prakash "Opacity Effect on Photon Emissivity Coefficients (PECs) of Neutral Helium Line Emissions and its Impact on Line Ratios" presented poster and **received best paper award** in the 25th National Symposium on Plasma Science and Technology (Plasma-2010) held at IASST, Guwahati during 8-11 Dec, 2010

[35] Goto M *J. Quant. Spectrosc. Radiat. Transfer* **76** (2003) 331

[36] Fujimoto T, Sawada K (1997) *NIFS-DATA-***39**

[37] Bishu Agarwal, Ram Prakash, Jalaj Jain, Vinay Kumar and P. Vasu Phys. Scr., VOL. 80 (2009), 055505

[38] Ram Prakash, P. Vasu, M.B. Chowdhury, Vinay Kumar & R. Manchanda (2003) IPR/RR-300/2003

[39] Ram Prakash, P. Vasu, V. Kumar, R. Manchanda, M. B. Chowdhuri & M. Goto, IPR/RR-324/2004 (2004) J. Appl. Phys. 97 (2005) p.43301

[40] Ram Prakash, et al. IPP (2006) 10/31

[41] Ram Prakash, Jalaj Jain, Vinay Kumar, R Manchanda, Bishu Agarwal, M. B. Chowdhari, Santanu Banerjee, and P J. Phys. B: At. Mol. Opt. Phys. 5 July 43 (2010) 144012